OPTICAL SPECTROSCOPY OF LANTHANIDES

Magnetic and Hyperfine Interactions

T0203965

OPTICAL SPECTROSCOPY OF LANTHANIDES

Magnetic and Hyperfine Interactions

Brian G. Wybourne

Lidia Smentek

CRC Press
Taylor & Francis Group
Boca Raton London New York

CRC Press is an imprint of the
Taylor & Francis Group, an **informa** business

CRC Press
Taylor & Francis Group
6000 Broken Sound Parkway NW, Suite 300
Boca Raton, FL 33487-2742

First issued in paperback 2019

© 2007 by Taylor & Francis Group, LLC
CRC Press is an imprint of Taylor & Francis Group, an Informa business

No claim to original U.S. Government works

ISBN-13: 978-0-8493-7264-3 (hbk)
ISBN-13: 978-0-367-38917-8 (pbk)

Library of Congress Cataloging-in-Publication Data

Smentek, Lidia.
    Optical spectroscopy of lanthanides : magnetic and hyperfine interactions / Lidia Smentek, Brian G. Wybourne. -- 1st ed.
        p. cm.
    Includes bibliographical references and index.
    ISBN-13: 978-1-4200-0693-3 (alk. paper)
    ISBN-10: 1-4200-0693-2 (alk. paper)
    1. Rare earth metals--Spectra. 2. Ions--Spectra. 3. Spectrum analysis. I. Wybourne, Brian G. II. Title.

QC462.R2S64 2007
546'.41--dc22   2006029158

Visit the Taylor & Francis Web site at
http://www.taylorandfrancis.com

and the CRC Press Web site at
http://www.crcpress.com

The authors' royalties are being donated to the St. Joseph's Indian School for Lakota (Sioux) Children in Chamberlain, South Dakota.

Contents

Foreword

Memories are a gift of the past,
That we hold in the present,
To create what can be a great future.
Treasure and keep memories,
For the sake of Life.

Mattie J. T. Stepanek, from "About Memories" (Hyperion/VSP, 2002)

This book is a result of an unplanned project which has been ruled by fate. Neither Brian Wybourne nor I were even thinking about writing a book. Being deeply involved in day-after-day fascinating research, we simply did not have time. There was no time to slow down, collect thoughts and results of our investigations, and to prepare the manuscript, since too many revolutionary and novel materials, properties and applications were continuously brought to the light, not only in the spectroscopy of lanthanides, the field of our research. For 13 years of our friendship and collaboration in the Physics Department, Nicolaus Copernicus University in Toruń, Poland, we were suggesting to each other to write a book, just teasing and treating this idea only as a joke, since we both realized the size of such a task and the nature of its responsibility. Brian knew this as an experienced author, who published his first book when I was still in high school, and I, without any experience, having been only the editor of special issues of scientific journals, was simply overwhelmed by such a project and even afraid to think about it.

2003 was very difficult year for me. I was a fresh cancer survivor who was trying to recover after my health crisis to gain the strength to go through a long and painful procedure of a promotion in my native institute. Unfortunately the procedure was based on the old-fashioned judgment of a "woman in physics" rather than on the evaluation of the scientific achievements of a physicist, and eventually I was denied promotion. Brian suffered with me, trying to give me all possible support. Being saved several times from health breakdowns, and being blessed in my life to be a scientist not for recognition but for the sake of fascination of my research, I have survived the final decision of my colleagues simply by focusing on my work. But I was crushed at the beginning of 2003, when my mother passed away after almost two years of struggles. My world shrank to the size of a dot. As the best therapy for my sorrow I started to teach at Vanderbilt University in Nashville (I have held an appointment in the Department of Chemistry since 1994) where a group of very talented, hard working and nice students from the Graduate School of the Physics Department helped me to go through the toughest time.

From Nashville I was keeping vivid contact with Brian Wybourne, and thanks to e-mail we were able to discuss science, exchange results and opinions, share the joy of teaching and also talk about our students. However I must admit that very often I was missing just chatting with Brian, the atmosphere of which cannot be reproduced by any electronic medium.

The first week of November 2003 I spent in Toruń to follow the Polish tradition of celebrating All Saints Day and the Day of the Dead. Because of this special holiday there was not enough time to go through all the interesting subjects and issues with Brian that we had kept for some months to talk about directly. After my return to Nashville I received an e-mail message from Brian, which was his reply to my question about the elegance of his lecture notes we discussed while in Toruń. Brian's letter was a reaction to the joke I made, that in spite of the promises he had given to himself, it looks as though he was getting ready to write a monograph.

Dear Lidia,

Pleased to know you managed to download the notes - I thought you might have trouble with postscript. I have no plans to publish - I attached the copyright symbol to avoid [some colleagues]! *I prepared the notes for my class but Andrjez (?) was the only survivor. He has been good at going through them and finding corrections hence I also gave him a complete copy. To complete the notes we need radial integrals etc to give some examples - not my forte! As a result the notes are left hanging. I don't think I have the energy to bring the notes to a publishable form. For that I would need a young dynamic collaborator who could knock them into shape. I understand that there is a person called Smentek who would fit the role and would, of course, become the first author. Maybe KBN could support such a project?*

Regards Brian

I have to explain that the student mentioned in this letter (his name is misspelt above, which is not difficult to do for a foreigner) is my graduate student. This is Andrzej Kędziorski, whom I left going to Nashville under the wings, guidance and care of one of the founders of the field of spectroscopy of rare earths, the outstanding and widely known Professor Brian Wybourne.

The tone of Brian's letter made me nervous especially when realizing that possibly we were approaching a dangerous point of talking seriously about writing a monograph. Being called to the blackboard by the Master, when the first astonishment went away, I was trying to imagine what it would have meant if we had indeed decided to write a book. We had co-authored several papers, we had spent hours on discussions, explaining to each other various issues and also arguing with each other. I knew that due to Brian's extensive knowledge and calm personality, writing a monograph with him would be not only a creative activity, but also it would provide pleasure and satisfaction. At the same time however I remembered so well Brian's negative attitude to writing a long manuscript and all the worries about problems he was aware of as a former author of several books. In order not to despair I was repeating to myself that it must be a joke....

It was not a joke. When I received the next e-mail message, I sensed that the idea had become a serious matter, but still the plan had all the aspects of a project for the future. At that moment in fact I was planning to go back to Toruń for an extended period of time, if my health would permit, and this move would make it easier to work together, if the decision was really made. Brian wrote:

Subj: Re: It is not easy....
Date: 11/12/03 9:21:04 AM Central Standard Time
From: bgw at phys.uni.torun.pl
To: mailto:Smentek1 at aol.com
Sent from the Internet (Details)
Dear Lidia,

I was pleased with Andrzej today. This morning he brought along two Masters students, one from Koepke and one from Wojtowicz, to my "lecture" on Spectro..... As Andrzej has up to today been the only student and I have simply taken the opportunity to talk with Andrzej I was totally unprepared. But it went well. I first got them to tell me what their thesis topics were, with Andrzej occasionally helping with English. One is working on Ce^{3+} and the other on energy transfer between Cr^{3+} and Nd^{3+} so we discussed the ground states of the ions - they didn't know the site symmetries so next week they are supposed to tell me. Anyhow it looks like next week I will discuss $3d^N$ and $4f^N$ and weak and strong crystal fields.

I didn't go to Jacek's group meeting though I did discuss some things with Jacek. I think his main thrust will be expecting everyone to compete for KBN grants etc with those who are producing being most favoured.

I said goodbye to Jacek as he was going to Warsaw after that to join Marta and fly to Katmandu.(...)

Unfortunately I did not have time to discuss a number of topics with Jacek. He had mentioned that KBN had a special fund for anyone writing a book. You mentioned Rudzikas's conference. I looked it up on the web. Unfortunately it was too late for me to discuss with Jacek. Possibly my KBN grant could cover your trip. However I need someone to check the status of my KBN grant. Clearly the grant must be totally expended by October next year.

I have no deadline for a completed manuscript. I had not thought of publishing until you raised the question. Maybe Springer Lecture Notes Series or locally like Physics as a Journey. I would expect to share authorship with my usual preference for alphabetical order. It would need a careful going through with you doing radial integrals etc etc. No rush. I'll see how I get on with the new students.

−4C this morning. Two Christmas cakes finished and I can smell the bread cooking! Too cloudy on Saturday to see the Lunar eclipse. I don't know how Copernicus became interested in such things. Likewise I have not seen Aurora Borealis which I understand was even seen in Florida!

Regards Brian

This was the last letter I received from Brian, ever; he was struck down by a stroke, and after ten days he passed away on November 26, 2003. These letters are like Brian's last will and testament left as instructions for what to do in the case of the worst. Therefore I did not have any choice but to start to think about the work which was dropped on my shoulders by fate. I wanted to be able to say what A. Einstein has said: *"Now he has departed from this strange world a little ahead of me. That means nothing. People like us, who believe in* **physics***, know that the distinction between past, present, and future is only a stubbornly persistent illusion"*; but I could not.

Brian left me alone with the task of adopting his lecture notes to the form of a monograph, of deriving new expressions and calculating radial integrals to continue our research and extend the knowledge about the f-electron systems, of deriving conclusions and presenting the future prospects of our field. I am left alone with the responsibility for a **BOOK**, we agreed to work on together...It is difficult, but *"Obstacles cannot crush me. Every obstacle yields to stern resolve. He who is fixed to a star does not change his mind"* (Leonardo da Vinci).

Here I am presenting a completed project, our book, hoping that Brian would be proud of it.

Since 2003 Andrzej has developed his skills and has performed interesting investigations that hopefully will be collected soon as his PhD thesis. Unfortunately, there is no Professor Wybourne around to advise, criticize, help with English and, most of all, judge the scientific merit of his work. Instead, in order to have a closure of Andrzej's lucky chapter of the beginning of his career, when he was given a chance to learn from a famous Professor about the beauty and the secrets of Nature, I am presenting here some results of his work. The numerical results and graphical illustration of the developed formalism that are presented in chapter 21 is a report we both, Andrzej and I, owe to Brian.

In the letters presented here Brian mentioned the possibility of getting financial support from the Science Foundation of Poland (KBN) for the project of writing a monograph; I did not apply for this aid. Instead I was awarded a one year stipend from the Nicolaus Copernicus University in Toruń which I would like to acknowledge. The financial aspect was not so important as the acceptance of my project that was granted by the authorities of the University together with the stipend; I do appreciate this moral support.

During the 13 years Brian lived in Toruń we shared professional and personal experiences. Those gained by him in his native New Zealand and exotic Poland; and my own, which I gained in my native Poland where I live with the imagination about exotic New Zealand. And today, on Brian's birthday, when I am facing a challenge of presenting the results of my work to a wide audience, I can hear his voice quoting Victoria Holt:

"Never regret. If it's good, it's wonderful. If it's bad, it's experience".

If so, having a positive attitude to the result of my mission, that is created by these words, I may follow the advice of Leonardo da Vinci:

"Every now and then go away, have a little relaxation, for when you come back to your work your judgment will be surer since to remain constantly at work will cause you to lose power of judgment. Go some distance away because then the work appears smaller, and more of it can be taken in at a glance, and a lack of harmony or portion is more readily seen".

Lidia Smentek
March 5th, 2006

Preface

The monograph is devoted to the theoretical description of the spectroscopic properties of the rare earth ions doped in various materials. These systems revolutionize modern science and technology at all possible scales, *nano, micro and macro*. The tensor operators and the concepts of the Racah algebra are the language of the presentation. The level of complexity in the course of this monograph begins with the basic knowledge of the theory of nuclei and theory of angular momentum, develops through the standard description of the fine and magnetic hyperfine interactions, and their impact upon the energy structure of the lanthanide ions, to culminate in the advanced description of $f \longleftrightarrow f$ electric and magnetic dipole transitions, including sensitized luminescence and its application in the detection and treatment of cancer in various tissues.

The standard description is understood as a model in which the interactions via a certain physical mechanism are limited to those within the $4f$ shell. The extension of the approach, which is analyzed through the book in the case of various physical interactions, includes the inter-shell interactions. The inclusion of the perturbing impact of the excited configurations breaks the limitations of the single configuration approximation. It leads very often not only to a better accuracy of the theoretical model, but also introduces new selection rules, which shed new light upon the understanding of the observed phenomena.

Thanks to the screening of the electrons of the optically active $4f$ shell by the closed shells of the $5s$ and $5p$ symmetry, the theory of the spectroscopic properties of the lanthanides in crystals is based on perturbation theory. In practical applications, the zeroth order problem describes the free ion within the single configuration approximation. The influence of important physical mechanisms upon the energy or the transition probability is included in a perturbative way taking into account corrections to the wave functions caused by various perturbations. The list of the perturbing operators included in the theoretical approaches presented in the monograph contains the crystal field potential, electron correlation effects, spin-orbit interaction and magnetic and electric multipole hyperfine interactions. In order to extend this list of various interactions, which are in most cases limited to interactions within the orbital space, an effectively relativistic model of various transitions is also presented.

Almost half of the book evolved from lectures prepared by Brian G. Wybourne for the graduate students of the Institute of Physics, Nicolaus Copernicus University in Toruń, Poland in 2002/2003. Wherever possible, the lecture format of presentation is kept (although the monograph is not prepared as a textbook). Therefore useful MAPLE routines are provided, several derivations of new expressions are presented and supported by numerical results, and some of the chapters are concluded by additional problems left for the reader to exercise. The advanced level of the course addressed to graduate students is elevated by the discussion of the results presented in the second part of the book; these results were not published previously. Among these new issues of the optical spectroscopy of lanthanides in various materials are the hyperfine induced $f \longleftrightarrow f$ electric dipole transitions. In their description the forcing role of the crystal field potential of the standard Judd-Ofelt theory is played by the electric dipole hyperfine interactions. This new source of contributions to the transition amplitude provides a potential explanation of the hypersensitivity of some electric dipole transitions to the environment. This is the property that causes serious problems with the theoretical reproduction of the intensities, when based on the standard parametrization of $f-$ spectra. As a consequence of this discussion, and also as a result of the analysis of the relativistic approach to the $f \longleftrightarrow f$ transitions, a new parametrization scheme is introduced. Namely, in order to include all possibly important physical interactions within the spin-orbital space when using the fitting procedure, the unit tensor operators of the Judd-Ofelt theory are replaced by the double tensor operators.

The chapter devoted to the radial integrals is unique in the sense that its contents may not be found in the literature of the field. This chapter is mainly addressed to all those who are interested in the direct calculations of the amplitude of the $f \longleftrightarrow f$ transitions. For this purpose the values of all radial integrals of the theory of $f-$spectra are presented for all lanthanide ions. They are evaluated within the perturbed function approach, due to which a complete radial basis sets of one electron functions of given symmetry are taken into account. The tables include the radial integrals of the Judd-Ofelt effective operators, their values modified by the electron correlation effects taken into account at the third order, as well as those associated with the effective operators originating from intra-shell and inter-shell interactions via the spin-orbit and hyperfine interactions.

As an illustration of the formalism developed throughout the monograph to describe the subtleties of the electronic structure of lanthanide ions, its very attractive and promising application is presented in the last chapter. The particular case of sensitized luminescence of the organic chelates with the lanthanide ion placed in the center of the cage that are the tissue selective markers of the cancerous cells are described. The architecture of the cage, its symmetry, and in turn, their impact upon the efficiency of the energy transfer and consequently the luminescence, require theoretical techniques developed and presented in the monograph. To construct adequate and reliable theoretical models, and through them to gain the understanding of the physical reality of these materials, is a challenging task. This research is also important in the light of existing strong demand for non-invasive diagnostic aids for early detection of cancer and for powerful tools to cure the attacked tissue.

I hope that the reader finds the presentation clear and the discussion interesting and useful for further research devoted to the lanthanides and their amazing properties.

Lidia Smentek

About the Authors

Brian Garner Wybourne, a fellow of the Royal Society of New Zealand, was an outstanding physicist. A native of New Zealand, he was widely recognized throughout the scientific world. Just after completing his PhD at the University of Canterbury in Christchurch, New Zealand, he moved in 1960 to Johns Hopkins University in Baltimore. In 1963 he continued his research at Argonne National Laboratory. Three years later he returned to his home country as the youngest professor of physics ever appointed to a chair of the Physics Department of Canterbury University in Christchurch, to become its Head in 1982, until 1989. In 1991 he arrived in Toruń, Poland, as a Visiting Professor at the Institute of Physics, Nicolaus Copernicus University, and he stayed there for 13 years teaching, doing research, collaborating with many in the world, and enjoying the gothic architecture of the medieval town, the birthplace of Copernicus. He died on November 26, 2003 leaving a rich legacy to fellow-physicists and mathematicians, as well as to all *secular* people who are honest, ready to accept the diversities of the world, and are tolerant of all differences.

His interests were broad and deep. On his still existing personal web page he classified the fields of his research: "Generally: applications of symmetry in physics. Specifically: spectroscopic properties of lanthanides and actinides, mathematical and physical applications of Lie groups, theory of symmetric functions, computing properties of groups, interacting boson model of nuclei, mesoscopic and many-body systems".

Brian G. Wybourne is the author of five books and of almost 190 scientific papers published in the best journals in the world. He is one of the pioneers and successful founders of the spectroscopy of rare earths. At the age of 30 he wrote the famous *Spectroscopic Properties of Rare Earths* (Wiley-Interscience, 1965), and then, almost every five years, a new book appeared, namely *Symmetry Principles and Atomic Spectroscopy* (Wiley-Interscience,1970) and *Classical Groups for Physicists* (Wiley-Interscience, 1974).

Lidia Smentek is a physicist. Educated at Nicolaus Copernicus University in Toruń, Poland, she became a faculty member of the Institute of Physics just after graduation in 1971, and since 1994 is a professor of chemistry at Vanderbilt University, Nashville, Tennessee. She had the privilege to collaborate with Brian Wybourne, and the pleasure of being his friend. She is the author of about 80 scientific papers, the majority of which is devoted to the theory of f-spectra and unique ab-initio type numerical calculations performed for various spectroscopic properties of lanthanides. Recently she participates in the inter-disciplinary research (funded by NSF) on the tissue selective organic chelates containing the lanthanide ions, which are being designed as probes for detection of cancer. In the past she has served as a guest editor of special issues of Molecular Physics (three times) and International Journal of Quantum Chemistry. In 2005 she edited a book, Brian Garner Wybourne; Memories and Memoirs (Adam Marszałek Pub., 2005), to commemorate Brian's scientific achievements and personal impact he had upon many in various places in the world.

1 Basic Facts of Nuclei

...when he imagined his education was completed, it had in fact not commenced; and that, although he had been at a public school and a university, he in fact knew nothing. To be conscious that you are ignorant is a great step to knowledge.

Benjamin Disraeli, *Sybil or The Two Nations* (1844)

The concept of quantum degeneracy and its controlled lifting by magnetic and electric fields is perhaps the most technologically important and practical development of the past century. It has lead to the development of lasers, NMR imaging, modern telecommunications, the realization of Bose-Einstein condensation, and the potential realization of quantum computing. These applications have required a detailed understanding of the quantum theory of angular momentum in all its manifestations, starting with the angular momentum of nuclear states, the angular momentum of electronic states and of the coupling of angular momentum of nuclear with electronic states.

1.1 NUCLEONS

The basic building blocks of nuclei (here we ignore the quark substructure of the nucleons) are neutrons and protons. Both are spin $\frac{1}{2}$ particles with positive intrinsic parity, i.e. $J_p = \frac{1}{2}^+$. Following Heisenberg[1], the proton and neutron can be regarded as different charge states of the *nucleon*. The respective masses are (we will generally put $c = 1$)

$$m_p = 938.3 MeV \qquad m_n = 939.6 MeV \tag{1-1}$$

In free space the neutron is unstable with a half-life of $t_{1/2} = 614.6s$ whereas the proton appears to be stable with $t_{1/2} > 10^{33} y$. Within the nucleus, as a consequence of the Pauli exclusion principle, the proton *and* the neutron are stable.

Remarkably, the proton *and* the neutron both possess magnetic moments.

$$\mu_p = 2.7928 \qquad \mu_n = -1.9128 \tag{1-2}$$

where the units are the Bohr nuclear magneton defined as

$$\mu_{B_N} = \frac{e\hbar}{2m_p} \tag{1-3}$$

An *isotope* of an element X having Z protons (Z is the *atomic number*) and N neutrons will be designated as

$$^A_Z X$$

where

$$A = N + Z \tag{1-4}$$

with A being the *nucleon number*.

The force between nucleons is, to a good approximation, charge independent and of a short range. Its origin is to be found in the quark model which we shall not explore here. The basic nuclear model we shall consider is the *nuclear shell model*. To a first approximation we can consider the nucleons as executing harmonic oscillations about the nuclear centre of mass and hence as nucleons in an isotropic 3-dimensional harmonic oscillator.

1.2   THE ISOTROPIC HARMONIC OSCILLATOR

Recall that the energy eigenvalues of a 3-dimensional isotropic harmonic oscillator potential containing a single particle are given by

$$E_n = (n + \tfrac{3}{2})\hbar\omega \quad n = 0, 1, 2, \ldots \tag{1-5}$$

corresponding to an infinite series of equally spaced levels. The $n-$th level has an orbital degeneracy of

$$\frac{(n + 1)(n + 2)}{2} \tag{1-6}$$

These are precisely the dimensions of the symmetric irreducible representations $\{n\}$ of the special unitary group $SU(3)$. The $n-$th level is associated with orbitals having the angular momentum quantum number ℓ such that

$$\ell = n, n - 2, \ldots, \begin{pmatrix} 1 \\ 0 \end{pmatrix} \tag{1-7}$$

Given that the nucleons have *even* intrinsic parity the states associated with the $n-$th level are all of the *same* parity which is *even* or *odd* as n is *even* or *odd*. The nucleons have spin $\frac{1}{2}$ and hence each orbital ℓ has a spin-orbital degeneracy of $4\ell + 2$. It is common in nuclear physics to prefix the orbital quantum number with the number of nodes in the single particle wavefunction. Thus the orbitals associated with $n-$level are in the sequence

$$1s; 1p; 2s, 1d; 2p, 1f; 3s, 2d, 1g; \ldots \tag{1-8}$$

Since nucleons have spin $\frac{1}{2}$, they are *fermions* and hence must be associated with wavefunctions that are *totally antisymmetric*. This means that in building up many-nucleon states the Pauli exclusion principle must be followed. Many-nucleon states may be constructed by filling the spin-orbitals with neutrons and protons up to their maximum allowed occupancy. The building-up principle is very similar to that for periodic table for atoms except one fills neutron and proton orbitals separately to construct nuclei with given A, N, Z numbers.

1.3 MAGIC NUCLEI NUMBERS

Nuclei in which the proton and neutron numbers Z,N belong to the *magic numbers*

$$\text{Nuclei Magic Numbers} \quad 2, 8, 20, 28, 50, 82, 126 \tag{1-9}$$

tend to be exceptionally stable and evidently associated with the closure of shells. The magic numbers 2, 8, 20 correspond to the closure of the shells corresponding to the first, second and third levels of an isotropic 3-dimensional harmonic oscillator. But closure of the fourth level would give the fourth magic number as 40 rather than 28. This constituted a puzzle until Goeppert Mayer introduced the spin-orbit interaction into the nuclear shell model. In her case the spin-orbit interaction has the *opposite sign* to the conventional spin-orbit interaction of electrons. This means, for example, that in the $1p$ shell the $1p_{3/2}$ level is *below* the $1p_{1/2}$ level whereas in atoms one has the opposite ordering. The effect of introducing the spin-orbit interaction is also to partially lift the single particle energy degeneracies so that, apart from the case of $s-$states each orbital ℓ becomes characterized by a total single particle angular momentum

$$j_{\pm} = \ell \pm s \tag{1-10}$$

Henceforth, let us use n as the *nodal* quantum number rather than as the harmonic oscillator level number so that a given spin-orbital is designated by the single particle quantum numbers $n\ell jm$. For a single nucleon moving in a nucleus we write the spin-orbit interaction as

$$V_{s.o} = \zeta(r)_{n\ell}(\mathbf{s} \cdot \ell) \tag{1-11}$$

where $\zeta(r)_{n\ell}$ is the *spin-orbit coupling constant* which is a *radial* function dependent on the nature of the nuclear field and upon the quantum numbers $n\ell$. For a single nucleon

$$(\mathbf{s} \cdot \ell) = \tfrac{1}{2}\left[j(j+1) - \ell(\ell+1) - \tfrac{3}{4}\right] \tag{1-12}$$

The energy separation between the two components of a spin-orbit split doublet characterized by the quantum numbers $n\ell$ becomes

$$\Delta E_{n\ell} = \left(\ell + \tfrac{1}{2}\right)\zeta(r)_{n\ell} \tag{1-13}$$

Thus the level with j_{+} is *lower* than the level with j_{-}. Furthermore, it is a practical observation that states with higher values of ℓ have larger doublet splittings.

Even with the introduction of the spin-orbit interaction the single nucleon degeneracy is only partially lifted. The degeneracy associated with the isotropic harmonic oscillators is partially lifted so that states of a given harmonic oscillator level are no longer degenerate with respect to ℓ while each set of states associated with a particular orbital angular momentum ℓ is split as a doublet of degenerate states labeled by the quantum numbers $n\ell jm$. The degeneracy with respect to the *total angular momentum projection quantum number m* remains, and hence each level with total angular momentum j is $(2j + 1)$−fold degenerate. The spin-orbit splitting leads to subshells with a given j−level accommodating up to $2j + 1$ protons or neutrons.

1.4   NUCLEAR PAIRING INTERACTIONS

It is remarkable that nuclei having *even* numbers of protons *and* neutrons, so called *even-even* nuclei, are always found to have zero nuclear angular momentum, i.e. $I = 0$. In nuclear physics the total angular momentum $J = L + S$ is nevertheless usually referred to as the *nuclear spin* and designated by the letter I. So called *odd-even* and *even-odd* nuclei always have *half-integer* nuclear spin while *odd-odd* nuclei always have *integer* nuclear spin. It would appear that *even* numbers of neutrons or protons *pair* to produce a lowest energy state, which has nuclear spin $I = 0$. Indeed, Racah, showed that a strong short range nuclear force, such as a delta type force, leads to such a pairing. It is this feature that is the key to predicting the nuclear spin of the ground states of nuclei. If there is an even number of protons or neutrons in a given orbital then those orbitals make no contribution to the nuclear spin of the ground state. Furthermore, there is no nuclear spin contribution from closed shells.

1.4.1 SENIORITY AND PAIRING INTERACTIONS

Racah introduced the concept of *seniority* in both atomic and nuclear physics (and indeed also in superconductivity where pairing is also important). He showed that for a strong pairing interaction such as occur in nuclei states of lowest seniority number v lie lowest. Thus in a configuration of identical nucleons, j^N, the integer, $[\frac{N-v}{2}]$, corresponds to the number of pairs of particles that are coupled to zero angular momentum where

$$v = \begin{cases} 0, 2, \ldots, N & N \ even \\ 1, 3, \ldots, N & N \ odd \end{cases} \qquad (1\text{-}14)$$

This has the consequence that in a configuration j^N if N is *even* then the lowest state will have zero angular momentum, whereas if N is *odd* the angular momentum of the lowest state will be $J = j$.

In the case of atoms however, where there is Coulomb repulsion between pairs of electrons, states of maximal seniority lie lowest and hence in some respects nuclear states are simpler than electronic states!

The angular momentum states J and seniority numbers for the identical particle configurations j^N are given for $j = \frac{1}{2}, \frac{3}{2}, \ldots, \frac{7}{2}$ in Table 1-1. For $j = \frac{5}{2}, \frac{7}{2}$ we list just the states up to $N = j + \frac{1}{2}$.

TABLE 1-1

Angular Momentum J and Seniority Numbers for Some j^N Configurations

j^N	J	v	$\frac{N-v}{2}$
$(\frac{1}{2})^0$	0	0	0
$(\frac{1}{2})^1$	$\frac{1}{2}$	1	0
$(\frac{1}{2})^2$	0	0	1
$(\frac{3}{2})^0$	0	0	0
$(\frac{3}{2})^1$	$\frac{3}{2}$	1	0
$(\frac{3}{2})^2$	0	0	1
	2	2	0
$(\frac{3}{2})^3$	$\frac{3}{2}$	1	1
$(\frac{3}{2})^4$	0	0	2
$(\frac{5}{2})^0$	0	0	0
$(\frac{5}{2})^1$	$\frac{5}{2}$	1	0
$(\frac{5}{2})^2$	0	0	1
	2, 4	2	0
$(\frac{5}{2})^3$	$\frac{5}{2}$	1	1
	$\frac{3}{2}, \frac{9}{2}$	3	0
$(\frac{7}{2})^0$	0	0	0
$(\frac{7}{2})^1$	$\frac{7}{2}$	1	0
$(\frac{7}{2})^2$	0	0	1
	2, 4, 6	2	0
$(\frac{7}{2})^3$	$\frac{7}{2}$	1	1
	$\frac{3}{2}, \frac{5}{2}, \frac{9}{2}, \frac{11}{2}, \frac{15}{2}$	3	0
$(\frac{7}{2})^4$	0	0	2
	2, 4, 6	2	1
	2, 4, 5, 8	4	0

1.5 NUCLEAR SPIN OF NUCLEI GROUND STATES

In atomic physics interest is almost entirely restricted to atomic effects involving nuclei in their ground states. A given isotope is characterized by the number of neutrons, N, and number of protons, Z. Starting with Table 1-1 giving the ordering of the $n\ell j$ quantum numbers for a single nucleon in an isotropic three-dimensional harmonic oscillator potential together with the spin-orbit interaction, we can determine the nuclear spin of the ground states of most nuclei. Let us now consider some examples.

1.5.1   NUCLEAR SPIN OF H AND He ISOTOPES

Hydrogen has three well-known isotopes, hydrogen, $_1^1H$, deuterium, $_1^2H$, and tritium, $_1^3H$. Hydrogen involves a single proton that may be assigned to the $(1s_{1/2})_p^1$ config-uration and hence $_1^1H$ has a nuclear spin $I = \frac{1}{2}$. Deuterium has a single proton and neutron, and hence the nuclear ground state configuration is $(1s_{1/2})_p^1(1s_{1/2})_n^1$. The nu-clear spin results from the coupling of two $\frac{1}{2}$ spins to produce $I = 0, 1$. Experiment shows that the ground state has $I = 1$. Finally, tritium has a single proton and two neutrons, and therefore the nuclear configuration $(1s_{1/2})_p^1(1s_{1/2})_n^2$. The two neutrons close the $1s_{1/2}$ shell and hence make no contribution to the nuclear spin, and thus the nuclear spin of the tritium nucleus is $I = \frac{1}{2}$, the same as for hydrogen. Note that in each case the ground state has *even* parity.

The two principal isotopes of helium are $_2^4He$ and $_2^3He$. For the more abundant isotope, $_2^4He$, we have the nuclear configuration $(1s_{1/2})_p^2(1s_{1/2})_n^2$, and both shells are closed. Hence the nuclear spin is $I = 0$, whereas for $_2^3He$ we have the nuclear configuration $(1s_{1/2})_p^2(1s_{1/2})_n^1$ and a nuclear spin of $I = \frac{1}{2}$.

1.5.2   SILICON ISOTOPES

Silicon has three stable isotopes, $_{14}^{28}Si$ (92.23%), $_{14}^{29}Si$ (4.67%), $_{14}^{30}Si$ (3.10%). Two of the isotopes are *even-even* nuclei, and both have nuclear spin and parity $I = 0^+$. All three isotopes have $Z = 14$, and therefore the 14 protons occur in the proton configuration $(1s_{1/2}^2 1p_{3/2}^4 1p_{1/2}^2 1d_{5/2}^6)$. In practice we normally omit all closed shells except for the highest and thus write the configuration as simply $(1d_{5/2}^6)_p$. Since the proton number is *even*, it follows that the protons make no contribution to the nuclear spin. Fourteen of the neutrons go into the same type of configuration as the protons with the fifteenth neutron occupying the $(2s_{1/2})_n$ orbital. Thus the nuclear spin of $_{14}^{29}Si$ is $I = \frac{1}{2}^+$.

The above observations are of significance in considering the possibility of us-ing silicon in quantum computing. Only the $_{14}^{29}Si$ nuclei will respond to an external magnetic field. Isotopically pure $_{14}^{28}Si$ with nuclear spin $I^p = 0^+$ gives no response to magnetic fields. Phosphorus has one stable isotope $_{15}^{31}P$ with one unpaired proton in the $2s_{1/2}$ shell, and hence nuclear spin $I^p = \frac{1}{2}^+$. Noting these facts, Kane[2] has proposed building a quantum computer using $_{14}^{28}Si$ doped with $_{15}^{31}P$.

1.5.3   RUBIDIUM ISOTOPES

Rubidium has two stable isotopes, $_{37}^{85}Rb$ (72.16%), $_{37}^{87}Rb$ (27.83%) (actually the isotope $_{37}^{87}Rb$ is slightly unstable with a half-life of $t_{1/2} = 4.75 \times 10^{10} y$). The two nuclei are *odd-even* with the neutrons making no contribution to the nuclear spin. The first 28 protons go into filling closed shells leaving a further nine protons to be distributed among the $2p_{3/2}$ and $1f_{5/2}$ orbitals. Eight of the protons will pair to produce no contribution to the nuclear spin leaving one unpaired proton. Experi-mentally it is found that for $_{37}^{85}Rb$ the nuclear spin is $I^p = \frac{5}{2}^-$ while for $_{37}^{87}Rb$

$I^p = \frac{3}{2}^-$. These results show some of the limitations of the simple nuclear shell model, but as always, final appeal must be made to experiment. For an excellent data base on properties of isotopes go to http://ie.lbl.gov/education/isotopes.htm or to http://www.webelements.com/webelements/elements/text/periodic-table/isot.html; the latter gives the Periodic Table and nuclear magnetic moments.

The rubidium isotopes play a key role in studies of *Bose-Einstein Condensation (BEC)*. Rubidium atoms behave as *bosons*, since their nuclear spins are half-integer. However, the number of their electrons is *odd*, and therefore the net electron angular momentum J is necessarily half-integer. The total angular momentum of the atom F comprises the vector addition of the nuclear spin I and electronic angular momentum J such that

$$\mathbf{F} = \mathbf{J} + \mathbf{I} \tag{1-15}$$

F is thus necessarily an integer and the rubidium atoms behave as *bosons*.

REFERENCES

1. Wybourne B G, (1965) *Spectroscopic Properties of Rare Earths* Wiley - Interscience Publication, John Wiley and Sons Inc., New York.
2. Kane B E, (1998) A silicon-based nuclear spin quantum computer. *Nature* **393**, 133.

2 Notes on the Quantum Theory of Angular Momentum

Pure mathematics is on the whole distinctly more useful than applied. For what is useful above all is technique, and mathematical technique is taught mainly through pure mathematics

G H Hardy

In the language, and within the understanding, of the quantum mechanics the symmetry of a many particle system is determined by the transformation properties of the Hamiltonian. Among the collection of all transformations those that commute with the Hamiltonian are *symmetry operations*. The commutation relations mean that the Hamiltonian remains invariant under the action of the symmetry operations. Keeping in mind our future applications of the tools of quantum mechanics for evaluation of various matrix elements, in the case of free atoms and ions, the most important are the symmetry operations of the rotation group in three dimensions, $SO(3)$. The angular momentum operators are the generators of all the rotations in three dimensional space. As a consequence, the Hamiltonian of spherically symmetric systems have to commute with the generators of the symmetry operators. In this particular case it means that Hamiltonian has to commute with the angular momentum operators. Consequently, the basic property of the hermitan operators states that if two operators commute, they have a common set of the eigenfunctions (but not the same eigenfunctions!) and simultaneously measured eigenvalues. As a result the energy states carry also information about the eigenvalues of angular momentum operators (only of these that commute with the Hamiltonian). These transformation properties, distinguished by the eigenvalues of appropriate angular momentum operators, are the formal source of the quantum numbers that identify the energy states of a many electron system. At the same time, the spherical harmonics form the basis sets for the irreducible representations of the three dimensional rotation group. In the case of the systems of symmetry lower than spherical, as in the case of crystals, the transformation properties are defined by the point groups, and the identification of the energy states is also based on the commutation relations of the Hamiltonian and the symmetry operators. In such cases, instead of quantum numbers, the irreducible representations of a given group directly identify the energy levels.

We review some basic aspects of the quantum theory of angular momentum, which is needed in the further discussion. In making practical calculations we must ultimately be able to evaluate the matrix elements of interactions in suitable angular momenta bases.

2.1   COUPLING AND UNCOUPLING OF ANGULAR MOMENTA

Consider the components, J_x, J_y and J_z of the angular momentum vector $\mathbf{J}$ that satisfy the commutation relations (we shall normally put $\hbar = 1$),

$$[J_x, J_y] = iJ_z, \quad [J_y, J_z] = iJ_x, \quad [J_z, J_x] = iJ_y \tag{2-1}$$

States that are simultaneous eigenfunctions of $\mathbf{J}^2$ and J_z are designated in Dirac's ket notation as $|JM\rangle$. Thus the standard angular momentum operator relations have the form

$$\mathbf{J}^2|JM\rangle = J(J+1)|JM\rangle \tag{2-2a}$$

$$J_z|JM\rangle = M|JM\rangle \tag{2-2b}$$

$$J_\pm|JM\rangle = [J(J+1) - M(M\pm 1)]^{\frac{1}{2}}|JM\pm 1\rangle \tag{2-2c}$$

where

$$J_\pm = J_x \pm iJ_y \tag{2-3}$$

are the usual angular momentum ladder operators.

For a given eigenvalue J there are $2J+1$ values of the M quantum number

$$M = -J, \; J-1, \ldots, \; J-1, \; J \tag{2-4}$$

and for the states with the maximum and minimum values of M

$$J_+|JJ\rangle = 0 \qquad J_-|J-J\rangle = 0 \tag{2-5}$$

It is a common problem in the quantum theory of angular momentum to couple together two ket states, say $|j_1m_1\rangle$, $|j_2m_2\rangle$ to produce coupled states $|(j_1j_2)jm\rangle$. Thus to have

$$|j_1m_1\rangle|j_2m_2\rangle = \sum_{j,m}\langle jm|m_1m_2\rangle|(j_1j_2)jm\rangle \tag{2-6}$$

or inversely to uncouple coupled states, the *coupling coefficients* or *Clebsch-Gordan coefficients* $\langle j_1m_1j_2m_2|j_1j_2jm\rangle$ are introduced. They represent the elements of a unitary transformation that couples the uncoupled states $|j_1m_1\rangle|j_2m_2\rangle$ to produce the coupled states $|(j_1j_2)jm\rangle$,

$$|(j_1j_2)jm\rangle = \sum_{m_1,m_2}\langle j_1m_1j_2m_2|j_1j_2jm\rangle|j_1m_1\rangle|j_2m_2\rangle \tag{2-7}$$

Such transformations arise, for example in relating basis states in the $|SM_SLM_L\rangle$ scheme to the coupled basis states $|SLJM\rangle$, where $M = M_S + M_L$. Thus,

$$|SLJM\rangle = \sum_{M_S,M_L}\langle M_SM_L|SLJM\rangle|SM_SLM_L\rangle \tag{2-8}$$

Note that we shall often abbreviate the Clebsch-Gordan coefficient $\langle j_1m_1j_2m_2|j_1j_2jm\rangle$ to just $\langle m_1m_2|j_1j_2jm\rangle$.

As an example we analyze the triply ionized thulium Tm^{3+}, which has as its ground state the spectroscopic term $4f^{12}\,^3H_6$. This means that the state is described by the quantum numbers $S = 1$, $L = 5$, and $J = 6$. The state is $(2J + 1) = 13$-fold degenerate, and the degenerate states are distinguished by the quantum number M. These states could be described by the kets $|SLJM\rangle$ in the so-called *Russell-Saunders* or LS-coupling, where S is coupled to L to give a total angular momentum J. Alternatively the states could be described by the kets defined in an uncoupled momenta scheme, $|SM_SLM_L\rangle$. These two sets of states correspond to two different bases that are linked by the Clebsch-Gordan coupling coefficients as in (2-8). For maximal M we expect for the ground state that

$$|(1, 5)66\rangle \equiv e^{i\phi}|(1, 1)(5, 5)6\rangle \tag{2-9}$$

where the left-hand ket is in the $|(SL)JM\rangle$ scheme, and the right-hand in the $|(SM_S)(LM_L)\rangle$ scheme, and $e^{i\phi}$ is a phase factor, which we choose as unity. In fact, at this point the choice of the phase factor is arbitrary, since the wave functions, by themselves, do not have any physical interpretation, and only their squared modulus, that cancels all signs, does. However, any phase factor convention once chosen has to be followed throughout the course of the calculations.

The other states in the $SLJM$ scheme that are assigned to the remaining values of the quantum numbers are expressed as linear combinations of those in the SM_SLM_L scheme. As an example we determine some of these linear combinations. First note that

$$J_\pm = L_\pm + S_\pm \tag{2-10}$$

Let us apply the lowering ladder operator to both sides of (2-9), using at the left-hand-side its form from the left hand-side of (2-10), and consequently, using its form from the right-hand-side of (2-10) to the right-hand-side of (2-9), namely

$$J_-|(1, 5)66\rangle = \sqrt{6(6 + 1) - 6(6 - 1)}|(1, 5)65\rangle = \sqrt{12}|(1, 5)65\rangle \tag{2-11a}$$

$$L_- + S_-|(1, 1)(5, 5)6\rangle = \sqrt{5(5 + 1) - 5(5 - 1)}|(1, 1)(5, 4)5\rangle$$
$$+ \sqrt{1(1 + 1) - 1(1 - 1)}|(1, 0)(5, 5)5\rangle$$
$$= \sqrt{10}|(1, 1)(5, 4)5\rangle + \sqrt{2}|(1, 0)(5, 5)5\rangle \tag{2-11b}$$

Equating (2-11a) and (2-11b) gives

$$|(1, 5)65\rangle = \frac{1}{\sqrt{6}}\left[\sqrt{5}|(1, 1)(5, 4)5\rangle + |(1, 0)(5, 5)5\rangle\right] \tag{2-12}$$

Using the ladder operators in the way presented in the example above it is possible to find the coefficients of all the other linear combinations that define the states of the $4f^{12}$ configuration of thulium ion. In order to make this procedure more straightforward, the subsequent application of (2-2c) to lower the values of the quantum numbers to desired values (in some cases the operators have to be applied several times), the Clebsch-Gordan coefficients are introduced. The concept of these coefficients as transformation matrices between coupled and uncoupled angular momenta schemes is presented in (2-6) and (2-7); their algebraic form is as follows

$$\langle m_1 m_2 | j_1 j_2 j m \rangle =$$

$$\delta_{m_1+m_2,m} \sqrt{\frac{(2j+1)(j_1+j_2-j)!\,(j_1-m_1)!\,(j_2-m_2)!\,(j+m)!\,(j-m)!}{(j_1+j_2+j+1)!\,(j+j_1-j_2)!\,(j-j_1+j_2)!\,(j_1+m_1)!\,(j_2+m_2)!}}$$

$$\times \sum_z (-1)^{j_1-m_1-z} \frac{(j_1+m_1+z)!\,(j+j_2-m_1-z)!}{z!\,(j-m-z)!\,(j_1-m_1-z)!\,(j_2-j+m_1+z)!} \qquad (2\text{-}13)$$

where the summation is performed over such values of z for which all the factorials are well defined (the arguments are positive).

2.2   THE 3j−SYMBOLS

While Clebsch-Gordan coefficients possesses considerable symmetry and from their structure it is possible to verify the coupling scheme of angular momenta, a more symmetrical object was defined by Wigner and is now commonly known as the $3j-$symbol or $3jm-$symbol. The $3j-$symbol is related to the Clebsch-Gordan coefficient by

$$\begin{pmatrix} j_1 & j_2 & j_3 \\ m_1 & m_2 & m_3 \end{pmatrix} = (-1)^{j_1-j_2-m_3} \frac{\langle m_1 m_2 | j_1 j_2 j_3 - m_3 \rangle}{\sqrt{(2j_3+1)}} \qquad (2\text{-}14)$$

The $3j-$symbol is invariant with respect to an *even* permutation of its columns, while for *odd* permutations of its columns is multiplied by a phase factor equal to the sum of the arguments in its top row,

$$\begin{pmatrix} j_1 & j_2 & j_3 \\ m_1 & m_2 & m_3 \end{pmatrix} = (-1)^{j_1+j_2+j_3} \begin{pmatrix} j_2 & j_1 & j_3 \\ m_2 & m_1 & m_3 \end{pmatrix} \qquad (2\text{-}15)$$

Furthermore, changing the sign of all three lower arguments results also in multiplication by a phase factor equal to the sum of the arguments in its top row,

$$\begin{pmatrix} j_1 & j_2 & j_3 \\ m_1 & m_2 & m_3 \end{pmatrix} = (-1)^{j_1+j_2+j_3} \begin{pmatrix} j_1 & j_2 & j_3 \\ -m_1 & -m_2 & -m_3 \end{pmatrix} \qquad (2\text{-}16)$$

A $3j-$symbol having all its m quantum numbers zero vanishes unless $j_1 + j_2 + j_3$ is *even*. Likewise any $3j-$symbol having two identical columns will vanish unless $j_1 + j_2 + j_3$ is *even*.

The unitarity property of the Clebsch-Gordan coefficients leads directly to the orthonormality conditions for the $3j-$symbols

$$\sum_{j_3,m_3} (2j_3+1) \begin{pmatrix} j_1 & j_2 & j_3 \\ m_1 & m_2 & m_3 \end{pmatrix} \begin{pmatrix} j_1 & j_2 & j_3 \\ m_1' & m_2' & m_3 \end{pmatrix} = \delta_{m_1,m_1'} \delta_{m_2,m_2'} \qquad (2\text{-}17a)$$

$$\sum_{m_1,m_2} \begin{pmatrix} j_1 & j_2 & j_3 \\ m_1 & m_2 & m_3 \end{pmatrix} \begin{pmatrix} j_1 & j_2 & j_3' \\ m_1 & m_2 & m_3' \end{pmatrix} = \frac{\delta_{j_3,j_3'} \delta_{m_3,m_3'}}{2j_3+1} \qquad (2\text{-}17b)$$

The values of the angular coupling coefficients are collected in various tables.

2.3 THE $6j$–SYMBOLS

The $6j$–symbol is defined by the relation

$$\langle(j_1 j_2)j_{12}, j_3; jm\,|\,j_1, (j_2 j_3)j_{23}; jm\rangle$$

$$= (-1)^{j_1+j_2+j_3+j}\sqrt{(2j_{12}+1)(2j_{23}+1)}\begin{Bmatrix} j_1 & j_2 & j_{12} \\ j_3 & j & j_{23} \end{Bmatrix} \qquad (2\text{-}18)$$

Similarly as in the case of the $3j$– symbols also here it is possible to analyze the coupling schemes of angular momenta on both sides of the transformation matrix. Indeed, it is seen that in general the $6j$– symbols connect two schemes of coupling of three momenta. The $6j$–symbol may be evaluated by first expressing it as a sum over a triple product of $3j$–symbols and then using the fact that the $6j$–symbol is independent of m to produce a sum involving a single variable to finally yield

$$\begin{Bmatrix} a & b & c \\ d & e & f \end{Bmatrix} = \sqrt{\Delta(abc)\Delta(aef)\Delta(dbf)\Delta(dec)} \times \sum_z (-1)^z (z+1)!$$

$$\times [(z-a-b-c)!(z-a-e-f)!(z-d-b-f)!$$
$$\times (z-d-e-c)!(a+b+d+e-z)!(b+c+e+f-z)!$$
$$\times (a+c+d+f-z)!]^{-1} \qquad (2\text{-}19)$$

where

$$\Delta(abc) = [(a+b-c)!(a-b+c)!(b+c-a)!/(a+b+c+1)!]^{\frac{1}{2}}$$

which also represents the triangular condition that has to be satisfied for its arguments. In particular, the $6j$–symbol vanishes unless the four triangular conditions portrayed below are satisfied (all four are represented by appropriate Δ's under the square root in the above expression),

$$\qquad (2\text{-}20)$$

where for example $|a-b| <= c <= a+b$.

The $6j$–symbol is invariant with respect to any interchange of columns and also with respect to the interchange of the upper and lower arguments of any two columns. The $6j$–symbols satisfy the orthogonality condition

$$\sum_{j_{12}}(2j_{12}+1)(2j_{23}+1)\begin{Bmatrix} j_3 & j & j_{12} \\ j_1 & j_2 & j_{23} \end{Bmatrix}\begin{Bmatrix} j_3 & j & j_{12} \\ j_1 & j_2 & j'_{23} \end{Bmatrix} = \delta_{j_{23},j'_{23}} \qquad (2\text{-}21)$$

2.4   THE $9j$–SYMBOL

The $6j$–symbol arose in discussing the coupling of three angular momenta. Clearly more complex nj–symbols arise for couplings that involve more than three angular momenta. The $9j$–symbol is defined as a transformation matrix between two coupling schemes of four angular momenta

$$\langle ((j_1 j_2) j_{12}, (j_3 j_4) j_{34}; j \,|\, (j_1 j_3) j_{13}, (j_2 j_4) j_{24}; j \rangle$$

$$= \sqrt{(2 j_{12} + 1)(2 j_{34} + 1)(2 j_{13} + 1)(2 j_{24} + 1)} \begin{Bmatrix} j_1 & j_2 & j_{12} \\ j_3 & j_4 & j_{34} \\ j_{13} & j_{24} & j \end{Bmatrix} \quad (2\text{-}22)$$

The $9j$–symbol may be expressed in terms of $6j$–symbols as

$$\begin{Bmatrix} a & b & c \\ d & e & f \\ g & h & i \end{Bmatrix} = \sum_z (-1)^{2z} [z] \begin{Bmatrix} a & d & g \\ h & i & z \end{Bmatrix} \begin{Bmatrix} b & e & h \\ d & z & f \end{Bmatrix} \begin{Bmatrix} c & f & i \\ z & a & b \end{Bmatrix} \quad (2\text{-}23)$$

The $9j$–symbol is left invariant with respect to any *even* permutation of its rows or columns or a transposition of rows and columns. Under an *odd* permutation of rows or columns the symbol is invariant but for a phase factor equal to the sum of its all arguments. If one argument of the $9j$–symbol is zero the symbol collapses to a single $6j$–symbol, namely

$$\begin{Bmatrix} a & b & c \\ d & e & f \\ g & h & 0 \end{Bmatrix} = \delta_{c,f} \delta_{g,h} \frac{(-1)^{b+d+f+g}}{\sqrt{(2c+1)(2g+1)}} \begin{Bmatrix} a & b & c \\ e & d & g \end{Bmatrix} \quad (2\text{-}24)$$

2.5   TENSOR OPERATORS

A fundamental problem of quantum mechanical description of many particle systems is to calculate matrix elements of relevant interactions. To do this in an elegant and efficient way one has to express the interactions in terms of tensor operators as pioneered by Racah[1–4] and outlined by Judd[5] and Edmonds[6]. The theory of tensor operators has a deep group theoretical basis, which is not considered here[5–8]. Here we follow Racah's original introduction of tensor operators[2]. An *irreducible tensor operator* $\mathbf{T}(k)$, of *rank k* has $(2k + 1)$ components $T(kq)$ where $q = -k, -k+1, \ldots, k-1, k$, which satisfy the commutation relations

$$[J_z, T(kq)] = q T(kq) \quad (2\text{-}25a)$$

$$[J_\pm, T(kq)] = \sqrt{k(k+1) - q(q \pm 1)} T(k, q \pm 1) \quad (2\text{-}25b)$$

Group theoretically this implies that the tensor operator components form a basis for the $(2k + 1)$–dimensional irreducible representation $[k]$ of the rotation group in three dimensions, $SO(3)$. Furthermore since the components satisfy the same commutation relations, they can be regarded as objects that transform themselves like angular momentum states $|kq\rangle$. As a result we can use standard angular momentum coupling techniques to form coupled products of tensor operators. For a rank $k = 1$ tensor operator we have, in terms of the Cartesian components (T_x, T_y, T_z)

$$T_{\pm 1}^{(1)} = \frac{\mp 1}{\sqrt{2}}(T_x \pm i T_y), \quad T_0^{(1)} = T_z \tag{2-26}$$

Thus $\mathbf{J}$ is a tensor operator of rank $k = 1$ with components

$$J_0^{(1)} = J_z, \quad J_{\pm 1}^{(1)} = \pm \frac{J_\mp}{\sqrt{2}} \tag{2-27}$$

A more complex example is the Coulomb interaction and its tensorial form. The matrix elements of the N-particle repulsive Coulomb interaction

$$H_c = \sum_{i<j}^{N} \frac{e^2}{r_{ij}} \tag{2-28}$$

play an important role in atomic physics. The interaction between each pair of electrons may be expanded in terms of Legendre polynomials of the cosine of the angle ω_{ij} between the vectors from the nucleus to the two electrons as

$$\frac{e^2}{r_{ij}} = e^2 \sum_k \frac{r_<^k}{r_>^{k+1}} P_k(\cos \omega_{ij}) \tag{2-29}$$

where $r_<$ indicates the distance from the nucleus of the nearer electron and $r_>$ the distance from the nucleus to the further away electron. Using the spherical harmonic addition theorem[9] it is possible to obtain the angular part of the Coulomb interaction operator in a tensorial form

$$
\begin{aligned}
P_k(\cos \omega_{ij}) &= \frac{4\pi}{2k+1} \sum_q Y_{kq}^*(\theta_i, \phi_i) Y_{kq}(\theta_j, \phi_j) \\
&= \sum_q (-1)^q \left(C_{-q}^{(k)} \right)_i \left(C_q^{(k)} \right)_j \\
&= \left(\mathbf{C}_i^{(k)} \cdot \mathbf{C}_j^{(k)} \right)
\end{aligned} \tag{2-30}
$$

where the *spherical tensors* $C_q^{(k)}$ are defined in terms of the usual spherical harmonics, Y_{kq} (that obviously satisfy (2-25a) and (2-25b), and therefore formally are also tensor operators) as

$$C_q^{(k)} = \left(\frac{4\pi}{2k+1} \right)^{1/2} Y_{kq} \tag{2-31}$$

The $C_q^{(k)}$ are the components of tensor operator $\mathbf{C}^{(k)}$ with a rank k, and $(\mathbf{C}_i^{(k)} \cdot \mathbf{C}_j^{(k)})$ denotes the *scalar product* of two spherical tensors.

2.6   THE WIGNER-ECKART THEOREM FOR *SO(3)*

The key for calculating the matrix elements of tensor operators that act between angular momentum states comes from the Wigner-Eckart theorem as applied to $SO(3)$.

The theorem is only introduced here, and its detailed derivations can be found elsewhere[5-8]. The m−dependence of the matrix elements of the tensor operator component $T_q^{(k)}$ in the jm scheme is given by the $3j−$ symbol that in addition is multiplied by a direction-independent factor, namely

$$\langle\alpha j m|T_q^{(k)}|\alpha' j'm'\rangle = (-1)^{j-m} \begin{pmatrix} j & k & j' \\ -m & q & m' \end{pmatrix} \langle\alpha j||T^{(k)}||\alpha' j'\rangle \quad (2\text{-}32)$$

The important point to note is that the entire dependence of the matrix element on the m projection quantum numbers is encased in a phase factor and a $3j$−symbol. The quantity $\langle\alpha j||T^{(k)}||\alpha' j'\rangle$ is a *reduced matrix element*, which is independent of the m quantum numbers. The numbers α and α' stand for any other descriptors required to complete the identification of the states. The theorem means that for all tensor operators with rank k the only $m−$ dependent part is as it is defined in (2-32), and the individual properties of each tensor operator are represented by the reduced matrix element.

Note that the $3j$−symbol vanishes and hence renders the vanishing matrix element unless

$$|j - j'| \le k \le j + j' \quad and \quad m - m' = q \quad (2\text{-}33)$$

Thus in order to evaluate the matrix element of any tensor operator the tables with the values of the $3j−$ symbols are required together with the expression for the reduced matrix elements. In fact, some reduced matrix elements may be readily determined. Consider the reduced matrix element $\langle\alpha j||J^{(1)}||\alpha' j'\rangle$. We first note that

$$\langle\alpha j m|J_z|\alpha' j'm'\rangle = m\delta(\alpha, \alpha')\delta(j, j')\delta(m, m') \quad (2\text{-}34)$$

Choosing for example $m = m' = \frac{1}{2}$ and $q = 0$ we have from (2-32)

$$\langle\alpha j \tfrac{1}{2}|J_0^{(1)}|\alpha' j' \tfrac{1}{2}\rangle = \tfrac{1}{2}\delta(\alpha, \alpha')\delta(j, j') = (-1)^{j-\frac{1}{2}} \begin{pmatrix} j & 1 & j \\ -\frac{1}{2} & 0 & \frac{1}{2} \end{pmatrix} \langle\alpha j||J^{(1)}||\alpha j\rangle$$

$$(2\text{-}35)$$

The $3j$−symbol may be expressed explicitly in terms of its arguments[5,6] to give a result

$$\langle\alpha j||J^{(1)}||\alpha j\rangle = \delta(\alpha, \alpha')\delta(j, j')\sqrt{j(j + 1)(2j + 1)} \quad (2\text{-}36)$$

In an exactly similar way

$$\langle\alpha L||L^{(1)}||\alpha' L'\rangle = \delta(\alpha, \alpha')\delta(L, L')\sqrt{L(L + 1)(2L + 1)} \quad (2\text{-}37a)$$

and

$$\langle\alpha S||S^{(1)}||\alpha' S'\rangle = \delta(\alpha, \alpha')\delta(S, S')\sqrt{S(S + 1)(2S + 1)} \quad (2\text{-}37b)$$

With somewhat greater difficulty[5] the reduced matrix element of the spherical tensor is obtained,

$$\langle\alpha\ell||C^{(k)}||\alpha'\ell'\rangle = \delta(\alpha, \alpha')(-1)^\ell[\ell, \ell']^{\frac{1}{2}} \begin{pmatrix} \ell & k & \ell' \\ 0 & 0 & 0 \end{pmatrix} \quad (2\text{-}37c)$$

where the abbreviation, customary for the atomic physics, is introduced

$$[\ell, \ell'] \equiv (2\ell + 1)(2\ell' + 1)$$

2.7   COUPLED TENSOR OPERATORS

We have noted the close connection between the transformation properties of tensor operators and angular momentum states. For two tensor operators $\mathbf{T}^{(k_1)}$ and $\mathbf{U}^{(k_2)}$ we define a coupled tensor operator $\mathbf{X}^{(k_1 k_2)K}$ via

$$\mathbf{X}_Q^{(k_1 k_2)K} \equiv \mathbf{X}_Q^K = \sum_{q_1, q_2} T_{q_1}^{(k_1)} U_{q_2}^{(k_2)} < k_1 q_1 k_2 q_2 |(k_1 k_2) K Q > \qquad (2\text{-}38)$$

which follows directly from the coupling scheme in (2-7). Indeed, $\mathbf{T}^{(k_1)}$ and $\mathbf{U}^{(k_2)}$ transform as $|k_1 q_1 >$ and $|k_2 q_2 >$, respectively, and consequently they are coupled together in the same manner as angular momenta. The coupled tensor operator $\mathbf{X}_Q^K$ is defined above as the tensorial product of two tensor operators, which in the literature is denoted as

$$\mathbf{X}_Q^K \equiv [\mathbf{T}^{(k_1)} \times \mathbf{U}^{(k_2)}]_Q^K \equiv [\mathbf{T}^{(k_1)} \mathbf{U}^{(k_2)}]_Q^K$$

Explicit evaluation of the Clebsch-Gordan coefficient for the case of $K = 0$ (and obviously $Q = 0$) leads to a simplified tensorial product of two tensor operators

$$[\mathbf{T}^{(k)} \mathbf{U}^{(k)}]_0^0 = \frac{(-1)^k}{\sqrt{(2k+1)}} \sum_q (-1)^{-q} T_q^{(k)} U_{-q}^{(k)} \qquad (2\text{-}39)$$

The scalar product of two tensor operators in general is defined as (see also (2-30))

$$(\mathbf{T}^{(k)} \cdot \mathbf{U}^{(k)}) = \sum_q (-1)^q T_q^{(k)} U_{-q}^{(k)} \qquad (2\text{-}40)$$

Therefore, it follows from (2-39) and (2-40) that

$$[\mathbf{T}^{(k)} \mathbf{U}^{(k)}]_0^0 = \frac{(-1)^k}{\sqrt{(2k+1)}} (\mathbf{T}^{(k)} \cdot \mathbf{U}^{(k)}) \qquad (2\text{-}41)$$

From the Wigner-Eckart theorem we have that

$$\langle \alpha j_1 j_2 J M | X_Q^{(K)} | \alpha' j_1' j_2' J' M' \rangle$$
$$= (-1)^{J-M} \begin{pmatrix} J & K & J' \\ -M & Q & M' \end{pmatrix} \langle \alpha j_1 j_2 J \| X^{(K)} \| \alpha' j_1' j_2' J' \rangle \qquad (2\text{-}42)$$

and the only problem is now to evaluate the reduced matrix element in (2-42). Basically this is done by an uncoupling of the bra and ket states and of the tensor operator followed by appropriate recouplings and summations (for the details see the books of Judd[5] and of Edmonds[6]).

If $\mathbf{T}^{(k)}$ and $\mathbf{U}^{(k)}$ act separately on parts 1 and 2 of a system such as in spin and orbit spaces or on different particles (like in the case of Coulomb interaction), or sets of particles, then we obtain the result

$$\langle \alpha j_1 j_2 J \| X^{(K)} \| \alpha' j_1' j_2' J' \rangle = \sum_{\alpha''} < \alpha j_1 \| T^{(k_1)} \| \alpha'' j_1' > \langle \alpha'' j_2 \| U^{(k_2)} \| \alpha' j_2' \rangle$$

$$\times [J, K, J']^{\frac{1}{2}} \begin{Bmatrix} j_1 & j_1' & k_1 \\ j_2 & j_2' & k_2 \\ J & J' & K \end{Bmatrix} \qquad (2\text{-}43)$$

We can specialize the above result for $K = 0$ to obtain the matrix element of a scalar product of tensor operators as

$$
\langle \alpha j_1 j_2 J M | (\mathbf{T}^{(k)} \cdot \mathbf{U}^{(k)}) | \alpha' j_1' j_2' J' M' \rangle
$$

$$
= \delta_{J,J'} \delta_{M,M'} (-1)^{j_1' + j_2 + J}
\begin{Bmatrix} j_1' & j_2' & J \\ j_2 & j_1 & k \end{Bmatrix}
$$

$$
\times \sum_{\alpha''} \langle \alpha j_1 \| T^{(k)} \| \alpha'' j_1' \rangle \langle \alpha'' j_2 \| U^{(k)} \| \alpha' j_2' \rangle
\tag{2-44}
$$

The result of an operator $\mathbf{T}^{(k)}$ acting on part 1 of a system can be found by putting $k_2 = 0$ in (2-43) to yield

$$
\langle \alpha j_1 j_2 J \| T^{(k)} \| \alpha' j_1' j_2' J' \rangle = \delta_{j_2, j_2'} (-1)^{j_1 + j_2 + J' + k} [J, J']^{\frac{1}{2}}
\begin{Bmatrix} J & k & J' \\ j_1' & j_2 & j_1 \end{Bmatrix}
$$

$$
\times < \alpha j_1 \| T^{(k)} \| \alpha' j_1' >
\tag{2-45}
$$

while the action on part 2 is found by putting $k_1 = 0$ in (2-43) to yield

$$
\langle \alpha j_1 j_2 J \| U^{(k)} \| \alpha' j_1' j_2' J' \rangle = \delta_{j_1, j_1'} (-1)^{j_1 + j_2' + J + k} [J, J']^{\frac{1}{2}}
\begin{Bmatrix} J & k & J' \\ j_2' & j_1 & j_2 \end{Bmatrix}
$$

$$
\times \langle \alpha j_2 \| U^{(k)} \| \alpha' j_2' \rangle
\tag{2-46}
$$

A weaker result applicable to both cases, where the operators act either on different parts of a system or indeed on the same system, may be derived to give

$$
\langle \alpha J \| X^{(K)} \| \alpha' J' \rangle = (-1)^{J + K + J'} [K]^{\frac{1}{2}} \sum_{\alpha'', J''}
\begin{Bmatrix} k_2 & K & k_1 \\ J & J'' & J' \end{Bmatrix}
$$

$$
\times \langle \alpha J \| T^{(k_1)} \| \alpha'' J'' \rangle \langle \alpha'' J'' \| U^{(k_2)} \| \alpha' J' \rangle
\tag{2-47}
$$

The results given in (2-42) to (2-47) form the basis for all subsequent applications of the theory of tensor operators.

2.8   SOME SPECIAL 3nj–SYMBOLS

For subsequent work it is useful to collect together a number of special cases of $3nj$–symbols, a much fuller set can be found in Edmonds[6].

$$
\begin{pmatrix} j & j & 0 \\ m & -m & 0 \end{pmatrix} = \frac{(-1)^{j-m}}{\sqrt{2j+1}}
\tag{2-48a}
$$

$$
\begin{pmatrix} j & k & j \\ -j & 0 & j \end{pmatrix} = \frac{(2j)!}{\sqrt{(2j-k)!(2j+k+1)!}}
\tag{2-48b}
$$

$$
\begin{pmatrix} j & 1 & j \\ -m & 0 & m \end{pmatrix} = (-1)^{j-m} \frac{m}{\sqrt{j(2j+1)(j+1)}}
\tag{2-48c}
$$

$$\begin{Bmatrix} j_1 & j_2 & j_3 \\ j_2 & j_1 & 1 \end{Bmatrix} = (-1)^{j_1+j_2+j_3+1} \frac{[j_1(j_1+1) + j_2(j_2+1) - j_3(j_3+1)]}{\sqrt{4j_1(j_1+1)(2j_1+1)j_2(j_2+1)(2j_2+1)}}$$

(2-49a)

$$\begin{Bmatrix} j_1 & j_2 & j_3 \\ j_2-1 & j_1 & 1 \end{Bmatrix} = (-1)^{j_1+j_2+j_3}$$

$$\times \sqrt{\frac{2(j_1+j_2+j_3+1)(j_1+j_2-j_3)(j_2+j_3-j_1)(j_1-j_2+j_3+1)}{2j_1(2j_1+1)(2j_1+2)(2j_2-1)2j_2(2j_2+1)}}$$

(2-49b)

Let us now turn to some practical applications of the preceding formalism.

2.9 THE ZEEMAN EFFECT - WEAK FIELD CASE

To illustrate the formalism of the tensor operators we analyze a case of a magnetic field B_z directed along the $z-$axis and a set of states $|\alpha SLJM >$ associated with a spectroscopic term ^{2S+1}L. The presence of the magnetic field adds to the Hamiltonian a term

$$H_{mag} = -B_z\mu_z = B_z\mu_0[L_z + g_s S_z] \tag{2-50}$$

where $g_s \cong 2.0023$. In terms of tensor operators we need to evaluate the matrix elements of the operator $L_0^{(1)} + g_s S_0^{(1)}$. Application of the Wigner-Eckart theorem to the diagonal matrix element gives

$$\langle \alpha SLJM|L_0^{(1)} + g_s S_0^{(1)}|\alpha SLJM\rangle$$

$$= (-1)^{J-M} \begin{pmatrix} J & 1 & J \\ -M & 0 & M \end{pmatrix} \langle \alpha SLJ\|L^{(1)} + g_s S^{(1)}\|\alpha SLJ\rangle$$

$$= \frac{M}{\sqrt{J(J+1)(2J+1)}} \langle \alpha SLJ\|L^{(1)} + g_s S^{(1)}\|\alpha SLJ\rangle \tag{2-51}$$

The $S^{(1)}$ operator acts on the first part of $|\alpha SLJ >$, therefore its matrix element has the form (use (2-45))

$$\langle \alpha SLJ\|g_s S^{(1)}\|\alpha SLJ\rangle$$

$$= g_s(-1)^{S+L+J+1}(2J+1)\begin{Bmatrix} J & 1 & J \\ S & L & S \end{Bmatrix} \langle \alpha S\|S^{(1)}\|\alpha S\rangle \tag{2-52a}$$

while $L^{(1)}$ acts on the second part (use (2-46)) to give

$$\langle \alpha SLJ\|L^{(1)}\|\alpha SLJ\rangle$$

$$= (-1)^{S+L+J+1}(2J+1)\begin{Bmatrix} J & 1 & J \\ L & S & L \end{Bmatrix} \langle \alpha L\|L^{(1)}\|\alpha L\rangle \tag{2-52b}$$

The appropriate reduced matrix elements are evaluated in (2-37a,b) and the $6j-$ symbols may be evaluated explicitly. Combining these results we finally obtain

$$\langle \alpha SLJM|H_{mag}|\alpha SLJM\rangle = B_z\mu_0 Mg(SLJ) \tag{2-53}$$

where

$$g(SLJ) = 1 + (g_s - 1)\frac{J(J+1) - L(L+1) + S(S+1)}{2J(J+1)} \qquad (2\text{-}54)$$

is the so-called Lande g−factor. This result shows that for a weak magnetic field with states of different J well separated, the magnetic field produces splittings linearly dependent on the M quantum number. This is the so-called weak field Zeeman effect. For a $J = \frac{1}{2}$ level we obtain the pattern

Note that we have not only determined the number of sublevels (two) but also the *magnitude* of splitting. For a $J = 1$ level we obtain the pattern

where there are three sublevels. In general we obtain $(2J + 1)$ sublevels. For a system having an *odd* number of electrons we obtain an *even* number of sublevels while for an *even* number of electrons we obtain an *odd* number of sublevels.

For a magnetic field in the z−direction the M−quantum number remains a good quantum number. This is because we have preserved SO_2 symmetry. However, H_{mag} does not preserve SO_3 symmetry, since a particular direction in 3−space is chosen. The total angular momentum J is no longer a good quantum number. Therefore there are non-zero matrix elements of H_{mag} that couple the states with $\Delta J = \pm 1$. We first note that $J_z = L_z + S_z$, and hence $L_z + g_s S_z = J_z + (g_s - 1)S_z$. But, at the same time the matrix elements of J_z are diagonal in J, and therefore to calculate the off-diagonal matrix elements of H_{mag} we need only to calculate the off-diagonal matrix element of S_z,

$$\langle \alpha SLJM|S_0^{(1)}|\alpha SLJ + 1M\rangle$$
$$= (-1)^{J-M}\begin{pmatrix} J & 1 & J+1 \\ -M & 0 & M \end{pmatrix}\langle \alpha SLJ\|S^{(1)}\|\alpha SLJ + 1\rangle \qquad (2\text{-}55)$$

Explicit evaluation of the $3j$−symbol gives

$$(-1)^{J-M}\begin{pmatrix} J & 1 & J+1 \\ -M & 0 & M \end{pmatrix} = -\sqrt{\frac{(J+M+1)(J-M+1)}{(2J+1)(J+1)(2J+3)}} \qquad (2\text{-}56)$$

Using subsequently (2-45) and (2-37b) the reduced matrix element in (2-55) has the form

$$\langle \alpha SLJ \| S^{(1)} \| \alpha SLJ + 1 \rangle$$

$$= (-1)^{S+L+J} \sqrt{(2J+1)(2J+3)} \begin{Bmatrix} J & 1 & J+1 \\ S & L & S \end{Bmatrix} \langle S \| S^{(1)} \| S \rangle$$

$$= -\sqrt{\frac{(S+L+J+2)(S+J+1-L)(J+1+L-S)(S-J+L)}{4(J+1)}} \quad (2\text{-}57)$$

and finally the off-diagonal matrix element is determined by the following expression

$$\langle \alpha SLJM | H_{mag} | \alpha SLJ + 1M \rangle$$

$$= B_z \mu_0 (g_s - 1) \sqrt{(J+1)^2 - M^2}$$

$$\times \sqrt{\frac{(S+L+J+2)(S+J+1-L)(J+1+L-S)(S-J+L)}{4(J+1)^2(2J+1)(2J+3)}} \quad (2\text{-}58)$$

In a particular case of a ^3P term, with $S = 1$ and $L = 1$, we can have $J = 0, 1$ and 2. In a free atom it is expected that the spin-orbit coupling gives rise to the three spectroscopic levels:

$$^3P_2 \qquad ^3P_1 \qquad ^3P_0$$

For simplicity of this illustrative calculations we assume $g_s = 2$ in (2-54) to evaluate

$$g(^3P_2) = \frac{3}{2} \qquad g(^3P_1) = \frac{3}{2}$$

The off-diagonal matrix elements are determined by (2-58) and we can obtain separate matrices, one for each value of M_J. The matrices for M_J and $-M_J$ differ only in the sign of the diagonal elements, which is in fact the sign of M_J. In units of $\mu_0 B_z$ we have the following matrices,

$$M_J = \pm 2 \quad \langle ^3P2 \pm 2 | \begin{pmatrix} |^3P2 \pm 2\rangle \\ \pm 3 \end{pmatrix} \quad (2\text{-}59a)$$

$$M_J = \pm 1 \quad \begin{matrix} \langle ^3P2 \pm 1| \\ \langle ^3P1 \pm 1| \end{matrix} \begin{pmatrix} |^3P2 \pm 1\rangle & |^3P1 \pm 1\rangle \\ \pm\frac{3}{2} & \frac{1}{2} \\ \frac{1}{2} & \pm\frac{3}{2} \end{pmatrix} \quad (2\text{-}59b)$$

$$M_J = 0 \quad \begin{matrix} \langle ^3P20| \\ \langle ^3P10| \\ \langle ^3P00| \end{matrix} \begin{pmatrix} |^3P20\rangle & |^3P10\rangle & |^3P00\rangle \\ 0 & \frac{\sqrt{3}}{3} & 0 \\ \frac{\sqrt{3}}{3} & 0 & \frac{\sqrt{6}}{3} \\ 0 & \frac{\sqrt{6}}{3} & 0 \end{pmatrix} \quad (2\text{-}59c)$$

The effect of the off-diagonal matrix elements is to mix states of different J and to lead to level shifts that are non-linear in M_J.

If the external magnetic field is strong and the energy separation of the different J states is small, then one has to expect a strong $J-$mixing, and H_{mag} is not regarded as a perturbation. So far we have considered states in an $|SLJM\rangle$ basis. The calculation of energy levels requires now to add to the above matrices contributions from the other interactions that are included in the Hamiltonian such as the Coulomb and spin-orbit interactions. In the event of a very strong magnetic field we may consider states in a $|SLM_SM_L\rangle$ basis (these are the uncoupled states $\equiv |SM_SLM_L\rangle$). In that case we have the matrix elements

$$\langle \alpha SLM_SM_L|H_{mag}|\alpha SLM_SM_L\rangle = \mu_0 B_z \langle \alpha SLM_SM_L|L_0^{(1)} + g_s S_0^{(1)}|\alpha SLM_SM_L\rangle$$
$$= \mu_0 B_z(M_L + g_s M_S) \tag{2-60}$$

For the states $|^3PM_SM_L\rangle$, taking $g_s = 2$ we have, again in units of $\mu_0 B_z$:

$$M_J = \pm 2 \quad \langle ^3P \pm 1 \pm 1| \left(\begin{array}{c} |^3P \pm 1 \pm 1\rangle \\ \pm 3 \end{array} \right) \tag{2-61a}$$

$$M_J = \pm 1 \quad \begin{array}{c} \langle ^3P \pm 10| \\ \langle ^3P0 \pm 1| \end{array} \left(\begin{array}{cc} |^3P \pm 10\rangle & |^3P0 \pm 1\rangle \\ \pm 2 & 0 \\ 0 & \pm 1 \end{array} \right) \tag{2-61b}$$

$$M_J = 0 \quad \begin{array}{c} \langle ^3P1 - 1| \\ \langle ^3P - 11| \\ \langle ^3P00| \end{array} \left(\begin{array}{ccc} |^3P1 - 1\rangle & |^3P - 11\rangle & |^3P00\rangle \\ 1 & 0 & 0 \\ 0 & -1 & 0 \\ 0 & 0 & 0 \end{array} \right) \tag{2-61c}$$

We note that as expected these matrices are diagonal. Their eigenvalues are precisely the eigenvalues that would be obtained if the matrices in (2-59) were diagonalized. This gives a method of checking Zeeman matrices calculated in the $|SLJM\rangle$ basis. Upon diagonalization we must obtain the corresponding values found in the $|SLM_SM_L\rangle$ basis.

2.10   EXERCISES

2-1. Obtain the expansion for $|(1,5)64\rangle$.

2-2. Use (2-13) to obtain the values of the Clebsch-Gordan coefficients to give an alternative derivation of (2-12) via (2-8).

REFERENCES

1. Racah G, (1941) Theory of Complex Spectra I *Phys. Rev.,***61** 186.
2. Racah G, (1942) Theory of Complex Spectra II *Phys. Rev.,***62** 438.
3. Racah G, (1943) Theory of Complex Spectra III *Phys. Rev.,***63** 367.
4. Racah G, (1949) Theory of Complex Spectra IV *Phys. Rev.,***76** 1352.
5. Judd B R, (1998) *Operator Techniques in Atomic Spectroscopy* Princeton: Princeton University Press.

6. Edmonds A R, (1960) *Angular Momentum in Quantum Mechanics* Princeton: Princeton University Press.
7. Wybourne B G, (1974) *Classical Groups for Physicists* Wiley-Interscience Publication, John Wiley and Sons Inc., New York.
8. Fano U and Racah G, (1959) *Irreducible Tensorial Sets* New York: Academic Press.
9. Condon E U and Shortley G H, (1935) *The Theory of Atomic Spectra* Cambridge: Cambridge University Press.

3 Interactions in One- and Two-electron Systems

He is rather a good mathematician, but he will never be as good as Schottky.

G Frobenius,
in a letter recommending the appointment of David Hilbert at Göttingen

We now try to use the ideas introduced in the previous chapter to the calculation of interactions in some one- and two-electron systems. Firstly, we consider the Coulomb repulsion in two-electron systems, secondly the spin-orbit interaction in one- and two-electron systems and thirdly, intermediate coupling and its effect on the Lande $g-$factors.

3.1 STATES OF TWO-ELECTRON SYSTEMS

Before starting to calculate matrix elements we need to choose a suitable angular momentum basis. The Coulomb repulsion operator for an $N-$electron system,

$$H_c = e^2 \sum_{i<j} \frac{1}{r_{ij}},$$ (3-1)

commutes with the angular momentum operators

$$\mathbf{S} = \sum_{i=1}^{N} \mathbf{s}_i \quad \text{and} \quad \mathbf{L} = \sum_{i}^{N} \mathbf{l}_i$$ (3-2)

with the consequence that the matrix elements of H_c are diagonal in the $SLJM$ and SM_SLM_L schemes, and independent of the M quantum numbers.

For a two-electron system say $n\ell n'\ell'$, the total orbital quantum number L takes on the range of values,

$$L = \ell + \ell', \ell + \ell' - 1, \dots, |\ell - \ell'|,$$ (3-3)

while the total spin S has just the two values,

$$S = 1, 0.$$ (3-4)

Thus for two electron system we obtain triplet $(S = 1)$ and singlet $(S = 0)$ spin states. For the special case of *equivalent electrons* where $n = n'$ and $\ell = \ell'$ we must apply the Pauli Exclusion Principle, which in this case amounts to excluding all values of S, L except those where $L + S$ is *even*. For example for the $4f^2$ configuration we have the spectroscopic terms

$$^3(PFH) + {}^1(SDGI) \tag{3-5}$$

whereas for the $4f5f$ configuration the allowed terms are

$$^{3,1}(SPDFGHI) \tag{3-6}$$

and in the $4f5d$ configuration

$$^{3,1}(PDFGH) \tag{3-7}$$

Recall that spectroscopists use historical letter notation for the orbital angular momentum integers, L,

$$L = \begin{array}{ccccccccc} 0 & 1 & 2 & 3 & 4 & 5 & 6 & 7 & \ldots \\ S & P & D & F & G & H & I & K & \ldots \end{array}$$

with the orbital angular momentum of single electrons being represented by the corresponding lower case letters. In some cases the principal quantum number n is suppressed in our notation. For example f^2 stands for a generic configuration that contains a pair of occupied f orbitals that possess the same principal quantum number, whereas ff' denotes a pair of f orbitals each with different principal quantum numbers.

3.2   THE CENTRAL FIELD APPROXIMATION

For N electrons moving about a point nucleus of charge Ze we have the approximate non-relativistic Hamiltonian

$$H = \sum_{i=1}^{N} \left(\frac{p_i^2}{2m} - \frac{Ze^2}{r_i} \right) + \sum_{i<j}^{N} \frac{e^2}{r_{ij}} \tag{3-8}$$

where the first part is a sum of one particle operators while the second, the Coulomb interaction, is a two particle object. In terms of the central field approximation we may consider each electron to be moving independently in the field of the nuclear charge and a spherically averaged potential field due to each of the other electrons. Hence each electron may be said to move in a spherically symmetric potential $U'(r_i)$. The Hamiltonian for the central field approximation, H_{cf}, becomes

$$H_{cf} = \sum_{i=1}^{N} \left[\frac{p_i^2}{2m} + U(r_i) \right]. \tag{3-9}$$

where the effective potential contains the interactions between the nucleus and the electrons together with the new term that represents the central field, namely

$$U(r_i) = \sum_i \left[-\frac{Ze^2}{r_i} + U'(r_i) \right]$$

The difference, $H - H_{cf}$ may now be treated as a perturbation potential and in fact it represents the non-central part of Coulomb interaction V_{corr} that is responsible for electron correlation effects,

$$H - H_{cf} \equiv V_{corr} = \sum_{i<j}^{N} \frac{e^2}{r_{ij}} - \sum_{i}^{N} U'(r_i) \qquad (3\text{-}10)$$

If the central field potential is well chosen, then V_{corr} gives rise to small corrections of energy and the operator might be treated as a perturbation. The Schrödinger equation for the central field becomes

$$\sum_{i=1}^{N} \left[\frac{p_i^2}{2m} + U(r_i) \right] \Psi = E_{cf} \Psi \qquad (3\text{-}11)$$

and the Hamiltonian has a simple form of a sum of one particle operators. This is why the wave function that describes the state of an N electron system is a product of one-electron functions, and the total energy is a sum of one-electron energies,

$$\Psi = \prod_{i=1}^{N} \psi_i(a^i) \quad \text{and} \quad E_{cf} = \sum_{i=1}^{N} E_i \qquad (3\text{-}12)$$

In this model each electron moves in the central field, $U(r_i)$, which satisfies the equation

$$\left[\frac{p^2}{2m} + U(r) \right] \psi(a^i) = E(a^i) \psi(a^i) \qquad (3\text{-}13)$$

where (a^i) represents a set of one-electron quantum numbers $(n\ell m_\ell)$, which specify the state of the electron in the central field. A further separation of variables can be made by introducing spherical coordinates (r, θ, ϕ). As a consequence, one-electron eigenfunctions, *the orbitals*, are expressed as a product of *angular* and *radial* parts to give for bound states

$$\psi(a^i) = r^{-1} R_{n\ell}(r) Y_{\ell m_\ell}(\theta, \phi) \qquad (3\text{-}14)$$

where $R_{n\ell}$ are the solutions of the radial equation (a similar as for the hydrogen atom but with the central field potential), and $Y_{\ell m_\ell}$ are the spherical harmonics (the solutions of the eigenvalue problem for the rotator).

In order to describe each electron in a proper way, a spin function $\chi(\sigma, m_s)$ has to be introduced, and instead of the orbitals, the spin-orbitals have to be constructed. These functions satisfy the orthonormality relation

$$\sum_{\sigma} \chi(\sigma, m_s) \chi(\sigma, m'_s) = \delta(m_s, m'_s) \qquad (3\text{-}15)$$

and instead of the orbital in (3-14), now the electron is described by

$$\varphi(n\ell m_\ell m_s) = \chi(\sigma, m_s) r^{-1} R_{n\ell}(r) Y_{\ell m_\ell}(\theta, \phi) \qquad (3\text{-}16)$$

which is a simple product of both functions, since the hamiltonian in (3-8) is spin-independent. The energy state of an N electron system is described by a product of spin-orbitals, each of which is a solution of (3-13)

$$\Psi = \prod_{i=1}^{N} \varphi(\alpha^i) \tag{3-17}$$

where α^i represents the quantum number quartet $(n\ell m_l m_s)$ associated with the $i-th$ electron. However, since the electrons are indistinguishable particles their numbering is totally arbitrary. From a physical point of view, a system of two electrons is described by $\Psi(1, 2)$ as well as by $\Psi(2, 1)$, since $|\Psi(1, 2)|^2 = |\Psi(2, 1)|^2$. This means that $\Psi(1, 2) = \pm\Psi(2, 1)$, and there are symmetric, for "$+$", and antisymmetric functions, for "$-$"; the latter are the functions of a proper symmetry in the case of a description of electrons (in fact of all fermions). Thus, instead of a simple product of one-electron functions in (3-17), a wave function that is totally antisymmetric with respect to all permutations of the coordinates of pairs of electrons has to be constructed

$$\Psi = \frac{1}{\sqrt{N!}} \sum_{p}^{N} (-1)^p P \varphi_1(\alpha^1) \varphi_2(\alpha^2) \dots \varphi_N(\alpha^N) \tag{3-18}$$

where P represents a permutation of the *spin* and *spatial* coordinates of a pair of electrons, and p is the *parity* of the permutation; the summation is over all $N!$ permutations of the N-electron coordinates. In determinantal form (3-18) becomes

$$\Psi = \frac{1}{\sqrt{N!}} \begin{vmatrix} \varphi_1(\alpha^1) & \varphi_2(\alpha^1) & \dots & \varphi_N(\alpha^1) \\ \varphi_1(\alpha^2) & \varphi_2(\alpha^2) & \dots & \varphi_N(\alpha^2) \\ \dots & \dots & \dots & \dots \\ \varphi_1(\alpha^N) & \varphi_2(\alpha^N) & \dots & \varphi_N(\alpha^N) \end{vmatrix} \tag{3-19}$$

Note that the properties of determinant guarantee that the Pauli exclusion principle is obeyed. Indeed, if two columns in (3-19) are identical, which means that two electrons are in the same state, then the function vanishes giving zero probability for such a situation to happen.

In addition to a requirement of an appropriate behavior with respect to the permutations, a many electron wave function has to be also a solution of the eigenvalue problems of all the operators that commute with the Hamiltonian. Thus a function of a proper symmetry is not only antisymmetric, and in the form of the Slater determinant, but it also has to be identified by appropriate quantum numbers that originate from the angular momenta operators commuting with the Hamiltonian. As a result, very often these requirements lead to a linear combination of determinants, and in order to construct a function of a proper symmetry, tools of coupling of angular momenta have to be applied.

Inspection of equation (3-13) shows that it is a rather easy task to describe a many electron system within the central field approximation, if it is required to solve a simple eigenvalue problem of each electron separately. This would indeed be the case if the potential that is one particle and reproduces all Coulomb interactions would be known. It is possible to postulate a potential that is appropriate for a given system and its physical properties; it is possible to apply Slater's potential which is in a simple analytical form, or to find the optimal description within the Hartree-Fock model based on the single configuration approximation and constructed by means of the variational principle. In the Hartree-Fock method the central field potential, the *Hartree-Fock potential*, which is felt by each electron, is simply a Coulomb interaction averaged over all $N - 1$ remaining electrons. From a formal point of view it means that a single integration in a matrix element of the Coulomb potential is performed leaving an object that depends only on coordinates of a particular electron. Since the orbitals and energy are unknown, and in addition the potential is built of the unknown functions, the Hartree-Fock equations are coupled integro-differential equations that to be solved must be treated by means of the iterative process, the so-called *Self Consistent Field - SCF.*

3.3 COULOMB INTERACTION IN TWO-ELECTRON SYSTEMS

Starting with (3-19) for $N = 2$ we may write the antisymmetrized eigenfunction for two electrons whose angular momenta are $LS-$coupled

$$\langle n_a\ell_a, n_b\ell_b; SL| = \frac{1}{\sqrt{2}}\left(\langle n_{a1}\ell_{a1}, n_{b2}\ell_{b2}; SL| - \langle n_{a2}\ell_{a2}, n_{b1}\ell_{b1}; SL|\right) \qquad (3\text{-}20)$$

where the indices 1 and 2 refer to the coordinates of the first and second electron respectively, and for brevity we have suppressed the spin angular momenta. The second eigenstate in (3-20) differs from that of the first by an *odd* permutation of the electron coordinates relative to the quantum numbers. We can recouple the angular momenta of the second eigenstate to produce

$$\langle n_a\ell_a, n_b\ell_b; SL|$$
$$= \frac{1}{\sqrt{2}}\left(\langle n_{a1}\ell_{a1}, n_{b2}\ell_{b2}; SL| - (-1)^{s_a+s_b+\ell_a+\ell_b+S+L}\langle n_{b1}\ell_{b1}, n_{a2}\ell_{a2}; SL|\right)$$

$$(3\text{-}21)$$

If $n_a\ell_a = n_b\ell_b = n\ell$ the antisymmetric bra vector may be taken as

$$\langle n\ell, n\ell; SL| = \langle (n\ell)^2; SL| \quad S+L \quad even \qquad (3\text{-}22)$$

where the phase factor of the second component originates from the odd permutation of two columns in two $3j-$ symbols representing the coupling of appropriate angular momenta (spin and orbital separately).

Likewise for the antisymmetrized ket vector of another two-electron configuration we may write the analogue of (3-21) as

$$|n_c\ell_c, n_d\ell_d; SL\rangle$$
$$= \frac{1}{\sqrt{2}}\left(|n_{c1}\ell_{c1}, n_{d2}\ell_{d2}; SL\rangle - (-1)^{s_c+s_d+\ell_c+\ell_d+S+L}|n_{d1}\ell_{d1}, n_{c2}\ell_{c2}; SL\rangle\right)$$

$$(3\text{-}23)$$

and of (3-22) as

$$|n\ell, n\ell; SL\rangle = |(n\ell)^2; SL\rangle \quad S + L \quad even \tag{3-24}$$

Using the wave functions from (3-21) and (3-23) it is straightforward to write the Coulomb interaction as a matrix element of the form

$$
\langle n_a\ell_a, n_b\ell_b; SL|e^2/r_{12}|n_c\ell_c, n_d\ell_d; SL\rangle
$$
$$
= \tfrac{1}{2}\big[\langle n_a\ell_a, n_b\ell_b; SL|e^2/r_{12}|n_c\ell_c, n_d\ell_d; SL\rangle
$$
$$
+ \langle n_b\ell_b, n_a\ell_a; SL|e^2/r_{12}|n_d\ell_d, n_c\ell_c; SL\rangle
$$
$$
+ (-1)^{\ell_a+\ell_b+S+L}\big(\langle n_a\ell_a, n_b\ell_b; SL|e^2/r_{12}|n_d\ell_d, n_c\ell_c; SL\rangle
$$
$$
+ \langle n_b\ell_b, n_a\ell_a; SL|e^2/r_{12}|n_c\ell_c, n_d\ell_d; SL\rangle\big)\big] \tag{3-25}
$$

Owing to the symmetry of e^2/r_{12}, since it is a scalar tensor operator, and consequently the parity of the functions on both sides of the matrix element must be the same (which means that $p(\ell_a + \ell_b) = p(\ell_c + \ell_d)$), (3-25) simplifies to

$$
\langle n_a\ell_a, n_b\ell_b; SL|e^2/r_{12}|n_c\ell_c, n_d\ell_d; SL\rangle
$$
$$
= \langle n_a\ell_a, n_b\ell_b; SL|e^2/r_{12}|n_c\ell_c, n_d\ell_d; SL\rangle
$$
$$
+ (-1)^{\ell_a+\ell_b+S+L}\langle n_a\ell_a, n_b\ell_b; SL|e^2/r_{12}|n_d\ell_d, n_c\ell_c; SL\rangle \tag{3-26}
$$

We can now write (3-26) in tensorial form as

$$
\langle n_a\ell_a, n_b\ell_b; SL|e^2/r_{12}|n_c\ell_c, n_d\ell_d; SL\rangle
$$
$$
= e^2 \sum_k \Big[\langle n_a\ell_a, n_b\ell_b; SL \Big| \frac{r_<^k}{r_>^{k+1}}(C_1^{(k)} \cdot C_2^{(k)}) \Big| n_c\ell_c, n_d\ell_d; SL\rangle
$$
$$
+ (-1)^{\ell_a+\ell_b+S+L}\langle n_a\ell_a, n_b\ell_b; SL \Big| \frac{r_<^k}{r_>^{k+1}}(C_1^{(k)} \cdot C_2^{(k)}) \Big| n_d\ell_d, n_c\ell_c; SL\rangle \Big]
$$
$$
\tag{3-27}
$$

which is expressed in the following way

$$
= \sum_k \big[f_k(\ell_a, \ell_b; \ell_c, \ell_d)R^k(n_a\ell_a, n_b\ell_b; n_d\ell_d, n_c\ell_c)
$$
$$
+ g_k(\ell_a, \ell_b; \ell_d, \ell_c)R^k(n_a\ell_a, n_b\ell_b; n_d\ell_d, n_c\ell_c)\big] \tag{3-28}
$$

where f_k and g_k represent the *angular* parts of the matrix elements and the R^k's are the *Slater radial integrals*. The angular factors can be evaluated by application of (2-44),

$$
\langle \alpha j_1 j_2 J M|(T^{(k)} \cdot U^{(k)})|\alpha' j_1' j_2' J' M'\rangle = \delta_{J,J'}\delta_{M,M'}(-1)^{j_1'+j_2+J}
\begin{Bmatrix} j_1' & j_2' & J \\ j_2 & j_1 & k \end{Bmatrix}
$$
$$
\times \sum_{\alpha''} \langle \alpha j_1 \| T^{(k)} \| \alpha'' j_1' \rangle \langle \alpha'' j_2 \| U^{(k)} \| \alpha' j_2' \rangle
$$

to give

$$
f_k(\ell_a, \ell_b; \ell_c, \ell_d) = (-1)^{\ell_a+\ell_b+L}\langle \ell_a \| C_1^{(k)} \| \ell_c \rangle \langle \ell_b \| C_2^{(k)} \| \ell_d \rangle
\begin{Bmatrix} \ell_a & \ell_c & k \\ \ell_d & \ell_b & L \end{Bmatrix} \tag{3-29a}
$$

and

$$g_k(\ell_a, \ell_b; \ell_d, \ell_c) = (-1)^S \langle \ell_a \| C_1^{(k)} \| \ell_d \rangle \langle \ell_b \| C_2^{(k)} \| \ell_c \rangle \begin{Bmatrix} \ell_a & \ell_d & k \\ \ell_c & \ell_b & L \end{Bmatrix} \quad (3\text{-}29b)$$

with the reduced matrix elements being given by (2-37c),

$$\langle \alpha \ell \| C^{(k)} \| \alpha' \ell' \rangle = \delta(\alpha, \alpha')(-1)^{\ell}[\ell, \ell']^{\frac{1}{2}} \begin{pmatrix} \ell & k & \ell' \\ 0 & 0 & 0 \end{pmatrix}$$

The Slater radial integrals are defined by

$$R^k(n_a \ell_a, n_b \ell_b; n_c \ell_c, n_d \ell_d)$$

$$= e^2 \int_0^\infty \int_0^\infty \frac{r_<^k}{r_>^{k+1}} R_{n_a \ell_a}(r_1) R_{n_b \ell_b}(r_2) R_{n_c \ell_c}(r_1) R_{n_d \ell_d}(r_2) dr_1 dr_2 \quad (3\text{-}30)$$

When $n_a \ell_a = n_c \ell_c$ and $n_b \ell_b = n_d \ell_d$ (3-28) simplifies to

$$\langle n_a \ell_a, n_b \ell_b; SL | e^2/r_{12} | n_a \ell_a, n_b \ell_b; SL \rangle$$

$$= \sum_k \left[f_k(\ell_a, \ell_b) F^{(k)}(n_a \ell_a, n_b \ell_b) + g_k(\ell_a, \ell_b) G^{(k)}(n_a \ell_a, n_b \ell_b) \right] \quad (3\text{-}31)$$

whereas

$$F^{(k)}(n_a \ell_a, n_b \ell_b) = R^k(n_a \ell_a, n_b \ell_b; n_a \ell_a, n_b \ell_b) \quad (3\text{-}32a)$$

and

$$G^{(k)}(n_a \ell_a, n_b \ell_b) = R^k(n_a \ell_a, n_b \ell_b; n_b \ell_b, n_a \ell_a) \quad (3\text{-}32b)$$

The $F^{(k)}$'s are the *direct integrals* and are necessarily *positive* and *decreasing* functions of k. These integrals represent the actual electrostatic interaction between two electronic clouds with the density determined by $e|n_a \ell_a|^2$ and $e|n_b \ell_b|^2$, respectively. The $G^{(k)}$'s are the *exchange integrals*, and they are *positive* and $G^{(k)}/(2k + 1)$ is necessarily a *decreasing function* of k. They result from the fact that electrons are indistinguishable particles, and they have a purely quantum mechanical origin. In the literature it is possible to find the so-called *exchange density* defined as $en_a \ell_a n_b \ell_b$ for the purpose of interpretation of $G^{(k)}$ as *the exchange interaction* or simply the interaction between the electronic clouds with the *exchange density*. Obviously such an interpretation is strictly theoretical and as one without physical background should be treated only formally.

To avoid large denominators appearing in explicit calculations Condon and Shortley[1] have redefined the radial $F^{(k)}$ and $G^{(k)}$ integrals in terms of reduced radial integrals F_k and G_k where

$$F_k = \frac{F^{(k)}}{D_k} \quad \text{and} \quad \frac{G^{(k)}}{D_k} \quad (3\text{-}33)$$

where the D_k's are given in their tables 1^6 and 2^6.

As seen from the present discussion, to evaluate the angular parts of the Coulomb interaction is a trivial task having in hand the tables of the angular coupling coefficients and defined reduced matrix elements. The only difficulty for their evaluation are the radial integrals, since the radial one-electron functions have to be found as solutions of the radial equation. This step of the calculations opens many possibilities for the construction of the potential that describes the interactions within a many electron system. Among these possibilities there is also the simplest procedure in which all the radial integrals are treated as parameters, which are adjusted to the measured values of the energy.

3.4   COULOMB MATRIX ELEMENTS FOR THE F^2 ELECTRON CONFIGURATION

In the case of all the electrons being equivalent we have for the two-electron configuration $(n\ell)^2$

$$\langle (n\ell)^2; SL|e^2/r_{12}|(n\ell)^2; SL\rangle = \sum_k f_k(\ell, \ell)F^{(k)}(n\ell, n\ell) \quad (S+L) \text{ even} \qquad (3\text{-}34)$$

with

$$f_k(\ell, \ell) = (-1)^L \langle \ell \| C^{(k)} \| \ell \rangle^2 \begin{Bmatrix} \ell & \ell & k \\ \ell & \ell & L \end{Bmatrix} \qquad (3\text{-}35)$$

For the f^2 configuration we have $\ell = 3$ and the triangular conditions on the relevant $3j-$ and $6j-$symbols limit k to 0, 2, 4, 6. For the particular case of f^N configurations we also have

$$F_0 = F^{(0)}$$

$$F_2 = \frac{F^{(2)}}{225}$$

$$F_4 = \frac{F^{(4)}}{1089}$$

$$F_6 = \frac{25 F^{(6)}}{184041}$$

The following short MAPLE programme f2.map can evaluate (3-34)

```
read"njsym";
fk:=proc()
local result,L;
for L from 0 to 6 do;
result:=CLS(3,0,L)*F0+CLS(3,2,L)*225*F2
+CLS(3,4,L)*1089*F4
+CLS(3,6,L)*184041/25*F6;
lprint('L =',L,result);
end do;
end;
```

which may be run in MAPLE as

```
> read"f2.map";
> fk();
```
'L =', 0, F0+60*F2+198*F4+1716*F6
'L =', 1, F0+45*F2+33*F4-1287*F6
'L =', 2, F0+19*F2-99*F4+715*F6
'L =', 3, F0-10*F2-33*F4-286*F6
'L =', 4, F0-30*F2+97*F4+78*F6
'L =', 5, F0-25*F2-51*F4-13*F6
'L =', 6, F0+25*F2+9*F4+F6

and MAPLE package njsym is available at http://www.fizyka.umk.pl/~bgw/njsym
Let us write our Coulomb matrix elements in the form

$$E = \sum_{k=0}^{6} f_k F^{(k)}(nf, nf) = \sum_{k=0}^{6} f^k F_k(nf, nf) \tag{3-36a}$$

Racah[2] has given a group-theoretical analysis of the Coulomb interaction for the f^N configurations rewriting the Coulomb matrix elements in the form

$$E = \sum_{k=0}^{3} e_k E^k \tag{3-36b}$$

where

$$e_0 = f^0$$

$$e_1 = \frac{9f^0}{7} + \frac{f^2}{42} + \frac{f^4}{77} + \frac{f^6}{462}$$

$$e_2 = \frac{143 f^2}{42} - \frac{130 f^4}{77} + \frac{35 f^6}{462}$$

$$e_3 = \frac{11 f^2}{42} + \frac{4 f^4}{77} - \frac{7 f^6}{462} \tag{3-37}$$

and

$$E^0 = F_0 - 10F_2 - 33F_4 - 286F_6$$

$$E^1 = \frac{70F_2 + 231F_4 + 2002F_6}{9}$$

$$E^2 = \frac{F_2 - 3F_4 + 7F_6}{9}$$

$$E^3 = \frac{5F_2 + 6F_4 - 91F_6}{3} \tag{3-38}$$

Conversely, we have

$$F_0 = \frac{7E^0 + 9E^1}{7}$$

$$F_2 = \frac{E^1 + 143E^2 + 11E^3}{42}$$

$$F_4 = \frac{E^1 - 130E^2 + 4E^3}{77}$$

$$F_6 = \frac{E^1 + 35E^2 - 7E^3}{462} \tag{3-39}$$

We can readily change the f2.map programme to

```
read"njsym";
fk:=proc()
local result,L,F0,F2,F4,F6;
F0:=(7*E0+9*E1)/7;
F2:=(E1+143*E2+11*E3)/42;
F4:=(E1-130*E2+4*E3)/77;
F6:=(E1+35*E2-7*E3)/462;
for L from 0 to 6 do;
result:=CLS(3,0,L)*F0+CLS(3,2,L)*225*F2+CLS(3,4,L)*1089*F4
+CLS(3,6,L)*184041/25*F6;
lprint('L =',L,result);
end do;
end;
```

which may be run to give the output:

```
> fk();
'L =', 0, E0+9*E1
'L =', 1, E0+33*E3
'L =', 2, E0+2*E1+286*E2-11*E3
'L =', 3, E0
'L =', 4, E0+2*E1-260*E2-4*E3
'L =', 5, E0-9*E3
'L =', 6, E0+2*E1+70*E2+7*E3
```

Thus for the spectroscopic terms of the f^2 configuration we have for the triplet states

$$E(^3H) = E^0 - 9E^3$$
$$E(^3F) = E^0$$
$$E(^3P) = E^0 + 33E^3 \tag{3-40a}$$

and for the singlets

$$E(^1G) = E^0 + 2E^1 - 260E^2 - 4E^3$$
$$E(^1D) = E^0 + 2E^1 + 286E^2 - 11E^3$$
$$E(^1I) = E^0 + 2E^1 + 70E^2 + 7E^3$$
$$E(^1S) = E^0 + 9E^1 \tag{3-40b}$$

Note that the relative Coulomb energies of the triplets depend only on the coefficient e_3 of E^3. This is indeed the case for the states of maximum multiplicity in all f^N configurations.

3.5 THE SPIN-ORBIT INTERACTION

The Coulomb interaction results in different SL terms having different energies but does not depend on the total angular momentum J of the electron states. The spin-orbit interaction, H_{s-o}, is a one-electron type operator of the form

$$H_{s-o} = \sum_{i=1}^{N} \xi(r_i)(\mathbf{s}_i \cdot \mathbf{l}_i) \tag{3-41}$$

where

$$\xi(r_i) = \frac{\hbar^2}{2m^2c^2r_i} \frac{dU(r_i)}{dr_i} \tag{3-42}$$

The spin-orbit interaction is diagonal in the one-electron orbital quantum number ℓ but *not* in the principal quantum number n. H_{s-o} commutes with $\mathbf{J}^2$ and J_z and is thus diagonal in J and independent of M_J. It does not commute with $\mathbf{L}^2$ or $\mathbf{S}^2$ and hence can couple states of different SL quantum numbers leading to a breakdown of LS−coupling.

The *spin-orbit coupling constant* $\zeta_{n\ell}$ is constant for the states of a given configuration and is defined as the radial integral,

$$\zeta_{n\ell} = \int_0^\infty R_{n\ell}^2 \xi(r) dr \tag{3-43}$$

For a single electron we have from (2-44)

$$\langle s\ell jm | \zeta_{n\ell}(\mathbf{s} \cdot \mathbf{l}) | s\ell jm \rangle = \zeta_{n\ell}(-1)^{j+l+s} \begin{Bmatrix} \ell & \ell & 1 \\ s & s & j \end{Bmatrix} \langle s\ell \| (\mathbf{s} \cdot \mathbf{l}) \| s\ell \rangle \tag{3-44}$$

The reduced matrix element can be evaluated by noting that

$$\begin{aligned} \langle s\ell \| (\mathbf{s} \cdot \mathbf{l}) \| s\ell \rangle &= \langle s \| s^{(1)} \| s \rangle \langle \ell \| l^{(1)} \| \ell \rangle \\ &= \sqrt{s(s+1)(2s+1)\ell(\ell+1)(2\ell+1)} \end{aligned} \tag{3-45}$$

Explicit evaluation of the $6j$−symbol in (3-44) combined with (3-45) leads to

$$\langle s\ell jm | \zeta_{n\ell}(\mathbf{s} \cdot \mathbf{l}) | s\ell jm \rangle = \zeta_{n\ell} \frac{j(j+1) - \ell(\ell+1) - s(s+1)}{2} \tag{3-46}$$

Noting that $j = \ell \pm s$ leads to

$$\langle s\ell jm | \zeta_{n\ell}(\mathbf{s} \cdot \mathbf{l}) | s\ell jm \rangle = \begin{cases} \zeta_{n\ell} \frac{\ell}{2} & j = \ell + s \\ -\zeta_{n\ell} \frac{\ell+1}{2} & j = \ell - s \end{cases} \tag{3-47}$$

Thus for a single electron we obtain a doublet and a splitting of the components ΔE of

$$\Delta E = \zeta_{n\ell}\frac{2\ell+1}{2} \tag{3-48}$$

For a two-electron configuration $(n\ell)^2$ we can write, making use of (2-44),

$$\langle (n\ell)^2 SLJM|H_{s-o}|(n\ell)^2 S'L'JM\rangle$$

$$= \left\langle (n\ell)^2 SLJM\left|\sum_{i=1}^{2}\zeta_{n\ell}\left(\mathbf{s}_i^{(1)}\cdot\mathbf{l}_i^{(1)}\right)\right|(n\ell)^2 S'L'JM\right\rangle$$

$$= \zeta_{n\ell}(-1)^{S'+L+J}\begin{Bmatrix} S & S' & 1 \\ L' & L & J \end{Bmatrix}\sum_{i=1}^{2}\langle s_1 s_2 S\|s_i^{(1)}\|s_1 s_2 S'\rangle\langle \ell_1 \ell_2 L\|l_i^{(1)}\|\ell_1 \ell_2 L'\rangle$$

$$\tag{3-49}$$

The reduced matrix elements may be evaluated by successive use of (2-45) and (2-46) followed by (2-37a,b). Remembering that for two equivalent electrons $S+L$ and $S'+L'$ must be *even*, we finally have

$$\langle (n\ell)^2 SLJM|H_{s-o}|(n\ell)^2 S'L'JM\rangle$$

$$= (-1)^{S'+L+J+1}2\zeta_{n\ell}$$

$$\times\sqrt{s(s+1)(2s+1)\ell(\ell+1)(2\ell+1)(2S+1)(2S'+1)(2L+1)(2L'+1)}$$

$$\times\begin{Bmatrix} S & S' & 1 \\ L' & L & J \end{Bmatrix}\begin{Bmatrix} S & 1 & S' \\ s & s & s \end{Bmatrix}\begin{Bmatrix} L & 1 & L' \\ \ell & \ell & \ell \end{Bmatrix} \tag{3-50}$$

3.6   SPIN-ORBIT MATRICES FOR f^2

We can implement (3-50) for the f^2 configuration by writing the MAPLE programme spinorbit.map as

```
read"njsym";
so:= proc(S, L, Sp, Lp, J)
local result;
result := combine(simplify(6*sqrt(14)*(-1)^
(Sp + L + J + 1)*sqrt((2*S + 1)*(2*Sp + 1)*
(2*L + 1)*(2*Lp + 1))*sixj(S, Sp, 1, Lp, L, J)*
sixj(S, Sp, 1, 1/2, 1/2, 1/2)*sixj(L, Lp, 1, 3, 3, 3)));
end proc:
```

Thus to compute the matrix element $\langle f^2\,{}^3P_2|H_{s-o}|f^2\,{}^1D_2\rangle$ we have, after reading in spinorbit.map,

```
> so(1,1,0,2,2);
                3/2 2^{1/2}
```

By repeatedly running the above programme we can construct the spin-orbit matrices for the f^2 configuration in terms of the spin-orbit coupling constant ζ_{nf} to obtain

$$
\begin{array}{cc}
 & \begin{array}{cc} ^3P & ^1S \end{array} \\
\begin{array}{c} J=0 \\ ^3P \\ ^1S \end{array} &
\begin{pmatrix} -1 & -2\sqrt{3} \\ -2\sqrt{3} & 0 \end{pmatrix}
\end{array}
\qquad
\begin{array}{cc}
 & \begin{array}{c} ^3P \\ ^3P \end{array} \\
\begin{array}{c} J=1 \\ ^3P \end{array} &
\begin{pmatrix} -\tfrac{1}{2} \end{pmatrix}
\end{array}
\qquad
\begin{array}{cccc}
 & \begin{array}{ccc} ^3P & ^1D & ^3F \end{array} \\
\begin{array}{c} J=2 \\ ^3P \\ ^1D \\ ^3F \end{array} &
\begin{pmatrix} \tfrac{1}{2} & \tfrac{3}{2}\sqrt{2} & 0 \\ \tfrac{3}{2}\sqrt{2} & 0 & -\sqrt{6} \\ 0 & -\sqrt{6} & -2 \end{pmatrix}
\end{array}
$$

$$(3\text{-}51)$$

$$
\begin{array}{cc}
 & \begin{array}{c} ^3F \end{array} \\
\begin{array}{c} J=3 \\ ^3F \end{array} &
\begin{pmatrix} -\tfrac{1}{2} \end{pmatrix}
\end{array}
\qquad
\begin{array}{cccc}
 & \begin{array}{ccc} ^3F & ^1G & ^3H \end{array} \\
\begin{array}{c} J=4 \\ ^3F \\ ^1G \\ ^3H \end{array} &
\begin{pmatrix} \tfrac{3}{2} & \frac{\sqrt{33}}{3} & 0 \\ \frac{\sqrt{33}}{3} & 0 & -\frac{\sqrt{30}}{3} \\ 0 & -\frac{\sqrt{30}}{3} & -3 \end{pmatrix}
\end{array}
\qquad
\begin{array}{cc}
 & \begin{array}{c} ^3H \end{array} \\
\begin{array}{c} J=5 \\ ^3H \end{array} &
\begin{pmatrix} -\tfrac{1}{2} \end{pmatrix}
\end{array}
$$

$$(3\text{-}52)$$

$$
\begin{array}{cc}
 & \begin{array}{cc} ^3H & ^1I \end{array} \\
\begin{array}{c} J=6 \\ ^3H \\ ^1I \end{array} &
\begin{pmatrix} \tfrac{5}{2} & \frac{\sqrt{6}}{2} \\ \frac{\sqrt{6}}{2} & 0 \end{pmatrix}
\end{array}
$$

3.7 INTERMEDIATE COUPLING

To compute the combined effect of the Coulomb and spin-orbit interactions we need to construct the energy matrices for $H_c + H_{s-o}$ using the results obtained for the Coulomb matrix elements and the above spin-orbit matrices to give,

$$
\begin{array}{cc}
 & \begin{array}{cc} ^3P & ^1S \end{array} \\
\begin{array}{c} J=0 \\ ^3P \\ ^1S \end{array} &
\begin{pmatrix} E^0 + 33E^3 - \zeta & -2\sqrt{3}\zeta \\ -2\sqrt{3}\zeta & E^0 + 9E^1 \end{pmatrix}
\end{array}
$$

$$
\begin{array}{cc}
 & \begin{array}{c} ^3P \end{array} \\
\begin{array}{c} J=1 \\ ^3P \end{array} &
\left(E^0 + 33E^3 - \tfrac{1}{2}\zeta \right)
\end{array}
$$

$$
\begin{array}{cccc}
 & \begin{array}{ccc} ^3P & ^1D & ^3F \end{array} \\
\begin{array}{c} J=2 \\ ^3P \\ ^1D \\ ^3F \end{array} &
\begin{pmatrix} E^0 + 33E^3 + \tfrac{1}{2}\zeta & \tfrac{3}{2}\sqrt{2}\zeta & 0 \\ \tfrac{3}{2}\sqrt{2}\zeta & E^0 + 2E^1 + 286E^2 - 11E^3 & -\sqrt{6}\zeta \\ 0 & -\sqrt{6}\zeta & E^0 + 2\zeta \end{pmatrix}
\end{array}
$$

$$
\begin{array}{cc}
 & \begin{array}{c} ^3F \end{array} \\
\begin{array}{c} J=3 \\ ^3F \end{array} &
\left(E^0 - \tfrac{1}{2}\zeta \right)
\end{array}
$$

$$
\begin{array}{cccc}
 & \begin{array}{ccc} ^3F & ^1G & ^3H \end{array} \\
\begin{array}{c} J=4 \\ ^3F \\ ^1G \\ ^3H \end{array} &
\begin{pmatrix} E^0 + \tfrac{3}{2}\zeta & \frac{\sqrt{33}}{3}\zeta & 0 \\ \frac{\sqrt{33}}{3}\zeta & E^0 + 2E^1 - 260E^2 - 4E^3 & -\frac{\sqrt{30}}{3}\zeta \\ 0 & -\frac{\sqrt{30}}{3}\zeta & E^0 - 9E^3 - 3\zeta \end{pmatrix}
\end{array}
$$

$J = 5$ ^3H

^3H $\left(E^0 - 9E^3 - \frac{1}{2}\zeta \right)$

$J = 6$ ^3H ^1I

$\begin{array}{c} ^3H \\ ^1I \end{array}$ $\begin{pmatrix} E^0 - 9E^3 + \frac{5}{2}\zeta & \frac{\sqrt{6}}{2}\zeta \\ \frac{\sqrt{6}}{2}\zeta & E^0 + 2E^1 + 70E^2 + 7E^3 \end{pmatrix}$

Let us choose the following values (in cm^{-1}) :=

$$E^0 = 6501, \quad E^1 = 4882, \quad E^2 = 21, \quad E^3 = 454, \quad \zeta = 737$$

and diagonalize the above matrices. We obtain the following eigenvalues and eigenvectors

-3	$-0.0282|^3F_4\rangle + 0.1523|^1G_4\rangle + 0.9879|^3H_4\rangle$
2047	$|^3H_5\rangle$
4209	$0.9985|^3H_6\rangle - 0.0540|^1I_6\rangle$
4764	$0.01321|^3P_2\rangle - 0.1444|^1D_2\rangle - 0.9894|^3F_2\rangle$
6132	$|^3F_3\rangle$
6799	$-0.8634|^3F_4\rangle + 0.4943|^1G_4\rangle - 0.1009|^3H_4\rangle$
10004	$0.5037|^3F_4\rangle + 0.8558|^1G_4\rangle - 0.1175|^3H_4\rangle$
17040	$-0.3059|^3P_2\rangle + 0.9415|^1D_2\rangle - 0.1415|^3F_2\rangle$
20528	$0.9964|^3P_0\rangle + 0.0850|^1S_0\rangle$
20962	$0.0540|^3H_6\rangle + 0.9985|^1I_6\rangle$
21114	$|^3P_1\rangle$
22351	$0.9520|^3P_2\rangle + 0.3046|^1D_2\rangle - 0.0317|^3F_2\rangle$
50656	$-0.0850|^3P_0\rangle + 0.9964|^1S_0\rangle$

The above energy levels are in reasonable agreement with experiment though the $^1I_6$ level is predicted to be significantly lower than observed. This observation led to the introduction of the effective operator $\alpha L(L+1)$ for the lanthanides[3]. The explanation of the origin of that operator was to come later[4-6].

3.8 EXERCISES

3-1. Show that the electrostatic interaction *between* the configurations $(n\ell)^2$ and $(n_a\ell_a, n_b\ell_b)$ is given by

$$\langle (n\ell)^2; SL|e^2/r_{12}|n_a\ell_a, n_b\ell_b; SL\rangle$$

$$= \sqrt{2}\sum_k \langle n\ell, n\ell; SL \left| \frac{r_<^k}{r_>^{k+1}}(C_1^{(k)} \cdot C_2^{(k)}) \right| n_a\ell_a, n_b\ell_b; SL\rangle$$

$$= \sum_k f_k(\ell, \ell; \ell_a, \ell_b)R^k(n\ell, n\ell; n_a\ell_a, n_b\ell_b) \qquad (3\text{-}53)$$

where

$$f_k(\ell, \ell; \ell_a, \ell_b) = (-1)^{\ell+\ell_a+L} \sqrt{2} \langle \ell \| C^{(k)} \| \ell_a \rangle \langle \ell \| C^{(k)} \| \ell_b \rangle \left\{ \begin{matrix} \ell & \ell_a & k \\ \ell_b & \ell & L \end{matrix} \right\}$$

(3-54)

3-2. Show that the electrostatic interaction *between* the configurations $(n\ell)^2$ and $(n'\ell')^2$ is given by

$$\langle (n\ell)^2; SL | e^2/r_{12} | (n'\ell')^2; SL \rangle = \sum_k f_k(\ell, \ell; \ell', \ell') R^k(n\ell, n\ell; n'\ell', n'\ell')$$

(3-55)

where

$$f_k(\ell, \ell; \ell', \ell') = (-1)^{\ell+\ell'+L} \langle \ell \| C^{(k)} \| \ell' \rangle^2 \left\{ \begin{matrix} \ell & \ell' & k \\ \ell' & \ell & L \end{matrix} \right\}$$ (3-56)

and

$$R^k(n\ell, n\ell; n'\ell', n'\ell') = G^{(k)}(n\ell, n'\ell')$$ (3-57)

3-3. Write out a full derivation of (3-48) starting with (3-44).

3-4. Give all the steps in deriving (3-50).

3-5. If you diagonalize the above matrices (3-51) and (3-52) you obtain just three distinct eigenvalues 3 (4), -4 (3), $-\frac{1}{2}$ (4) where the bracketed numbers are the number of times each eigenvalue occurs. Explain!
Hint: You will need to think about (3-48) and jj—coupling.

3-6. Calculate the correction to the ground state Lande g—factor for Pr^{3+} due to intermediate coupling.

3-7. Likewise calculate the correction for the level at 10004 cm^{-1}.

REFERENCES

1. Condon E U and Shortley G H, (1935) *The Theory of Atomic Spectra* Cambridge:Cambridge University Press.
2. Racah G, (1949) Theory of Complex Spectra IV *Phys. Rev.,* **76** 1352.
3. Runciman W A and Wybourne B G (1959) Spectra of Trivalent Praseodymium and Thulium Ions *J. Chem. Phys.,* **31** 1149.
4. Rajnak K and Wybourne B G (1963) Configuration Interaction Effects in ℓ^n Configurations *Phys. Rev.,* **132** 280.
5. Wybourne B G (1964) Orbit-orbit Interactions and the 'Linear Theory' of Configuration Interaction *J. Chem. Phys.,* **40** 1457.
6. Wybourne B G (1965) Generalization of the 'Linear Theory' of Configuration Interaction *Phys. Rev.,* **137** A 364.

4 Coupling Schemes of Angular Momenta

I find television very educating. Every time somebody turns on the set, I go into the other room and read a book.

Groucho Marx

In this chapter several coupling schemes that are relevant for subsequent discussions on hyperfine structure are discussed. The choice of a coupling scheme amounts to a choice of a particular basis in which the description of a structure of particular atomic system is simple and efficient. Different coupling schemes are often appropriate for different regions of the periodic table, since their choice is based on the physical reality of each system. Thus for light elements, where the Coulomb interactions are dominant over the spin-orbit interaction, the $LS-$coupling scheme (or Russell-Saunders coupling) is favored. In such a scheme the Coulomb interactions are diagonal, whereas the spin-orbit interactions can couple different LS terms. For heavy elements the spin-orbit interaction becomes comparable to, or is even greater than, the Coulomb interaction, and $jj-$coupling may be relevant. In other situations the relevant strengths of various interactions may favor other coupling schemes which we shall shortly explore.

4.1 NOTES ON *jj–COUPLING*

For a single electron, ℓ, the spin-orbit interaction is evaluated in (3-47), and for $\ell = 3$ we have, in terms of ζ_{nf}, the two values

$$\begin{cases} \frac{3}{2} & j = \frac{7}{2} \\ -2 & j = \frac{5}{2} \end{cases} \tag{4-1}$$

The spin-orbit interaction operator commutes with $\mathbf{j}^2$ and j_z, and hence is diagonal in any $N-$electron $jj-$coupled configuration j^N. If $N = N_+ + N_-$ and $N_\pm$ is the number of electrons described by $j_\pm = \ell \pm \frac{1}{2}$, it is rather straightforward to evaluate

$$\langle j_+^{N_+} j_-^{N_-} JM|H_{s-o}|j_+^{N_+} j_-^{N_-} J'M'\rangle = \delta_{J,J'}\delta_{M,M'}\zeta_{n\ell}\left(N_+\frac{\ell}{2} - N_-\frac{\ell+1}{2}\right) \tag{4-2}$$

For two $f-$electrons we have the three possible $jj-$coupled configurations

$$\tfrac{7}{2}^2 \; (+3), \qquad \tfrac{5}{2}^2 \; (-4), \qquad \tfrac{7}{2}\tfrac{5}{2} \; (-\tfrac{1}{2}) \tag{4-3}$$

where we have enclosed the corresponding spin-orbit interaction matrix elements, in units of ζ_{nf}, in parentheses (,).

Any jj−coupling ket $|(s_1\ell_1)j_1(s_2\ell_2)j_2JM\rangle$ may be expanded as a linear combination of LS−coupled states $|(s_1s_2)S(\ell_1\ell_2)LJM\rangle$ by noting that

$$|(s_1\ell_1)j_1(s_2\ell_2)j_2JM\rangle = p$$
$$\times \sum_{S,L} \langle(s_1s_2)S(\ell_1\ell_2)LJ|(s_1\ell_1)j_1(s_2\ell_2)j_2J\rangle|(s_1s_2)S(\ell_1\ell_2)LJM\rangle \quad (4\text{-}4)$$

where the recoupling coefficients are given by

$$\langle(s_1s_2)S(\ell_1\ell_2)LJ|(s_1\ell_1)j_1(s_2\ell_2)j_2J\rangle = \sqrt{[S,L,j_1,j_2]}\begin{Bmatrix} s_1 & s_2 & S \\ \ell_1 & \ell_2 & L \\ j_1 & j_2 & J \end{Bmatrix} \quad (4\text{-}5)$$

and

$$p = \begin{cases} 1 & \text{if } j_1 = j_2 \text{ and } \ell_1 = \ell_2 \\ \sqrt{2} & \text{otherwise} \end{cases} \quad (4\text{-}6)$$

which is the modification of (3-22) and (3-23) for jj−coupling. The expansion coefficients appearing on the right-hand-side of (4-4), and defined by (4-5), may be readily evaluated using the Maple programme below

```
read"njsym";
jjSL:=proc(j1,j2,S,L,J)
local result,p;
p:=1;
if (j1<>j2) then p:=sqrt(2);
end if;
result:=simplify(p*sqrt((2*j1+1)*(2*j2+1)*
(2*S+1)*(2*L+1))*
ninej(1/2,1/2,S,3,3,L,j1,j2,J));
end proc:
```

This procedure leads to the transformation coefficients

$$
\begin{array}{cc}
J = 0 & |^3P_0\rangle \quad |^1S_0\rangle \\
|(\tfrac{7}{2})^2_0\rangle & \left(-\tfrac{\sqrt{21}}{7} \quad \tfrac{2\sqrt{7}}{7}\right. \\
|(\tfrac{5}{2})^2_0\rangle & \left.\tfrac{2\sqrt{7}}{7} \quad \tfrac{\sqrt{21}}{7}\right)
\end{array}
$$

$$
\begin{array}{cc}
J = 1 & |^3P_1\rangle \\
|(\tfrac{7}{2}\tfrac{5}{2})_1\rangle & (\ 1\ )
\end{array}
$$

$$
\begin{array}{cccc}
J = 2 & |^3P_2\rangle & |^1D_2\rangle & |^3F_2\rangle \\
|(\tfrac{7}{2})^2_2\rangle & \left(\tfrac{3\sqrt{2}}{7}\right. & \tfrac{5}{7} & \left.-\tfrac{\sqrt{6}}{7}\right. \\
|(\tfrac{5}{2})^2_2\rangle & -\tfrac{2}{7} & \tfrac{3\sqrt{2}}{7} & \tfrac{3\sqrt{3}}{7} \\
|(\tfrac{7}{2}\tfrac{5}{2})_2\rangle & \left.\tfrac{3\sqrt{3}}{7}\right. & -\tfrac{\sqrt{6}}{7} & \left.\tfrac{4}{7}\right)
\end{array}
$$

$$
\begin{array}{cc}
J = 3 & |^3F_3\rangle \\
|(\tfrac{7}{2}\tfrac{5}{2})_3\rangle & (\ 1\ )
\end{array}
$$

$$
\begin{array}{cccc}
J = 4 & |^3F_4\rangle & |^1G_4\rangle & |^3H_4\rangle \\
|(\tfrac{7}{2})^2_4\rangle & \left(\tfrac{2\sqrt{66}}{21}\right. & \tfrac{3\sqrt{2}}{7} & \left.-\tfrac{\sqrt{15}}{21}\right. \\
|(\tfrac{5}{2})^2_4\rangle & -\tfrac{2\sqrt{3}}{21} & \tfrac{\sqrt{11}}{7} & \tfrac{\sqrt{330}}{21} \\
|(\tfrac{7}{2}\tfrac{5}{2})_4\rangle & \left.\tfrac{\sqrt{165}}{21}\right. & -\tfrac{2\sqrt{5}}{7} & \left.\tfrac{4\sqrt{6}}{21}\right)
\end{array}
$$

$$
\begin{array}{cc}
J = 5 & |^3H_5\rangle \\
|(\tfrac{7}{2}\tfrac{5}{2})_5\rangle & (\ 1\ )
\end{array}
$$

$$
\begin{array}{ccc}
J = 6 & |^3H_6\rangle & |^1I_6\rangle \\
|(\tfrac{7}{2})^2_6\rangle & \left(\tfrac{\sqrt{42}}{7}\right. & \left.\tfrac{\sqrt{7}}{7}\right. \\
|(\tfrac{7}{2}\tfrac{5}{2})_6\rangle & \left.\tfrac{\sqrt{7}}{7}\right. & \left.-\tfrac{\sqrt{42}}{7}\right)
\end{array}
$$

Suppose A_J is the $jj \rightarrow LS$ transformation matrix for a given J and B_J is a matrix corresponding to some interaction calculated in a $LS-$coupling basis, then the corresponding matrix C_J in the $jj-$coupling basis is given by the matrix multiplication

$$C_J = A_J \times B_J \times A_J^{-1} \qquad (4\text{-}7)$$

For example, consider the spin-orbit matrix for the $J = 2$ states of f^2 that we found earlier,

$$B_2 = \begin{array}{c} \\ {}^3P \\ {}^1D \\ {}^3F \end{array} \begin{array}{ccc} {}^3P & {}^1D & {}^3F \end{array} \left(\begin{array}{ccc} \frac{1}{2} & \frac{3}{2}\sqrt{2} & 0 \\ \frac{3}{2}\sqrt{2} & 0 & -\sqrt{6} \\ 0 & -\sqrt{6} & -2 \end{array} \right)$$

with $J = 2$ at top.

and the $jj \rightarrow LS$ transformation matrix A_2 for $J = 2$,

$$A_2 = \begin{array}{c} \langle(\frac{7}{2})^2_2| \\ \langle(\frac{5}{2})^2_2| \\ \langle(\frac{7}{2}\frac{5}{2})_2| \end{array} \begin{array}{ccc} |^3P_2\rangle & |^1D_2\rangle & |^3F_2\rangle \end{array} \left(\begin{array}{ccc} \frac{3\sqrt{2}}{7} & \frac{5}{7} & -\frac{\sqrt{6}}{7} \\ -\frac{2}{7} & \frac{3\sqrt{2}}{7} & \frac{3\sqrt{3}}{7} \\ \frac{3\sqrt{3}}{7} & -\frac{\sqrt{6}}{7} & \frac{4}{7} \end{array} \right)$$

The triple product given in (4-7) may be readily evaluated in Maple using the command "simplify(evalm(A&* C&* A^ (-1)))" to give

$$\left(\begin{array}{ccc} \frac{3\sqrt{2}}{7} & \frac{5}{7} & -\frac{\sqrt{6}}{7} \\ -\frac{2}{7} & \frac{3\sqrt{2}}{7} & \frac{3\sqrt{3}}{7} \\ \frac{3\sqrt{3}}{7} & -\frac{\sqrt{6}}{7} & \frac{4}{7} \end{array} \right) \times \left(\begin{array}{ccc} \frac{1}{2} & \frac{3}{2}\sqrt{2} & 0 \\ \frac{3}{2}\sqrt{2} & 0 & -\sqrt{6} \\ 0 & -\sqrt{6} & -2 \end{array} \right) \times \left(\begin{array}{ccc} \frac{3\sqrt{2}}{7} & -\frac{2}{7} & \frac{3\sqrt{3}}{7} \\ \frac{5}{7} & \frac{3\sqrt{2}}{7} & \frac{-\sqrt{6}}{7} \\ \frac{-\sqrt{6}}{7} & \frac{3\sqrt{3}}{7} & \frac{4}{7} \end{array} \right)$$

$$= \left(\begin{array}{ccc} 3 & 0 & 0 \\ 0 & -4 & 0 \\ 0 & 0 & -\frac{1}{2} \end{array} \right)$$

which is exactly the result we found earlier by diagonalizing the $J = 2$ spin-orbit matrix.

It is perhaps interesting to consider the $J = 2$ energy matrix constructed in chapter 3 to evaluate the contributions due to the Coulomb and spin-orbit interactions within the f^2 configuration,

$$B_2 = \begin{array}{c} \\ {}^3P \\ {}^1D \\ {}^3F \end{array} \begin{array}{ccc} {}^3P & {}^1D & {}^3F \end{array} \left(\begin{array}{ccc} E^0 + 33E^3 + \frac{1}{2}\zeta & \frac{3}{2}\sqrt{2}\zeta & 0 \\ \frac{3}{2}\sqrt{2}\zeta & E^0 + 2E^1 + 286E^2 - 11E^3 & -\sqrt{6}\zeta \\ 0 & -\sqrt{6}\zeta & E^0 - 2\zeta \end{array} \right)$$

with $J = 2$ header above.

If we now repeat the transformation (4-7) we obtain the transformation of the energy matrix into the jj−coupling basis as

$$
J = 2 \qquad |(\tfrac{7}{2}\tfrac{3}{2})^2_2 \qquad\qquad |(\tfrac{5}{2}\tfrac{3}{2})^2_2 \qquad\qquad |(\tfrac{7}{2}\tfrac{5}{2})_2
$$

$$
\begin{array}{c}
(\tfrac{7}{2}\tfrac{3}{2})^2_2| \\[4pt]
\\[4pt]
(\tfrac{5}{2}\tfrac{3}{2})^2_2| \\[4pt]
\\[4pt]
(\tfrac{7}{2}\tfrac{5}{2})_2|
\end{array}
\left(
\begin{array}{ccc}
E^0 + \tfrac{1}{49}(-50E^1 & \tfrac{3\sqrt{2}}{49}(-10E^1 + 1430E^2 & \tfrac{2\sqrt{6}}{49}(5E^1 - 715E^2 \\
+7150E^2 + 319E^3) & -121E^3) & +176E^3) \\
+3\zeta & & \\[8pt]
\tfrac{3\sqrt{2}}{49}(-10E^1 & E^0 + \tfrac{6}{49}(-6E^1 & \tfrac{12\sqrt{3}}{49}(E^1 + 143E^2 \\
+1430E^2 - 121E^3) & +858E^2 - 11E^3) & -11E^3) \\
& -4\zeta & \\[8pt]
\tfrac{2\sqrt{6}}{49}(5E^1 - 715E^2 & \tfrac{12\sqrt{3}}{49}(E^1 + 143E^2 & E^0 + \tfrac{3}{49}(-4E^1 \\
+176E^3) & -11E^3) & +572E^2 + 275E^3) \\
& & -\tfrac{1}{2}\zeta
\end{array}
\right)
$$

Note that whereas in the LS−basis the energy matrix had the electrostatic interaction in diagonal form and the spin-orbit interaction had off-diagonal matrix elements, in the jj−coupling basis we have the opposite situation.

4.2 J_1j−COUPLING

While most are familiar with the LS−coupling of Russell-Saunders and to a lesser extent jj−coupling, there are other important coupling schemes that find significant applications in atomic physics. Here we consider the case of J_1j−coupling. This scheme is particulary relevant for electron configurations that involve a core of the generic type ℓ^N to which an inequivalent electron in a state described by the orbital ℓ' is weakly coupled. In this case the core is coupled to form states characterized by a set of quantum numbers, say $S_1L_1J_1$. The spin and orbital angular momentum quantum numbers of the inequivalent electron $(s\ell')$ are coupled together to form states characterized by a total angular momentum j. Finally the angular momenta J_1 and j are coupled together to form states of total angular momentum J. The manner in which the electrons of the core are coupled to form J_1 need not be restricted to LS−coupling, although for performing the calculations in the J_1j−coupling scheme it is usually simplest to consider the states of the core in theLS−coupling basis.

For the purpose of labeling energy levels it is usual to give the designation of the core level and the quantum numbers J_1jJ. This coupling scheme may be considered as a direct consequence of the strong binding of the core electrons and the weak electrostatic interaction between the added electron and the core electrons. The separation of the levels of a given J_1j is a measure of the closeness of the physical coupling scheme to that of J_1jJ−coupling.

4.3   Nd I AND Nd II ENERGY LEVELS AND J_1j−COUPLING.

The low-lying energy levels of neutral neodymium, NdI, and singly ionized neodymium, NdII, provide a good illustration of J_1j−coupling. The ground

TABLE 4-1

The Energies of the Ground Term ^5I
of NdI(4 $f^4 6s^2$)

Term	J_1	Level (cm^{-1})
^5I	4	0.000
	5	1128.056
	6	2366.597
	7	3681.696
	8	5048.602

configurations are $4f^4 6s^2$ and $4f^4 6s$, respectively. The ground term of the $4f^4 6s^2$ configuration is ^5I with $J_1 = 4, \ldots, 8$. The energies, in cm^{-1}, are given in Table 4-1.

Table 4-2 has been arranged to display the levels grouped into two separate LS−coupled multiplets, which clearly overlap. An alternative would be to note the levels appearing in Table 4-1 and to consider the weak coupling of a $6s$ electron to the $4f^4(^5I_{J1})$ core to give the arrangement given in Table 4-3. This appears to give strong evidence that for these energy levels the description in terms of $J_1 j$−coupling is closer to the physical situation than LS−coupling. To confirm this suspicion let us try to make an approximate, though realistic calculation. To do this we need to be able to calculate the relevant matrix elements of the electrostatic and spin-orbit

TABLE 4-2

The Energies of the Ground Terms
^6I and ^4I of NdII(4 $f^4 6s$)

Term	J_1	Level (cm^{-1})
^6I	$\frac{7}{2}$	0.000
	$\frac{9}{2}$	513.322
	$\frac{11}{2}$	1470.097
	$\frac{13}{2}$	2585.453
	$\frac{15}{2}$	3801.917
	$\frac{17}{2}$	5085.619
^4I	$\frac{9}{2}$	1650.199
	$\frac{11}{2}$	3066.750
	$\frac{13}{2}$	4512.481
	$\frac{15}{2}$	5985.572

interactions in a $J_1 j$−coupling basis. Many technical details of such calculations can be found in B R Judd.[1]

In $J_1 j$−coupling the treatment of the spin-orbit interaction is much simpler than in LS−coupling. Making the abbreviations

$$\psi = \ell^N \alpha SL$$

we can write the spin-orbit interaction matrix elements for the states of the configuration $\ell^N \ell'$ in $J_1 j$−coupling as

$$\langle \psi_1 J_1, s\ell', j; JM | H_{s-o} | \psi_1' J_1', s\ell' j'; JM \rangle$$

$$= \delta_{J_1, J_1'} \delta_{j,j'} \left[\delta_{\psi_1, \psi_1'} \zeta_{\ell'} \langle s\ell' j | (\mathbf{s} \cdot \mathbf{l}) | s\ell' j' \rangle + \zeta_\ell \langle \psi_1 J_1 | \sum_{i=1}^{N} ((\mathbf{s}_i \cdot \mathbf{l}_i)) | \psi_1' J_1' \rangle \right] \quad (4\text{-}8)$$

where

$$\langle s\ell' j | (\mathbf{s} \cdot \mathbf{l}) | s\ell' j \rangle = \tfrac{1}{2} \left[j(j+1) - \ell'(\ell'+1) - s(s+1) \right] \quad (4\text{-}9)$$

and for the second part of the right-hand-side of (4-8) the spin-orbit matrix elements are just those calculated for the ℓ^N core.

TABLE 4-3
Low-Lying Levels of NdII($4 f^4 6s$)
Ordered in $J_1 j$ Coupling

$^5I_{J1}$	$J_1 s\, J$	Level (cm^{-1})
$^5I_4$	$4s\,\frac{7}{2}$	0.000
	$4s\,\frac{9}{2}$	513.322
$^5I_5$	$5s\,\frac{11}{2}$	1470.097
	$5s\,\frac{9}{2}$	1650.199
$^5I_6$	$6s\,\frac{13}{2}$	2585.453
	$6s\,\frac{11}{2}$	3066.750
$^5I_7$	$7s\,\frac{15}{2}$	3801.917
	$7s\,\frac{13}{2}$	4512.481
$^5I_8$	$8s\,\frac{17}{2}$	5085.619
	$8s\,\frac{15}{2}$	5985.572

The treatment of the spin-orbit interaction in $J_1 j$−coupling has the considerable advantage of being diagonal in J_1 and j. Thus the spin-orbit interaction is taken into account by simply adding the first term in the right-hand-side of (4-8) to the diagonal of the spin-orbit matrices calculated for the ℓ^N core.

In general the simplicity of calculations of the spin-orbit matrix elements is offset by the difficulty of evaluation of the electrostatic matrix elements. Fortunately that is not a problem for the special case of the $\ell^N s$ configurations. Let us calculate the matrix elements for the electrostatic interaction between the core and the s-electron. For $J_1 = J_1'$ we obtain

$$\langle \psi_1 J_1 s; J | H_c | \psi_1' J_1 s; J \rangle = \pm \delta_{\psi_1,\psi_1'} G^\ell(\ell,s) \frac{L_1(L_1+1) - S_1(S_1+1) - J_1(J_1+1)}{(2\ell+1)(2J+1)}$$

(4-10)

The plus sign is taken for $J = J_1 + \frac{1}{2}$ and the minus sign for $J = J_1 - \frac{1}{2}$. The off-diagonal matrix elements are given by

$$\langle \psi_1 J \pm \tfrac{1}{2}, s; J | H_c | \psi' J \mp \tfrac{1}{2}, s; J \rangle = \frac{\delta_{\psi_1,\psi_1'} G^\ell(\ell,s)}{(2\ell+1)(2J+1)}$$

$$\times \left[(S_1 + L_1 + J + \tfrac{3}{2})(S_1 + L_1 + \tfrac{1}{2} - J)(L_1 + J + \tfrac{1}{2} - S_1)(S_1 + J + \tfrac{1}{2} - L_1) \right]^{\frac{1}{2}}$$

(4-11)

From an explicit calculation for the case of NdII $4f^4(^5I_{J_1})6s; J$ we obtain the matrices

$$J = \tfrac{7}{2} \qquad |(^5I_4)s; \tfrac{7}{2}\rangle$$

$$\langle (^5I_4)s; \tfrac{7}{2}| \left(\tfrac{2}{7} G^3(f,s) - \tfrac{7}{2}\zeta_f \right)$$

$$J = \tfrac{9}{2} \qquad |(^5I_4)s; \tfrac{9}{2}\rangle \qquad\qquad |(^5I_5)s; \tfrac{9}{2}\rangle$$

$$\begin{array}{c} \langle (^5I_4)s; \tfrac{9}{2}| \\ \langle (^5I_5)s; \tfrac{9}{2}| \end{array} \left(\begin{array}{cc} \tfrac{8}{35} G^3(f,s) - \tfrac{7}{2}\zeta_f & \tfrac{4\sqrt{7}}{35} G^3(f,s) \\ \tfrac{4\sqrt{7}}{35} G^3(f,s) & \tfrac{-3}{35} G^3(f,s) - \tfrac{9}{4}\zeta_f \end{array} \right)$$

$$J = \tfrac{11}{2} \qquad |(^5I_5)s; \tfrac{11}{2}\rangle \qquad\qquad |(^5I_6)s; \tfrac{11}{2}\rangle$$

$$\begin{array}{c} \langle (^5I_5)s; \tfrac{11}{2}| \\ \langle (^5I_6)s; \tfrac{11}{2}| \end{array} \left(\begin{array}{cc} \tfrac{1}{14} G^3(f,s) - \tfrac{9}{4}\zeta_f & \tfrac{5}{14} G^3(f,s) \\ \tfrac{5}{14} G^3(f,s) & \tfrac{1}{14} G^3(f,s) - \tfrac{3}{4}\zeta_f \end{array} \right)$$

$$J = \tfrac{13}{2} \qquad |(^5I_6)s; \tfrac{13}{2}\rangle \qquad\qquad |(^5I_7)s; \tfrac{13}{2}\rangle$$

$$\begin{array}{c} \langle (^5I_6)s; \tfrac{13}{2}| \\ \langle (^5I_7)s; \tfrac{13}{2}| \end{array} \left(\begin{array}{cc} \tfrac{-3}{49} G^3(f,s) - \tfrac{3}{4}\zeta_f & \tfrac{4\sqrt{15}}{49} G^3(f,s) \\ \tfrac{4\sqrt{15}}{49} G^3(f,s) & \tfrac{10}{49} G^3(f,s) + \zeta_f \end{array} \right)$$

$$J = \tfrac{15}{2} \qquad |(^5I_7)s; \tfrac{15}{2}\rangle \qquad\qquad |(^5I_8)s; \tfrac{15}{2}\rangle$$

$$\begin{array}{c} \langle (^5I_7)s; \tfrac{15}{2}| \\ \langle (^5I_8)s; \tfrac{15}{2}| \end{array} \left(\begin{array}{cc} \tfrac{-5}{28} G^3(f,s) + \zeta_f & \tfrac{\sqrt{187}}{56} G^3(f,s) \\ \tfrac{\sqrt{187}}{56} G^3(f,s) & \tfrac{9}{28} G^3(f,s) + 3\zeta_f \end{array} \right)$$

$$J = \tfrac{17}{2} \qquad |(^5I_8)s; \tfrac{17}{2}\rangle$$

$$\langle (^5I_8)s; \tfrac{17}{2}| \left(\tfrac{-2}{7} G^3(f,s) + 3\zeta_f \right)$$

where the spin-orbit interaction for the f^4 core is included. We can estimate the value of the spin-orbit coupling constant ζ_{4f} from the width of the ^5I multiplet of NdI as $\sim 782cm^{-1}$. Then it is possible to adjust the value of $G^3(f, s)$ to optimize the separations of the pairs of levels for each value of J. Finally we add to the matrices a common term to match the ground state. The matrices may be readily set up as a Maple programme and the matrices diagonalized to give their eigenvalues and eigenvectors. We have made no attempt to include intermediate coupling in the f^4 core, configuration interaction etc. Nevertheless with our very simple calculation with no sophisticated fitting procedure we obtain a reasonably satisfying result as shown in Table 4-4, using $G^3(f, s) = 1300cm^{-1}$, $\zeta_{4f} = 800cm^{-1}$ and a constant term of $3130cm^{-1}$.

TABLE 4-4
Calculated and Experimental Low-Lying Energy Levels of NdII

$J_1 s\, J$	Experimental	Calculated	Eigenvector
$4s\frac{7}{2}$	0	29	$\lvert(^5I_4)s; \frac{7}{2}\rangle$
$4s\frac{9}{2}$	513	496	$0.890\lvert(^5I_4)s; \frac{9}{2}\rangle - 0.456\lvert(^5I_5)s; \frac{9}{2}\rangle$
$5s\frac{11}{2}$	1470	1321	$0.948\lvert(^5I_5)s; \frac{11}{2}\rangle - 0.317\lvert(^5I_6)s; \frac{11}{2}\rangle$
$5s\frac{9}{2}$	1650	1464	$0.456\lvert(^5I_4)s; \frac{9}{2}\rangle + 0.890\lvert(^5I_4)s; \frac{9}{2}\rangle$
$6s\frac{13}{2}$	2585	2363	$0.970\lvert(^5I_6)s; \frac{13}{2}\rangle - 0.243\lvert(^5I_7)s; \frac{13}{2}\rangle$
$6s\frac{11}{2}$	3066	2785	$0.317\lvert(^5I_5)s; \frac{11}{2}\rangle + 0.948\lvert(^5I_6)s; \frac{11}{2}\rangle$
$7s\frac{15}{2}$	3802	3633	$0.990\lvert(^5I_7)s; \frac{15}{2}\rangle - 0.139\lvert(^5I_8)s; \frac{15}{2}\rangle$
$7s\frac{13}{2}$	4512	4106	$0.243\lvert(^5I_6)s; \frac{13}{2}\rangle + 0.970\lvert(^5I_7)s; \frac{13}{2}\rangle$
$8s\frac{17}{2}$	5086	5099	$\lvert(^5I_8)s; \frac{17}{2}\rangle$
$8s\frac{15}{2}$	5985	5933	$0.139\lvert(^5I_7)s; \frac{15}{2}\rangle + 0.990\lvert(^5I_8)s; \frac{15}{2}\rangle$

4.4 $J_1 j$–COUPLING IN Gd III LEVELS OF $4f^7$ ($^8S_{7/2}$)6p

A bra state $\langle(S_1L_1)J_1, (s\ell)j; J\rvert$ may be expanded as a sum of LS–coupled states by writing

$$\langle(S_1L_1)J_1, (s\ell)j; J\rvert = \sum_{S,L}\langle(S_1L_1)J_1, (s\ell)j; J\lvert(S_1s)S, (L_1\ell)L; J\rangle\langle(S_1s)S, (L_1\ell)L; J\rvert$$

$$(4\text{-}12)$$

The transformation coefficients follow from (2-22). Using this result we can transform matrix elements of an operator, H_o calculated in a LS–coupling basis into matrix elements appropriate to the $J_1 j$—coupling basis, in particular

$$\langle \psi_1, J_1, (s\ell)j; J | H_o | \psi_1', J_1', (s\ell)j'; J' \rangle$$

$$= ([J_1, J_1', j, j'])^{\frac{1}{2}} \sum_{S,L,S',L'} ([S, L, S', L'])^{\frac{1}{2}} \begin{Bmatrix} S_1 & s & S \\ L_1 & \ell & L \\ J_1 & j & J \end{Bmatrix} \begin{Bmatrix} S_1' & s & S' \\ L_1' & \ell & L' \\ J_1' & j' & J' \end{Bmatrix}$$

$$\times \langle \psi_1, s\ell; SLJ | H_o | \psi_1', s\ell; S'L'J' \rangle \tag{4-13}$$

In the particular case of the Coulomb interactions within the states of $4f^7(^8S_{7/2})6p$ configuration, a considerable simplification is possible that leads to the $J_1 j$—coupling result

$$\langle f^7(^8S_{\frac{7}{2}})6p_j; J | H_c | f^7(^8S_{\frac{7}{2}})6p_{j'}; J \rangle$$

$$= -56([j, j'])^{\frac{1}{2}} \begin{Bmatrix} \frac{7}{2} & J & j' \\ \frac{1}{2} & j & 1 \\ 3 & \frac{7}{2} & \frac{1}{2} \end{Bmatrix} \sum_k \begin{pmatrix} 3 & k & 1 \\ 0 & 0 & 0 \end{pmatrix}^2 G^k(f, p) \tag{4-14}$$

This expression may be further simplified by defining

$$G(f, p) = 5G_2(f, p) + 4G_4(f, p) \tag{4-15}$$

where the Condon and Shortley denominator factors have been used,

$$175G_2(f, p) = G^2(f, p) \qquad 189G_4(f, p) = G^4(f, p) \tag{4-16}$$

As a result, the electrostatic interactions are expressed in the terms of a single parameter $G(f, p)$ (note that the radial integrals are not evaluated here directly but they are treated as adjustable parameters). The relevant $J_1 j$—coupled matrix elements may be evaluated using a simple Maple programme to yield

$$J = 5 \qquad |(\tfrac{7}{2}, \tfrac{3}{2})5\rangle$$
$$\langle(\tfrac{7}{2}, \tfrac{3}{2})5| \left(F - 21G + \tfrac{\zeta_p}{2} \right)$$

$$J = 4 \qquad |(\tfrac{7}{2}, \tfrac{3}{2})4\rangle \qquad |(\tfrac{7}{2}, \tfrac{1}{2})4\rangle$$
$$\begin{matrix} \langle(\tfrac{7}{2}, \tfrac{3}{2})4| \\ \langle(\tfrac{7}{2}, \tfrac{1}{2})4| \end{matrix} \begin{pmatrix} F - 11G + \tfrac{\zeta_p}{2} & -2\sqrt{35}G \\ -2\sqrt{35}G & F - 7G - \zeta_p \end{pmatrix}$$

$$J = 3 \qquad |(\tfrac{7}{2}, \tfrac{3}{2})3\rangle \qquad |(\tfrac{7}{2}, \tfrac{1}{2})3\rangle$$
$$\begin{matrix} \langle(\tfrac{7}{2}, \tfrac{3}{2})3| \\ \langle(\tfrac{7}{2}, \tfrac{1}{2})3| \end{matrix} \begin{pmatrix} F - 3X + \tfrac{\zeta_p}{2} & -6\sqrt{3}G \\ -6\sqrt{3}G & F - 15G - \zeta_p \end{pmatrix}$$

$$J = 2 \qquad |(\tfrac{7}{2}, \tfrac{3}{2})2\rangle$$
$$\langle(\tfrac{7}{2}, \tfrac{3}{2})2| \left(F + 3G + \tfrac{\zeta_p}{2} \right)$$

where we have added a constant term F to the diagonal of each matrix. The parameter G may be fixed by noting that independently of the coupling

$$^7P_2 - {}^9P_5 = 24G$$

leading to the trial value of $G = 82cm^{-1}$. Then to fix the center-of-gravity of the states we fix $F = 47424cm^{-1}$ and adjust the spin-orbit coupling constant ζ_p to give a best fit to the levels leading to $\zeta_p = 3050cm^{-1}$. Diagonalization of the matrices yields the results collected in Table 4-5.

Inspection of the eigenvectors shows clearly that in this case $J_1 j$—coupling gives an excellent account of the complete set of states.

TABLE 4-5
Calculated and Experimental Energy Levels for $4f^7(^8S_{7/2})6p$

J_1jJ	Experimental	Calculated	Eigenvector
$(\frac{7}{2}, \frac{1}{2})3$	43019	43022	$-0.1477\vert(\frac{7}{2}, \frac{3}{2})3\rangle - 0.9890\vert(\frac{7}{2}, \frac{1}{2})3\rangle$
$(\frac{7}{2}, \frac{1}{2})4$	43612	43593	$-0.2118\vert(\frac{7}{2}, \frac{3}{2})4\rangle - 0.9773\vert(\frac{7}{2}, \frac{1}{2})4\rangle$
$(\frac{7}{2}, \frac{3}{2})5$	47234	47234	$\vert(\frac{7}{2}, \frac{3}{2})5\rangle$
$(\frac{7}{2}, \frac{3}{2})4$	48339	48260	$-0.9773\vert(\frac{7}{2}, \frac{3}{2})4\rangle + 0.2118\vert(\frac{7}{2}, \frac{1}{2})4\rangle$
$(\frac{7}{2}, \frac{3}{2})3$	48860	48831	$-0.9890\vert(\frac{7}{2}, \frac{3}{2})3\rangle + 0.1477\vert(\frac{7}{2}, \frac{1}{2})3\rangle$
$(\frac{7}{2}, \frac{3}{2})2$	49195	49194	$\vert(\frac{7}{2}, \frac{3}{2})2\rangle$

4.5 $J_1\ell$—COUPLING

The $J_1\ell$—coupling scheme has had considerable success in the interpretation of noble gas spectra and in lanthanides and actinides where configurations such as $f^N g$ arise. $J_1\ell$—coupling arises in $\ell^N \ell'$ configurations when the electrostatic interaction of the outer ℓ' electron with the ℓ^N core is weak compared to the spin-orbit interaction of the external electron ℓ'. Here the orbital angular momentum ℓ' is first coupled to the total angular momentum J_1 of the core to give a resultant angular momentum K; then K is coupled to the spin of the external electron ℓ' to yield the total angular momentum J.

For the angular momentum K to be a good quantum number it is necessary that *both* the electrostatic and spin-orbit interactions of the external electron ℓ' be very weak. In the absence of spin-dependent interactions with the outer electron, each level, classified according to its K—value, will be two-fold degenerate. The effect of weak spin-dependent interactions is to remove this two-fold degeneracy, giving rise to the appearance of pairs of levels.

A $J_1\ell$ bra vector $\langle(S_1L_1)J_1\ell'; Ks; JM|$ may be expanded as a sum of LS states by writing

$$
\langle(S_1L_1)J_1\ell'; Ks; JM|
$$

$$
= \sum_{S,L} \langle(S_1L_1)J_1\ell'; Ks; JM|(S_1s)S, (L_1\ell')L; JM\rangle\langle(S_1s)S, (L_1\ell')L; JM|
$$

$$
= \sum_{S,L} (-1)^{S_1+L_1+s+\ell'+S+L+2K} \, ([J_1, S, L, K])^{\frac{1}{2}}
$$

$$
\times \left\{\begin{matrix} S_1 & L_1 & J_1 \\ \ell' & K & L \end{matrix}\right\}\left\{\begin{matrix} S & L & J \\ K & s & S_1 \end{matrix}\right\} \langle(S_1s)S, (L_1\ell')L; JM| \qquad (4\text{-}17)
$$

The matrix elements for the electrostatic interaction of an electron $n'\ell'$ with a $n\ell^N$ core becomes

$$
\langle\ell^N(S_1L_1)J_1\ell'; Ks; J|H_c|\ell^N(S_1'L_1')J_1'\ell'; K's; J\rangle
$$

$$
= \sum_k \left[\delta_{S_1,S_1'}\delta_{K,K'} f_k(J_1, J_1', K)F^k(n\ell, n'\ell') + g_k(J_1, J_1', K', K, J)G^k(n\ell, n'\ell')\right]
$$

$$
(4\text{-}18)
$$

where

$$
f_k(J_1, J_1', K) = (-1)^{J_1'+K+\ell'} \langle\ell\|C^{(k)}\|\ell\rangle\langle\ell'\|C^{(k)}\|\ell'\rangle
$$

$$
\times \left\{\begin{matrix} J_1 & J_1' & k \\ \ell' & \ell' & K \end{matrix}\right\} \langle\ell^N S_1 L_1 J_1\|U^{(k)}\|\ell^N S_1 L_1' J_1'\rangle \quad (4\text{-}19)
$$

and

$$
g_k(J_1, J_1', K', K, J) = -N \left([J_1, J_1', S_1, S_1', L_1, L_1', K, K']\right)^{\frac{1}{2}} \langle\ell\|C^{(k)}\|\ell'\rangle^2(-1)^{L_1+L_1'}
$$

$$
\times \sum_{\bar\psi}(\psi\{|\bar\psi)(\bar\psi|\}\psi') \sum_t [t] \left\{\begin{matrix} S_1 & L_1 & J_1 \\ \ell' & K & t \end{matrix}\right\}\left\{\begin{matrix} S_1' & L_1' & J_1' \\ \ell' & K' & t \end{matrix}\right\}
$$

$$
\times \left\{\begin{matrix} \bar S & s & S_1 \\ s & J & K \\ S_1' & K' & t \end{matrix}\right\}\left\{\begin{matrix} \bar L & \ell & L_1 \\ \ell & k & \ell' \\ L_1' & \ell' & t \end{matrix}\right\} \qquad (4\text{-}20)
$$

Note that the genealogy of the states in (4-19) is involved within the matrix element of $N-$ electron unit tensor operator $U^{(k)}$ while in (4-20) it is presented in an explicit way. The difference is caused by the fact that in the case of the direct interactions one coordinate of H_{el} is fixed to the position of the $(N+1)^{th}$ electron, that is described by ℓ', and the second coordinate runs over all the electrons of the core $4f^N$. Therefore for this particular electron configuration two particle object of its effective tensorial form is reduced, and only one particle operator determines the whole interaction. The situation is different in the case of the exchange term, since both coordinates of H_{el} are inter-changed, and additional re-coupling coefficients are required. The expression for the exchange term (*interaction*) is rather formidable. Fortunately, where $J_1\ell$−coupling is most appropriate the exchange interactions are negligible and frequently need not be evaluated.

The matrix elements of the spin-orbit interaction within the ℓ^N core are just those computed in the absence of the added electron and are diagonal in J_1. The corresponding matrix elements for the added electron $n'\ell'$ are

$$\langle \ell^N(S_1L_1)J_1\ell'; Ks; J|H_{s-o}|\ell^N(S_1'L_1')J_1'\ell'; K's; J\rangle$$

$$= (-1)^{K+K'+J_1+J+\ell'+s}\zeta_{n'\ell'}\left([K, K', s, \ell']\right)^{\frac{1}{2}}\left[s(s+1)\ell'(\ell'+1)\right]^{\frac{1}{2}}$$

$$\times \left\{\begin{array}{ccc} K & K' & 1 \\ s & s & J \end{array}\right\}\left\{\begin{array}{ccc} K & K' & 1 \\ \ell' & \ell' & J_1 \end{array}\right\} \qquad (4\text{-}21)$$

Again, where $J_1\ell$−coupling is valid, these matrix elements are negligible. In fact, for f^N configurations containing an outer electron with $\ell' >= 3$ the spin-orbit interaction, to a very good approximation, may be neglected, and only the spin-dependent electrostatic interactions need be considered to evaluate departures from $J_1\ell$−coupling.

The direct electrostatic interaction leaves K as a good quantum number. Therefore it is possible to calculate the energies of the two-fold degenerate levels that are characterized by different K values in *pure* $J_1\ell$−coupling, by simply adding to the energy matrices of the $n\ell^N$ core, the matrix elements of the direct electrostatic interaction between the added electron, $n'\ell'$, and the $n\ell^N$ core. The direct interactions are normally very weak because of the external nature of the added electron. Consequently if the terms in the $n\ell^N$ core are well separated we may, to a good approximation, neglect the matrix elements that couple the different terms. For g−electrons the direct interaction is non-negligible only for the $F^2(n\ell, n'g)$ integral, and as an additional approximation we may neglect all other terms. Adopting this approximation, writing out explicitly the $6 - j$ symbol of (4-19), it is straightforward to obtain the coefficients of $F^2(n\ell, n'\ell')$ as

$$f_2(J_1, K) = \langle\ell\|C^{(2)}\|\ell\rangle\langle\ell'\|C^{(2)}\|\ell'\rangle\langle\ell^N S_1 L_1 J_1\|U^{(2)}\|\ell^N S_1 L_1' J_1'\rangle$$

$$\times (J_1(2J_1 - 1)(J_1 + 1)(2J_1 + 1)(2J_1 + 3)$$

$$\times (2\ell' - 1)\ell'(\ell' + 1)(2\ell' + 1)(2\ell' + 3))^{-\frac{1}{2}}$$

$$\times \left[3h(2h + 1) - 2J_1(J_1 + 1)\ell'(\ell' + 1)\right] \qquad (4\text{-}22)$$

where

$$h = \frac{K(K+1) - J_1(J_1+1) - \ell'(\ell'+1)}{2} \tag{4-23}$$

This means that for a particular term $\alpha_1 J_1$ of the $n\ell^N$ core the energies $E(\alpha_1 J_1 K)$ of the levels formed by adding an electron ℓ' in pure $J_1\ell$–coupling are given by

$$E(\alpha_1 J_1 K) = a_{J_1} h(2h+1) + b_{J_1} \tag{4-24}$$

where b_{J_1} is a constant and a_{J_1} is the coefficient of the quantity $h(2h+1)$ in (4-22). Notice that (4-24) is quadratic in h and hence for perfect $J_1\ell$–coupling the plot of $E(\alpha_1 J_1 K)$ against h is represented by a parabola whose minimum is at $h = -\frac{1}{4}$, regardless of the term being studied. The levels of different J_1 lie on different parabolas.

$J_1\ell$–coupling is well seen in the spectra of the noble gases and in rare earth spectra such as in the $4f5g$ configuration of doubly ionized cerium $CeIII$. The energy levels of the $4f5g$ configuration are given in Table 4-6. Note the occurrence of close *pairs* of almost degenerate levels, which is exactly what one expects in $J_1\ell$–coupling. We give the mean energy for each pair. The $4f$ gives rise to two terms, $^2F_{5/2}$ and $^2F_{7/2}$ with, as expected, the $J_1 = \frac{5}{2}$ being of lower energy than the $J_1 = \frac{7}{2}$ due to the spin-orbit interaction given by (3-53) as $\frac{7}{2}\zeta_{4f}$. The angular momentum ℓ of the $5g$ electron is coupled to J_1, in each case, to yield the various values of the quantum number K and we designate a particular $J_1\ell$–coupling term by the quantum numbers $J_1\ell K$. Coupling the spin s of the added $5g$ electron then gives rise to the total angular momentum J given in the second column. In pure $J_1\ell$–coupling the mean energy levels, $E(J_1 K)$, are given by (4-24), in terms of the parameters a_{J_1} and b_{J_1}. It follows from (4-22) that a_{J_1} is directly related to the radial integral $F^2(4f5g)$. Furthermore, explicit calculation gives the ratio

$$\frac{a_{5/2}}{a_{7/2}} = \frac{9}{5} \tag{4-25}$$

Using (4-24) we obtain the values of $E(J_1 K)$ for the $4f5g$ $J_1\ell$–coupling levels as shown in Table 4-7.

Noting that the position of the $(\frac{7}{2}5g)\frac{11}{2}$ level depends only upon $b_{7/2}$, that the separation of the two levels $(\frac{7}{2}5g)\frac{1}{2}$ and $(\frac{7}{2}5g)\frac{11}{2}$ is $595a_{7/2}$, and having the ratio (4-25) it is possible to conclude that (in cm^{-1})

$$a_{5/2} = 0.419, \quad a_{7/2} = 0.233, \quad b_{5/2} = 122903.8, \quad b_{7/2} = 125157.4 \tag{4-26}$$

Using these values in Table 4-7 leads to the results displayed in Table 4-8.

Considering that we have totally ignored configuration interaction and have assumed pure $J_1\ell$–coupling the agreement between experiment and theory is surprisingly good.

TABLE 4-6
Experimental Levels for the CeIII $4f\,5g$
Configuration in $J_1\ell$–Coupling

Term	J	Level	Mean
$(\frac{5}{2}5g)\frac{9}{2}$	4	122905.69	
			122907.3
	5	122908.89	
$(\frac{5}{2}5g)\frac{11}{2}$	6	122919.83	
			122921.1
	5	122922.37	
$(\frac{5}{2}5g)\frac{7}{2}$	4	122932.21	
			122932.8
	3	122933.38	
$(\frac{5}{2}5g)\frac{5}{2}$	2	122976.30	
			122977.3
	3	122978.36	
$(\frac{5}{2}5g)\frac{13}{2}$	6	123010.29	
			123013.7
	7	123017.02	
$(\frac{5}{2}5g)\frac{3}{2}$	2	123028.39	
			123028.7
	1	123029.01	
$(\frac{7}{2}5g)\frac{11}{2}$	6	125155.89	
			125157.4
	5	125158.97	
$(\frac{7}{2}5g)\frac{9}{2}$	4	126164.86	
			125166.6
	5	125168.37	
$(\frac{7}{2}5g)\frac{13}{2}$	6	125181.54	
			125184.1
	7	125186.61	
$(\frac{7}{2}5g)\frac{7}{2}$	4	125193.91	
			125195.0
	3	125196.03	
$(\frac{7}{2}5g)\frac{5}{2}$	2	125230.90	
			125123.8
	3	125232.67	
$(\frac{7}{2}5g)\frac{3}{2}$	2	125268.40	
			125268.9
	1	125269.29	
$(\frac{7}{2}5g)\frac{15}{2}$	8	125270.97	
			125275.3
	7	125279.58	
$(\frac{7}{2}5g)\frac{1}{2}$	0	125295.21	
	1	125296.65	
			125295.9

TABLE 4-7
Values of $E(J_1 K)$ for the $4f\,5g$
$J_1\ell$-Coupling Levels

Term	h	$(E(J_1 K))$
$(\frac{5}{2}5g)\frac{13}{2}$	10	$210a_{\frac{5}{2}} + b_{\frac{5}{2}}$
$(\frac{5}{2}5g)\frac{11}{2}$	$\frac{7}{2}$	$28a_{\frac{5}{2}} + b_{\frac{5}{2}}$
$(\frac{5}{2}5g)\frac{9}{2}$	-2	$6a_{\frac{5}{2}} + b_{\frac{5}{2}}$
$(\frac{5}{2}5g)\frac{7}{2}$	$-\frac{13}{2}$	$78a_{\frac{5}{2}} + b_{\frac{5}{2}}$
$(\frac{5}{2}5g)\frac{5}{2}$	-10	$190a_{\frac{5}{2}} + b_{\frac{5}{2}}$
$(\frac{5}{2}5g)\frac{3}{2}$	$-\frac{25}{2}$	$300a_{\frac{5}{2}} + b_{\frac{5}{2}}$
$(\frac{7}{2}5g)\frac{15}{2}$	14	$406a_{\frac{7}{2}} + b_{\frac{7}{2}}$
$(\frac{7}{2}5g)\frac{13}{2}$	$\frac{13}{2}$	$91a_{\frac{7}{2}} + b_{\frac{7}{2}}$
$(\frac{7}{2}5g)\frac{11}{2}$	0	$b_{\frac{7}{2}}$
$(\frac{7}{2}5g)\frac{9}{2}$	$-\frac{11}{2}$	$55a_{\frac{7}{2}} + b_{\frac{7}{2}}$
$(\frac{7}{2}5g)\frac{7}{2}$	-10	$190a_{\frac{7}{2}} + b_{\frac{7}{2}}$
$(\frac{7}{2}5g)\frac{5}{2}$	$-\frac{27}{2}$	$351a_{\frac{7}{2}} + b_{\frac{7}{2}}$
$(\frac{7}{2}5g)\frac{3}{2}$	-16	$496a_{\frac{7}{2}} + b_{\frac{7}{2}}$
$(\frac{7}{2}5g)\frac{1}{2}$	$-\frac{35}{2}$	$595a_{\frac{7}{2}} + b_{\frac{7}{2}}$

4.6　EXERCISES

4-1. Given that

$$\begin{pmatrix} j_1 & j_2 & j_3 \\ 0 & 0 & 0 \end{pmatrix} = (-1)^{\frac{1}{2}J} \left[\frac{(J - 2j_1)!(J - 2j_2)!(J - 2j_3)!}{(\frac{1}{2}J - j_1)!(\frac{1}{2}J - j_2)!(\frac{1}{2}J - j_3)!} \right]^{\frac{1}{2}}$$

(4-27)

where $J = j_1 + j_2 + j_3$, show that

$$\begin{pmatrix} a & a & 2 \\ 0 & 0 & 0 \end{pmatrix} = (-1)^{a+1} \left[\frac{a(a + 1)}{(2a + 1)(2a - 1)(2a + 3)} \right]^{\frac{1}{2}}$$

(4-28)

and

$$\langle a \| C^{(2)} \| a \rangle = - \left[\frac{a(a + 1)(2a + 1)}{(2a + 3)(2a - 1)} \right]^{\frac{1}{2}}$$

(4-29)

TABLE 4-8

Experimental and Calculated $4f5g$

Mean Energy Levels in $J_1\ell$-Coupling

Term	Experimental	Calculated
$(\frac{5}{2}5g)\frac{9}{2}$	122907.3	122906.3
$(\frac{5}{2}5g)\frac{11}{2}$	122921.1	122915.5
$(\frac{5}{2}5g)\frac{7}{2}$	122932.8	122936.5
$(\frac{5}{2}5g)\frac{5}{2}$	122977.3	122983.4
$(\frac{5}{2}5g)\frac{13}{2}$	123013.7	122991.8
$(\frac{5}{2}5g)\frac{3}{2}$	123028.7	123029.5
$(\frac{7}{2}5g)\frac{11}{2}$	125157.4	125157.4
$(\frac{7}{2}5g)\frac{9}{2}$	125166.6	125170.2
$(\frac{7}{2}5g)\frac{13}{2}$	125184.1	125178.6
$(\frac{7}{2}5g)\frac{7}{2}$	125195.0	125201.7
$(\frac{7}{2}5g)\frac{5}{2}$	125231.8	125239.2
$(\frac{7}{2}5g)\frac{3}{2}$	125268.9	125273.0
$(\frac{7}{2}5g)\frac{15}{2}$	125275.3	125252.0
$(\frac{7}{2}5g)\frac{1}{2}$	125296.0	125296.0

4-2. Show that for $N = 1$ or $N = 4\ell + 1$

$$\langle s\ell j\|U^{(2)}\|s\ell j\rangle = \pm(-1)^{s+\ell+j}(2j+1)\begin{Bmatrix} 2 & j & j \\ \frac{1}{2} & \ell & \ell \end{Bmatrix} \quad (4\text{-}30)$$

where the sign is $+$ if $N = 1$ or $-$ if $N = 4\ell + 1$.

4-3. Use your results to put (4-22) into as simple a form as possible.

4-4. Use the NIST (National Institute of Standards and Technology) data base to obtain a list of the energy levels of neutral neon Ne I and make a list of configurations that could be best described in

(a) LS-coupling,
(b) $J_1 j$-coupling and
(c) $J_1\ell$-coupling.

REFERENCES

1. Judd B R, (1962) Low-Lying Levels in Certain Actinide Atoms, *Phys. Rev.,* **125**, 613.

5 Fine and Magnetic Hyperfine Structure

Some people think that physics is over once the equation is found which governs some phenomena. To me this seems as foolish as somebody who says English is over once he has learned the words and the grammar and never goes on to read and understand Shakespeare. Physics is not the equation but the multitude of phenomena which result from it. To know the equation is not the end but the beginning and to deduce from it the physics is an unending quest.

W. E. Thirring (1987)

5.1 INTERMEDIATE COUPLING, g–FACTORS AND g–SUM RULE

The effect of diagonalizing the combined Coulomb and spin-orbit interaction energy matrices is to yield eigenstates that are independent of the M quantum number and for which the total electronic angular momentum J remains a good quantum number. A typical eigenstate expressed in the intermediate coupling scheme is of the form

$$|aJ\rangle = \sum_{\alpha,S,L} \langle \alpha SLJ|aJ\rangle |\alpha SLJ\rangle \qquad (5\text{-}1)$$

where the $\langle \alpha SLJ|aJ\rangle$ are the eigenvector components. For example, we found for the ground state of Pr^{3+} the linear combination

$$|a4\rangle = -0.0282|^3F_4\rangle + 0.1523|^1G_4\rangle + 0.9879|^3H_4\rangle \qquad (5\text{-}2)$$

where each coefficient determines the weight with which a particular component contributes to the final state. Since in (5-2) the maximum coefficient is for the term ^3H it is possible to identify the function for $J = 4$ in the intermediate coupling scheme by the symmetry of this pure atomic term, replacing $a4$ by a whole specification $[^3H]4$, denoting at the same time the fact that S and L are not any longer good quantum numbers, and therefore they are enclosed in the brackets.

The diagonal matrix elements of the Zeeman Hamiltonian, H_{mag} for a state $|\alpha SLJM\rangle$ are given by (2-53),

$$\langle \alpha SLJM|H_{mag}|\alpha SLJM\rangle = B_z\mu_0 M g(SLJ)$$

In intermediate coupling, we have from (5-1)

$$\langle aJM|H_{mag}|aJM\rangle = \sum_{\alpha,S,L} \langle aJ|\alpha SLJ\rangle \langle \alpha SLJM|H_{mag}|\alpha SLJM\rangle \langle \alpha SLJ|aJ\rangle$$

$$= B_z\mu_0 M \sum_{\alpha,S,L} \langle aJ|\alpha SLJ\rangle g(SLJ)\langle \alpha SLJ|aJ\rangle \qquad (5\text{-}3)$$

Thus, using the linear combination from (5-2) we have for the ground state of Pr^{3+}

$$\langle a4M|H_{mag}|a4M\rangle = B_z\mu_0 M$$
$$\times \left[(-0.0282)^2 g(^3F_4) + (0.1523)^2 g(^1G_4) + (0.9879)^2 g(^3H_4)\right]$$
$$= B_z\mu_0 M(0.8045) \tag{5-4}$$

where

$$g(^3F_4) = 1.2506 \quad g(^1G_4) = 1.0000 \quad g(^3H_4) = 0.79954 \tag{5-5}$$

As a result of intermediate coupling the value of g−factor is changed from 0.7995 for the pure $^3H_4$ to 0.8045 for the intermediate coupling functions. Here the correction is rather small. As a second example consider the level calculated at $10,004cm^{-1}$ with the eigenstate

$$|b4\rangle = 0.5037|^3F_4\rangle + 0.8558|^1G_4\rangle - 0.1175|^3H_4\rangle \tag{5-6}$$

with the dominant component of ^1G symmetry. The g−factor evaluated with this function has the value of 1.0607, compared with that of the pure $^1G_4$ g−factor of 1.0000. Finally, for the third $J = 4$ eigenstate

$$|c4\rangle = -0.8634|^3F_4\rangle + 0.4943|^1G_4\rangle - 0.1009|^3H_4\rangle \tag{5-7}$$

which in its major part is of the ^3F symmetry. The intermediate coupling corrected g−factor is 1.1847 compared with that for a pure $^3F_4$ g−factor of 1.2506. The sum of intermediate coupling corrected g−factors gives

$$\sum_{i=1}^{3} g_i = 0.8045 + 1.0607 + 1.1847 = 3.0499$$

whereas the sum of the three LS g−factors gives

$$\sum_{i=1}^{3} g_i = 0.7995 + 1.0000 + 1.2506 = 3.0501$$

Although these two results are the same to within the precision of the calculation, they illustrate the *g-sum rule*.

In all cases where the states differing in J are well separated, and as a consequence only the diagonal elements need to be calculated, the intermediate coupling g−value becomes

$$g(\gamma J) = \sum_{\alpha SL} g(SLJ)\langle\alpha SLJ|\gamma J\rangle^2 \tag{5-8}$$

If we sum over the variable γ for a given J−value then

$$\sum_{\gamma} g(\gamma J) = \sum_{\alpha SL} g(SLJ) \tag{5-9}$$

since the transformation is necessarily unitary *for a complete set of states* and therefore

$$\sum_{\alpha SL} \langle \alpha SLJ | \gamma J \rangle^2 = 1 \qquad (5\text{-}10)$$

Equation (5-9) is the statement of the *g-sum rule*, but note the important qualification that it is valid *for a complete set of states*.

5.2   FINE STRUCTURE IN ALKALI ATOMS AND ZEEMAN EFFECT

The alkali elements appear, along with hydrogen, in the first column of the Periodic Table. The ground state of the n-th alkali element is, apart from closed shells, $ns(^2S_{1/2})$. The first excited multiplet is $np(^2P)$. There is no spin-orbit interaction associated with the ground state whereas in the first excited multiplet the effect of the spin-orbit interaction is to split the ^2P into two states $^2P_{1/2}$ and $^2P_{3/2}$, with the $^2P_{1/2}$ state lying below the $^2P_{3/2}$ as follows from (3-53).

The splitting

$$\Delta E = E(^2P_{3/2}) - E(^2P_{1/2}) = \tfrac{3}{2}\zeta_{np} \qquad (5\text{-}11)$$

is commonly referred to as the *fine structure* and is seen, for example, in the Sodium D lines, which at sufficient spectroscopic resolution appear as the famous yellow Sodium D doublet. Knowing the value of the fine structure splitting we can deduce a value for the spin-orbit coupling constant $\zeta_{n\ell}$, as shown in Table 5-1.

The alkali atoms, particularly Rubidium (Rb), play an important role in Bose Einstein Condensation (BEC), and the interplay of the fine structure, Zeeman effect and hyperfine structure are of major significance. For the ground state $ns(^2S_{1/2})$ the effect of an external magnetic field B_z, which disturbs the spherical symmetry of an atom, is to lift the two-fold degeneracy to produce a doublet with a Zeeman splitting of

$$\Delta E = E(^2S_{1/2,1/2}) - E(^2S_{1/2,-1/2}) = g_s \mu_0 B_z \qquad (5\text{-}12)$$

The behavior of the $np(^2P)$ term is more complicated, especially if the magnetic field is sufficiently strong as to produce Zeeman splittings comparable with those of the fine structure. It is necessary then to include the off-diagonal Zeeman matrix elements introduced in (2-58) in addition to the diagonal matrix elements given by (2-53).

For the states with $M = \pm\tfrac{3}{2}$ (and $J = \tfrac{3}{2}$) there are no off-diagonal matrix elements, and for the Hamiltonian

$$H = H_{s-o} + H_{mag} \qquad (5\text{-}13)$$

there is a single sum of distinct contributions arising from the spin orbit interaction and the perturbing influence of the applied magnetic field, namely

$$\langle np\,^2P\tfrac{3}{2}, \pm\tfrac{3}{2} | H | np\,^2P\tfrac{3}{2}, \pm\tfrac{3}{2} \rangle = \tfrac{1}{2}\zeta_{np} \pm \tfrac{3}{2}g(^2P_{3/2})\mu_0 B_z \qquad (5\text{-}14)$$

TABLE 5-1

Energy Levels of the Lowest States of Alkali Atoms

Element	State	Energy (cm^{-1})	ΔE	ζ_{np}
H	$1s\ ^2S^e_{\frac{1}{2}}$	0.000		
	$2p\ ^2P^o_{\frac{1}{2}}$	82258.9206		
	$2p\ ^2P^o_{\frac{3}{2}}$	82259.2865	0.3659	0.244
	$2s\ ^2S^e_{\frac{1}{2}}$	82258.9559		
Li	$2s\ ^2S^e_{\frac{1}{2}}$	0.000		
	$2p\ ^2P^o_{\frac{1}{2}}$	14903.660		
	$2p\ ^2P^o_{\frac{3}{2}}$	14904.000	0.34	0.226
Na	$3s\ ^2S^e_{\frac{1}{2}}$	0.000		
	$3p\ ^2P^o_{\frac{1}{2}}$	16956.172		
	$3p\ ^2P^o_{\frac{3}{2}}$	16973.368	17.20	11.47
K	$4s\ ^2S^e_{\frac{1}{2}}$	0.000		
	$4p\ ^2P^o_{\frac{1}{2}}$	12985.170		
	$4p\ ^2P^o_{\frac{3}{2}}$	13042.876	57.72	38.48
Rb	$5s\ ^2S^e_{\frac{1}{2}}$	0.000		
	$5p\ ^2P^o_{\frac{1}{2}}$	12578.960		
	$5p\ ^2P^o_{\frac{3}{2}}$	12816.560	237.60	158.38
Cs	$6s\ ^2S^e_{\frac{1}{2}}$	0.000		
	$6p\ ^2P^o_{\frac{1}{2}}$	11178.200		
	$6p\ ^2P^o_{\frac{3}{2}}$	11732.300	554.1	369.4

In this case of the excited state the Zeeman splitting is still linear in B_z, as in the case of the ground state in (5-12). For the states with $M = \pm\frac{1}{2}$, since this is a case of $J = \frac{1}{2}$ and also $J = \frac{3}{2}$, the matrices of rank 2 have to be considered,

$$
\begin{array}{c}
\begin{array}{cc}
\quad\quad |^2P\tfrac{1}{2}, \pm\tfrac{1}{2}\rangle & \quad\quad |^2P\tfrac{3}{2}, \pm\tfrac{1}{2}\rangle
\end{array}\\
\begin{array}{c}
\langle^2P\tfrac{1}{2}, \pm\tfrac{1}{2}| \\
\langle^2P\tfrac{3}{2}, \pm\tfrac{1}{2}|
\end{array}
\begin{pmatrix}
-\zeta_{np} \pm \tfrac{1}{2}g(^2P_{\frac{1}{2}})\mu_0 B_z & 0.4714(g_s - 1)\mu_0 B_z \\
0.4714(g_s - 1)\mu_0 B_z & \tfrac{1}{2}\zeta_{np} \pm \tfrac{3}{2}g(^2P_{\frac{3}{2}})\mu_0 B_z
\end{pmatrix}
\end{array}
\tag{5-15}
$$

where

$$g(^2 P_{1/2}) = 0.6659 \qquad \text{and} \qquad g(^2 P_{3/2}) = 1.3341 \qquad (5\text{-}16)$$

As it is seen in (5-15), the spin orbit matrix elements are obviously diagonal in J. Note also that contrary to the ground state, which exhibits a Zeeman splitting that is *linear* in the magnetic field B_z as do the states $np(^2 P \frac{3}{2}, \frac{3}{2})$, the states $np(^2 P \frac{3}{2}, \pm \frac{1}{2})$ and $np(^2 P \frac{1}{2}, \pm \frac{1}{2})$ are mixed, and the splittings of those states are no longer linear in B_z. The amount of mixing depends on the size of the fine structure splitting. However, the situation is changed when the hyperfine splitting is taken into account.

5.3 INTRODUCTORY REMARKS ON MAGNETIC HYPERFINE STRUCTURE

The nucleus possesses a total angular momentum **I**, often referred to as *nuclear spin*, and the electron system is identified by the total angular momentum **J**. The total angular momentum of the atom **F** is a result of these two angular momenta coupled together,

$$\mathbf{F} = \mathbf{I} + \mathbf{J}$$

where the magnitude of **F** is given by

$$F = I + J, \ I + J - 1, \ \dots, \ |I - J| \qquad (5\text{-}17)$$

Thus the number of hyperfine sublevels arising from a level of a given J is the minimum of $\{(2J + 1), (2I + 1)\}$. States of a given F are $2F + 1$-fold degenerate with respect to the quantum number M_F

$$M_F = F, \ F - 1, \ \dots, \ -F + 1, \ -F \qquad (5\text{-}18)$$

This degeneracy may be lifted by an applied external magnetic field or partially or completely by external electric fields.

For example, in the case of the ground state $1s(^2 S_{1/2})$ of atomic hydrogen $^1_1 H$ the nuclear angular momentum is $I = \frac{1}{2}$, and the electronic angular momentum is $J = \frac{1}{2}$. Hence $F = 0, 1$ and we get two hyperfine sublevels: a singlet with $F = 0$ and a triplet with $F = 1$. Experimentally, the energy separation between the states is

$$\Delta E = E(F = 1) - E(F = 0) = 5.9 \times 10^{-6} eV \qquad (5\text{-}19)$$

The frequency ν and wavelength λ associated with this transition are

$$\nu = 1420.4057517667(10) MHz \quad \text{and} \quad \lambda = 21.1 cm \qquad (5\text{-}20)$$

This is the origin of the well-known $21 cm$ line of radio astronomy associated with interstellar hydrogen. The transition is too slow to be seen in the laboratory by spontaneous emission but can be measured by stimulated emission to an extraordinary degree of accuracy. In the case of the interstellar medium the enormous amount of hydrogen makes the transition readily observable.

The situation is also interesting for the rubidium isotopes, $^{85}_{37} Rb$ and $^{87}_{37} Rb$, which have nuclear spins of $I = \frac{5}{2}$ and $I = \frac{3}{2}$, respectively. In both cases the electronic ground state is $5s(^2S_{1/2})$, hence $J = \frac{1}{2}$, and the two allowed values of F for each isotope in the ground state are

$$^{85}_{37}Rb \quad 5s(^2S_{1/2}) \quad F = 2, 3 \qquad ^{87}_{37}Rb \quad 5s(^2S_{1/2}) \quad F = 1, 2 \tag{5-21}$$

The first two excited electronic states of rubidium are the two fine structure levels $5p(^2P_{1/2})$ and $5p(^2P_{3/2})$, separated by $158.38cm^{-1}$. For these excited states the hyperfine sublevels occur with the following values of the total angular momentum F

$$^{85}_{37}Rb \quad 5p(^2P_{1/2}) \quad F = 2, 3 \qquad 5p(^2P_{3/2}) \quad F = 1, 2, 3, 4 \tag{5-22a}$$
$$^{87}_{37}Rb \quad 5p(^2P_{1/2}) \quad F = 1, 2 \qquad 5p(^2P_{3/2}) \quad F = 0, 1, 2, 3 \tag{5-22b}$$

That gives a qualitative description of the magnetic hyperfine structure but tells us nothing about the ordering of the hyperfine sublevels and of their separations.

5.4 MAGNETIC HYPERFINE STRUCTURE

The nuclear magnetic-dipole moment vector μ_I for a nucleus may be written as

$$\mu_I = g_I \mu_N \mathbf{I} \tag{5-23}$$

where g_I is the nuclear g−factor and β_N is the nuclear magneton, which is defined as

$$\mu_N = \frac{e\hbar}{2m_p} = \frac{m_e \beta}{m_p} \tag{5-24}$$

where m_p is the proton mass, m_e the electron mass and β is the Bohr magneton. The magnetic-moment vector of the nucleus can be taken as proportional to its angular momentum I and written as

$$\mu_I = \frac{\mu_I \mathbf{I}}{I} \tag{5-25}$$

where μ_I is the *nuclear magnetic moment* expressed in units of nuclear magnetons.

Each electron i in an unfilled shell produces a magnetic field B_i at the nucleus resulting in an interaction with the nuclear magnetic moment vector μ_I and adding to the Hamiltonian a term

$$H_{hfs} = -\sum_{i=1}^{N} \mathbf{B}_i \cdot \mu_I \tag{5-26}$$

For an electron in an orbital with ($\ell \neq 0$) the magnetic field $\mathbf{B}$ produced at the nucleus is[1-3]

$$\mathbf{B} = -2\mu_0 \frac{[\mathbf{l} - \mathbf{s} + 3\mathbf{r}(\mathbf{s} \cdot \mathbf{r})/r^2]}{r^3} \tag{5-27}$$

For N electrons we obtain[4-6]

$$H_{hfs} = 2\mu_0\mu_n g_I \sum_{i=1}^{N} \frac{\mathbf{N}_i \cdot \mathbf{I}}{r_i^3} \tag{5-28}$$

where

$$\mathbf{N}_i = \mathbf{l}_i - \mathbf{s}_i + \frac{3\mathbf{r}_i(\mathbf{s}_i \cdot \mathbf{r}_i)}{r_i^2} \tag{5-29}$$

$\mathbf{N}_i$ may be put into tensor operator form to yield[4]

$$\mathbf{N}_i = \mathbf{l}_i - \sqrt{10}(\mathbf{s}\mathbf{C}^{(2)})_i^{(1)} \tag{5-30}$$

and finally

$$H_{hfs} = a_\ell \sum_{i=1}^{N}[\mathbf{l}_i^{(1)} - \sqrt{10}(\mathbf{s}^{(1)}\mathbf{C}^{(2)})_i^{(1)}] \cdot \mathbf{I} \tag{5-31}$$

with

$$a_\ell = 2\mu_0\mu_n g_I <r^{-3}> = \frac{2\mu_0\mu_n\mu_I <r^{-3}>}{I} \tag{5-32}$$

where $<r^{-3}>$ is the expectation value of the inverse-cube radius in the one-electron state of an electron from an open shell.

Note that the matrix elements of $\mathbf{N}_i$ vanish for s-orbitals. The first part vanishes because of an obvious reason, since $\mathbf{l}_i = 0$; the second part does not contribute because of the triangular conditions for the non-vanishing matrix elements of double tensor operators. Indeed, the second term at the right-hand side of (5-31) acts as a tensor operator with rank 1 in the spin part of the space (as each vector), and the spherical tensor of rank 2 in the orbital part, while coupled together, it behaves as a tensor of rank 1 in the spin-orbital space. Thus, for the non-zero matrix elements $\langle 0|w^{(12)k}|0\rangle$ the triangular condition requires $k = 0$. However, for unpaired electrons in the s-orbitals Fermi[7] showed that there is, what is now known as the *Fermi contact term*, an additional contribution, and as a result the modified, and more general operator, has the form

$$\mathbf{N}_i' = \mathbf{N}_i + \frac{8}{3}\pi |\Psi_s(0)|^2 \mathbf{s}_i \tag{5-33}$$

where $\Psi_s(0)$ is the value of the normalized Schrödinger eigenfunction of the s-orbital at the nucleus. This means that for ($\ell >= 1$) only the first term, $\mathbf{N}_i$, is non-zero, while for ($\ell = 0$) only the second term contributes to the magnetic field generated at the nucleus.

Matrix elements of H_{hfs} may be taken as diagonal in the nuclear spin I, since states of different nuclear spin are usually separated by MeV compared with eV separations of electronic states. In the absence of external electric or magnetic fields the matrix elements are diagonal in F and independent of M_F. In such a situation it is convenient to work in a $JIFM_F$ scheme whereas if large external fields are present, it is more realistic to work in a $JM_JIM_IM_F$ scheme. If the fine structure is small, it may be necessary to consider matrix elements that are off-diagonal in J. In the case of intermediate coupling it may be necessary to consider matrix elements that couple different SL terms.

For simplicity let us first consider the matrix elements of H_{hfs}, which are diagonal in J though not necessarily diagonal in other quantum numbers α, α'. We have, after recalling (2-44) and that $\mathbf{I}$ is, as every vector, a tensor operator with a rank $k = 1$

$$\langle \alpha J I F M | H_{hfs} | \alpha' J I F M \rangle = (-1)^{J+I+F} a_\ell \begin{Bmatrix} J & J & 1 \\ I & I & F \end{Bmatrix} \langle I \| I^{(1)} \| I \rangle$$

$$\times \langle \alpha J \| \sum_{i=1}^{N} N_i^{(1)} \| \alpha' J \rangle \qquad (5\text{-}34)$$

The $6j$−symbol can be evaluated explicitly to yield

$$\langle \alpha J I F M | H_{hfs} | \alpha' J I F M \rangle = \tfrac{1}{2}[F(F+1) - J(J+1) - I(I+1)]$$

$$\times a_\ell \frac{\langle \alpha J \| \sum_{i=1}^{N} N_i^{(1)} \| \alpha' J \rangle}{\sqrt{J(J+1)(2J+1)}} \qquad (5\text{-}35\text{a})$$

$$= \tfrac{1}{2} A K \qquad (5\text{-}35\text{b})$$

where

$$K = F(F+1) - J(J+1) - I(I+1) \qquad (5\text{-}35\text{c})$$

and

$$A = a_\ell \frac{\langle \alpha J \| \sum_{i=1}^{N} N_i^{(1)} \| \alpha' J \rangle}{\sqrt{J(J+1)(2J+1)}} \qquad (5\text{-}35\text{d})$$

The *measured* value of A is known as the *magnetic hyperfine structure constant* and may be positive or negative. The quantity a_ℓ, defined in (5-32), is constant for all the states of a given electron configuration. The second factor in A is different for each of the states of the configuration and constant within the hyperfine splittings of a given level. If $A > 0$, then the lowest F−value lies lowest in the energy scale, whereas if $A < 0$, it lies highest. Knowing A from experiment it is possible to deduce a value for the nuclear magnetic moment, if we can calculate $< r^{-3} >$ and the angular part of (5-35d). Note that the reduced matrix element $\langle I \| I^{(1)} \| I \rangle$ vanishes unless $I > 0$, since for $I = 0$, $F = J$, and as a result $K = 0$, which gives the zero correction to the energy.

5.5   EXERCISES

5-1. Show that

$$\frac{\langle SLJ \| L^{(1)} \| SLJ \rangle}{\sqrt{J(J+1)(2J+1)}} = \frac{J(J+1) + L(L+1) - S(S+1)}{2J(J+1)}$$

$$= 2 - g(SLJ) \qquad (5\text{-}36)$$

5-2. Show that

$$\langle \ell \| C^{(2)} \| \ell \rangle = -\sqrt{\frac{\ell(\ell+1)(2\ell+1)}{(2\ell-1)(2\ell+3)}} \qquad (5\text{-}37)$$

5-3. Show that for a single electron in an orbital with ($\ell >= 1$)

$$-\frac{\sqrt{10}\langle s\ell j\|(s^{(1)}C^{(2)})^{(1)}\|s\ell j\rangle}{\sqrt{j(j+1)(2j+1)}} = (2j+1)\sqrt{\frac{30s(s+1)(2s+1)\ell(\ell+1)(2\ell+1)}{j(j+1)(2j+1)(2\ell-1)(2\ell+3)}}$$

$$\times \begin{Bmatrix} s & s & 1 \\ \ell & \ell & 2 \\ j & j & 1 \end{Bmatrix} \tag{5-38}$$

5-4. Show that

$$\begin{Bmatrix} \ell+\frac{1}{2} & \ell+\frac{1}{2} & 1 \\ \ell & \ell & 2 \\ \frac{1}{2} & \frac{1}{2} & 1 \end{Bmatrix} = -\frac{\sqrt{10\ell(2\ell-1)}}{30(\ell+1)(2\ell+1)} \tag{5-39a}$$

and

$$\begin{Bmatrix} \ell-\frac{1}{2} & \ell-\frac{1}{2} & 1 \\ \ell & \ell & 2 \\ \frac{1}{2} & \frac{1}{2} & 1 \end{Bmatrix} = +\frac{\sqrt{10(2\ell+3)(\ell+1)}}{30\ell(2\ell+1)} \tag{5-39b}$$

5-5. Show that for a single electron in an orbital with ($\ell >= 1$),

$$A = \begin{cases} a_\ell \frac{4\ell(\ell+1)}{(2\ell+1)(2\ell+3)} & j = \ell+\frac{1}{2} \\ a_\ell \frac{4\ell(\ell+1)}{(2\ell+1)(2\ell-1)} & j = \ell-\frac{1}{2} \end{cases} \tag{5-40}$$

5-6. Show that for a $J = \frac{1}{2}$ electronic level of an atom with a nuclear spin $I >= \frac{1}{2}$ and magnetic hyperfine structure constant A a pair of sublevels is formed with an energy separation of

$$\Delta E = \frac{1}{2}(2I+1)A \tag{5-41}$$

REFERENCES

1. Ramsey N F (1953) *Nuclear Moments* Wiley - Interscience Publication, John Wiley and Sons Inc., New York.
2. Kopfermann H (1958) *Nuclear Moments* 2edn New York: Academic Press.
3. Weissbluth M (1978) *Atoms and Molecules* New York: Academic Press.
4. Judd B R (1998) *Operator Techniques in Atomic Spectroscopy* Princeton: Princeton University Press.
5. Edmonds A R (1960) *Angular Momentum in Quantum Mechanics* Princeton: Princeton University Press.
6. Fano U and Racah G (1959) *Irreducible Tensorial Sets* New York: Academic Press.
7. Fermi E, (1930) *Über die Magnetischen Momente der Atomkerne Z Physik,* **60** 320.

6 Magnetic Dipole and Electric Quadrupole Hyperfine Structures

The road ahead can hardly help being strewn with many a mistake. The main point is to get those mistakes made and recognized as fast as possible!

John A Wheeler

In this chapter we first consider the magnetic hyperfine structure in the $J M_J I M_I$ scheme, which later is needed when the impact of the external magnetic or electric fields is discussed. The latter case is of particular importance when considering ions in crystal fields. The analysis here is concluded by consideration of hyperfine structure associated with nuclei possessing electric quadrupole moments.

An excellent compilation of data on the Sodium D lines and on *Caesium D* lines has been given by Steck[1]. He has also given a similar database on the *Rubidium* 87 *D* lines[2] (and the collection includes also an excellent set of references). From Table 5-1 we have for $^{133}_{55}Cs$ the magnetic dipole hyperfine structure constants

$$A(6p\,^2P_{1/2}) = 291.920(19)MHz$$
$$A(6p\,^2P_{3/2}) = 50.275(3)MHz$$

and hence for their ratio

$$\frac{A(6p\,^2P_{1/2})}{A(6p\,^2P_{3/2})} = 5.8064$$

which may be compared with the ratio 5 coming from (5-40) in Exercise 5-5 at the end of the previous chapter. In a similar way it is possible to find for $^{23}_{11}Na$ and $^{87}_{37}Rb$ the respective ratios

$$\frac{A(3p\,^2P_{1/2})}{A(3p\,^2P_{3/2})} = 5.005 \quad \text{and} \quad \frac{A(5p\,^2P_{1/2})}{A(5p\,^2P_{3/2})} = 4.82$$

6.1 MAGNETIC HYPERFINE STRUCTURE IN THE $J M_J I M_I$ BASIS

In the $J I F M_F$ scheme we used the operator

$$H_{hfs} = a_\ell(\mathbf{N}^{(1)} \cdot \mathbf{I}^{(1)}) \tag{6-1}$$

Noting (2-40) we can write the scalar product in terms of tensor operator components to give

$$H_{hfs} = a_\ell \sum_{q=-1}^{1} (-1)^q N_q^{(1)} I_{-q}^{(1)} \tag{6-2}$$

The matrix elements of H_{hfs} in the $JM_J IM_I$ scheme then become

$$\langle \alpha SLJM_J IM_I | H_{hfs} | \alpha' S'L'JM_J' IM_I' \rangle = a_\ell \sum_{q=-1}^{1} (-1)^{J-M_J} \begin{pmatrix} J & 1 & J \\ -M_J & q & M_J' \end{pmatrix}$$

$$\times (-1)^{I-M_I} \begin{pmatrix} I & 1 & I \\ -M_I & -q & M_I' \end{pmatrix} \langle \alpha SLJ \| N^{(1)} \| \alpha' S'L'J \rangle \langle I \| I^{(1)} \| I \rangle \tag{6-3}$$

Explicit evaluation of the $3j$–symbols allows one to write the matrix elements diagonal in M_J and M_I (which implies that $q = 0$) as

$$\langle \alpha SLJM_J IM_I | H_{hfs} | \alpha' S'L'JM_J IM_I \rangle = AM_J M_I \tag{6-4a}$$

The off-diagonal matrix elements (for $q = \pm 1$) are determined by the following expression

$$\langle \alpha SLJM_J IM_I | H_{hfs} | \alpha' S'L'JM_J \pm 1 IM_I \mp 1 \rangle$$
$$= \tfrac{1}{2} A \sqrt{(J \mp M_J)(J \pm M_J + 1)(I \pm M_I)(I \mp M_I + 1)} \tag{6-4b}$$

where A is the magnetic hyperfine structure constant defined in (5-35d).

6.2   ZEEMAN EFFECT IN THE $JIFM_F$ AND $JM_J IM_I M_F$ BASES

As a first example, a weak external magnetic field B_z acting on hyperfine levels in a $JIFM_F$ (electronic and nuclear momenta are coupled) basis is analyzed. The additional term in the Hamiltonian is written in terms of the electron spin, orbital and nuclear g–factors as

$$H_{mag} = \mu_B B_z \left(L_0^{(1)} + g_s S_0^{(1)} + g_I' I_0^{(1)} \right) \tag{6-5}$$

where $\mu_B \equiv \mu_0$ and $g_I' = -\frac{\mu_n}{\mu_B} g_I$.

In general the nuclear Zeeman effect, represented by the third term, is more than three orders of magnitude weaker than that of the electronic Zeeman effect, and as such is often ignored. However, we retain it, since it can lead to a direct method of determining nuclear magnetic moments.

The diagonal matrix elements of H_{mag} are determined using the tensor operator formalism developed in chapter 2 (see (2-42) - (2-46)) together with the explicit forms for the relevant $3j$– and $6j$–symbols. The Wigner-Eckart theorem from (2-42) applied together with (2-48c) gives

$$\langle \alpha JIFM_F | H_{mag} | \alpha' JIFM_F \rangle = \mu_B B_z \frac{M_F}{\sqrt{F(F+1)(2F+1)}}$$
$$\times [\langle \alpha JIF \| (L^{(1)} + g_s S^{(1)}) \| \alpha' JIF \rangle + \langle \alpha JIF \| g_I' I^{(1)} \| \alpha' JIF \rangle] \tag{6-6}$$

The first reduced matrix element can be evaluated using (2-45), since in this case the operators act within the electronic part of the wave function. The second matrix element is an example of the operator that acts within the second, nuclear part of the wave function, and therefore it is evaluated using (2-46). Finally, using the explicit form of the $6j$−symbol we have

$$\langle \alpha J I F M_F | H_{mag} | \alpha' J I F M_F \rangle = \mu_B B_z M_F \left[g_J \frac{[F(F+1) - I(I+1) + J(J+1)]}{2F(F+1)} \right.$$

$$\left. + g_I' \frac{[F(F+1) + I(I+1) - J(J+1)]}{2F(F+1)} \right] \qquad (6\text{-}7)$$

In the case of intermediate coupling the electronic Lande g−factor g_J may be replaced by its intermediate coupling corrected value evaluated in the previous chapter. For experimental convenience (6-7) is often rewritten as

$$\langle \alpha J I F M_F | H_{mag} | \alpha' J I F M_F \rangle = \mu_B B_z M_F g_F \qquad (6\text{-}8)$$

with

$$g_F = g_J \frac{[F(F+1) - I(I+1) + J(J+1)]}{2F(F+1)}$$

$$+ g_I' \frac{[F(F+1) + I(I+1) - J(J+1)]}{2F(F+1)} \qquad (6\text{-}9)$$

In the case of a weak magnetic field the energy shifts is linear in the magnetic field B_z. For fields that produce splitting comparable with the hyperfine splitting, or greater, it is necessary to consider matrix elements that are off-diagonal in F. In that case it is usually simpler to work in the $J M_J I M_I M_F$ scheme in which the electronic and nuclear momenta are uncoupled.

Assuming that the fine structure splitting is sufficiently large it is possible to ignore J−mixing and just consider the matrix elements that couple states differing in F by one unit.

$$\langle \alpha J I F M_F | H_{mag} | \alpha' J I F + 1 M_F \rangle$$

$$= \mu_B B_z (-1)^{F-M_F} \begin{pmatrix} F & 1 & F+1 \\ -M_F & 0 & M_F \end{pmatrix} (-1)^{J+I+F+1} \sqrt{(2F+1)(2F+3)}$$

$$\times \left[\begin{Bmatrix} F & 1 & F+1 \\ J & I & J \end{Bmatrix} \langle \alpha J \| (L^{(1)} + g_s S^{(1)}) \| \alpha' J \rangle \right.$$

$$\left. - \begin{Bmatrix} F & 1 & F+1 \\ I & J & I \end{Bmatrix} \langle \alpha I \| g_I' I^{(1)} \| \alpha' I \rangle \right] \qquad (6\text{-}10)$$

Explicit evaluation of the $3j−$ and $6j−$symbols using (2-56) and (2-49b) respectively leads to

$$\langle \alpha J I F M_F | H_{mag} | \alpha' J I F + 1 M_F \rangle$$

$$= \mu_B B_z (g_J - g_I') \sqrt{(F+1)^2 - M_F^2}$$

$$\times \sqrt{\frac{(F+J+I+2)(F+J-I+1)(F-J+I+1)(I+J-F)}{4(F+1)^2(2F+1)(2F+3)}} \qquad (6\text{-}11)$$

In the case of strong Zeeman effect the functions are expressed in the uncoupled angular momenta scheme $J M_J I M_I M_F$. In this particular basis the Zeeman matrix elements are especially simple. Further discussion is presented with the assumption that the fine structure is very much larger than the hyperfine structure, which is a good approximation for the heavier alkali atoms such as Rb and Cs. In that case we can ignore matrix elements that are non-diagonal in J and obtain

$$\langle \alpha J M_J I M_I M_F | H_{mag} | \alpha' J M_J I M_I M_F \rangle = \mu_B B_z (g_J M_J + g'_I M_I)$$

$$\text{with } M_J + M_I = M_F \qquad (6\text{-}12)$$

Note that M_F remains a conserved quantum number.

6.3   EXAMPLE OF A $J = \frac{1}{2}$ ELECTRONIC LEVEL

The particular case of a $J = \frac{1}{2}$ electronic level is relatively simple and of great practical importance. Of considerable interest are the ground and first excited states of the alkali atoms[1,3]. To construct a very specific example we consider an atom with nuclear spin $I = \frac{7}{2}$ and $J = \frac{1}{2}$ as is indeed the case for the ground state of $^{133}_{55}Cs$. Ultimately we want to compare our calculations with known experimental data. Clearly, there are two hyperfine sublevels with

$$F = 3 \quad \text{with} \quad M_F = \pm 3,\ \pm 2,\ \pm 1,\ 0 \qquad (6\text{-}13a)$$
$$F = 4 \quad \text{with} \quad M_F = \pm 4,\ \pm 3,\ \pm 2,\ \pm 1,\ 0 \qquad (6\text{-}13b)$$

To simplify matters it is further assumed that the $J = \frac{1}{2}$ level has a Lande g−factor g_J, and that the nuclear g−factor g'_I is small enough, compared with g_J, to be ignored. The magnetic field matrix elements are evaluated in the $J I F M_F$ scheme in terms of $\mu_B B_z$. Using (6-7) for the diagonal elements, and (6-11) for the elements that couple the two possible F values, we find the matrices

$$M_F = \pm 4 \quad |\tfrac{1}{2}\tfrac{7}{2}4, \pm 4\rangle$$
$$\langle \tfrac{1}{2}\tfrac{7}{2}4, \pm 4| \left(\pm \tfrac{g_J}{2} \right)$$

$$M_F = \pm 3 \quad |\tfrac{1}{2}\tfrac{7}{2}3, \pm 3\rangle \quad |\tfrac{1}{2}\tfrac{7}{2}4, \pm 3\rangle$$
$$\begin{array}{c} \langle \tfrac{1}{2}\tfrac{7}{2}3, \pm 3| \\ \langle \tfrac{1}{2}\tfrac{7}{2}4, \pm 3| \end{array} \left(\begin{array}{cc} \mp \tfrac{3g_J}{8} & \tfrac{\sqrt{7}g_J}{8} \\ \tfrac{\sqrt{7}g_J}{8} & \pm \tfrac{3g_J}{8} \end{array} \right)$$

$$M_F = \pm 2 \quad |\tfrac{1}{2}\tfrac{7}{2}3, \pm 2\rangle \quad |\tfrac{1}{2}\tfrac{7}{2}4, \pm 2\rangle$$
$$\begin{array}{c} \langle \tfrac{1}{2}\tfrac{7}{2}3, \pm 2| \\ \langle \tfrac{1}{2}\tfrac{7}{2}4, \pm 2| \end{array} \left(\begin{array}{cc} \mp \tfrac{g_J}{4} & \tfrac{g_J\sqrt{3}}{4} \\ \tfrac{g_J\sqrt{3}}{4} & \pm \tfrac{g_J}{4} \end{array} \right)$$

$$M_F = \pm 1 \quad |\tfrac{1}{2}\tfrac{7}{2}3, \pm 1\rangle \quad |\tfrac{1}{2}\tfrac{7}{2}4, \pm 1\rangle$$
$$\begin{array}{c} \langle \tfrac{1}{2}\tfrac{7}{2}3, \pm 1| \\ \langle \tfrac{1}{2}\tfrac{7}{2}4, \pm 1| \end{array} \left(\begin{array}{cc} \mp \tfrac{g_J}{8} & \tfrac{g_J\sqrt{15}}{8} \\ \tfrac{g_J\sqrt{15}}{8} & \pm \tfrac{g_J}{8} \end{array} \right)$$

$$M_F = 0 \quad |\tfrac{1}{2}\tfrac{7}{2}3, 0\rangle \quad |\tfrac{1}{2}\tfrac{7}{2}4, 0\rangle$$
$$\begin{array}{c} \langle \tfrac{1}{2}\tfrac{7}{2}3, 0| \\ \langle \tfrac{1}{2}\tfrac{7}{2}4, 0| \end{array} \left(\begin{array}{cc} 0 & \mp \tfrac{g_J}{2} \\ \tfrac{g_J}{2} & 0 \end{array} \right)$$

Diagonalization of the above matrices yields just the two eigenvalues $\pm \tfrac{g_J}{2}$ each with a degeneracy of 5. This is exactly what would be expected if the calculations would have been performed in the $J M_J I M_I M_F$ scheme.

However, to complete the calculations the magnetic hyperfine matrix elements evaluated in the same basis should be included. These follow from (5-35b) giving

$$\langle \tfrac{1}{2}\tfrac{7}{2}4M_F | H_{hfs} | \tfrac{1}{2}\tfrac{7}{2}4M_F \rangle = \frac{7}{4}A \qquad (6\text{-}14a)$$

$$\langle \tfrac{1}{2}\tfrac{7}{2}3M_F | H_{hfs} | \tfrac{1}{2}\tfrac{7}{2}3M_F \rangle = -\frac{9}{4}A \qquad (6\text{-}14b)$$

where A is the magnetic hyperfine structure constant. To find the Zeeman effect on the two hyperfine levels associated with a $J = \tfrac{1}{2}$ level the matrices of H_{mag} and H_{hfs}, expressed in consistent units, must be combined and then diagonalized. Experimentalists commonly express their measurements of hyperfine constants in terms of MHz, since many of their measurements involve microwave techniques. Thus in the data tables[1] for caesium we find the Bohr magneton given as

$$\mu_B = h \cdot 1.399\ 624\ 624(56) MHz/G \qquad (6\text{-}15)$$

which implies the magnetic field is measured in Gauss. The magnetic hyperfine structure constant A for the $6p\ ^2P_{\frac{1}{2}}$ state is given as

$$A_{6p^2 P_{\frac{1}{2}}} = h \cdot 291.920(19) MHz \qquad (6\text{-}16)$$

If $J = \tfrac{1}{2}$, then necessarily $F = I \pm \tfrac{1}{2}$, and for $M_F = \pm(I + \tfrac{1}{2})$ we have

$$
\begin{array}{c}
|\tfrac{1}{2}II + \tfrac{1}{2}, \pm(I + \tfrac{1}{2})\rangle \\
\langle \tfrac{1}{2}II + \tfrac{1}{2}, \pm(I + \tfrac{1}{2})| \left(\ \tfrac{1}{2}IA \pm k(2I + 1) \ \right)
\end{array}
\qquad (6\text{-}17)
$$

where

$$k = \frac{\mu_B B_z g_J}{2I + 1} \qquad (6\text{-}18)$$

For all other values of M_F we have the matrices of rank two,

$$
\begin{array}{cc}
|M_F| \le I - \tfrac{1}{2} & |\tfrac{1}{2}II - \tfrac{1}{2}, M_F\rangle \qquad |\tfrac{1}{2}II + \tfrac{1}{2}, M_F\rangle \\
\begin{array}{c}\langle \tfrac{1}{2}II - \tfrac{1}{2}, M_F| \\ \langle \tfrac{1}{2}II + \tfrac{1}{2}, M_F| \end{array} &
\begin{pmatrix}
-\tfrac{1}{2}(I+1)A - kM_F & k\sqrt{(I + \tfrac{1}{2})^2 - M_F^2} \\
k\sqrt{(I + \tfrac{1}{2})^2 - M_F^2} & \tfrac{1}{2}IA + kM_F
\end{pmatrix}
\end{array}
\qquad (6\text{-}19)
$$

Taking the determinant of (6-19) it is possible to evaluate the eigenvalues $\lambda_\pm$ by solving the secular equation

$$\lambda^2 + \tfrac{1}{2}A\lambda - \tfrac{1}{4}\left[A^2 I(I + 1) + 2kAM_F(2I + 1) + k^2(2I + 1)^2 \right] \qquad (6\text{-}20)$$

The roots of the quadratic equation give

$$\lambda_\pm = -\frac{A}{4} \pm \frac{A}{4}\sqrt{1 + 4I(I + 1) + \frac{8}{A}kM_F(2I + 1) + 4\frac{k^2}{A^2}(2I + 1)^2} \qquad (6\text{-}21)$$

Denoting

$$x = \frac{k}{A} \tag{6-22}$$

the solutions have a simpler form

$$\lambda_{\pm} = \frac{1}{4}A\left(-1 \pm (2I+1)\sqrt{1 + \frac{8x}{2I+1}M_F + 4x^2}\right) \tag{6-23}$$

This expression, together with the definitions in (6-18) and (6-22), allows one to describe the behavior of any $J = \frac{1}{2}$ level for any nuclear spin I having magnetic hyperfine structure in a magnetic field; expression (6-23) is related to the celebrated Breit-Rabi equation[4].

6.4 EXAMPLE OF $^{133}_{55}Cs$

In the case of $^{133}_{55}Cs$ we have $I = \frac{7}{2}$ and (6-23) becomes

$$\lambda_{\pm} = \frac{1}{4}A\left(-1 \pm 8\sqrt{1 + xM_F + 4x^2}\right) \tag{6-24}$$

The case $x = 0$ corresponds to $B_z = 0$ and gives just the two hyperfine levels found in (6-14a,b). Given that the experimental value of A is positive we have the $F = 3$ state of $6p(^2P_{1/2})$ below that of $F = 4$ with a separation of $4A$. For small x, i.e $1 \gg x \gg x^2$, we have from (6-24)

$$\lambda_{\pm} = \frac{1}{4}A\left(-1 \pm 8\sqrt{1 + xM_F}\right) \tag{6-25}$$

Expanding the square root to first-order in x, noting (6-22) and (6-18), using the value of $g_J = \frac{2}{3}$ for $6p(^2P_{1/2})$, we find that the states with $F = 3$ and $F = 4$ split in a magnetic field B_z as

$$E(F = 4) = \frac{7}{4}A + \frac{\mu_B}{12}B_z M_F \quad MHz \tag{6-26a}$$

$$E(F = 3) = -\frac{9}{4}A - \frac{\mu_B}{12}B_z M_F \quad MHz \tag{6-26b}$$

relative to the center of gravity of the $6p(^2P_{1/2})$ state. Thus in a weak magnetic field the two hyperfine levels determined in (6-14a) and (6-14b) split into $2F + 1$ equi-spaced Zeeman sublevels (as indicated by the number of values of M_F). The separation of consecutive Zeeman sublevels being

$$\Delta E = \frac{\mu_B}{12}B_z = 0.1166B_z \tag{6-27}$$

where in (6-26a,b) and (6-27) we have put $h = 1$ and B_z is in Gauss to give measurements of E and ΔE in MHz as is common in such experiments[1,3].

In the case of a very strong magnetic field and $g_J \gg g_I$, such that the Zeeman splittings are very much greater than the hyperfine splitting, the Zeeman levels coalesce to give two levels corresponding to $M_J = \pm\frac{1}{2}$, each having a degeneracy of $2I + 1$.

Using the results presented in (6-23) it is interesting to discuss the clock correction equation. The ground state of an alkali atom has $J = \frac{1}{2}$, and just the two hyperfine levels with $F = I \pm \frac{1}{2}$. These two hyperfine levels are separated by

$$\Delta E_{hfs} = \tfrac{1}{2}A(2I + 1) \tag{6-28}$$

It follows from (6-23) that there can be no first-order Zeeman shift for the Zeeman sublevels with $M_F = 0$. In such a case (6-23) simplifies to

$$\lambda_\pm = \frac{A}{4}\left(-1 \pm (2I + 1)\sqrt{1 + 4x^2}\right) \quad M_F = 0 \tag{6-29}$$

The difference in the two levels with $M_F = 0$ becomes

$$\Delta\lambda = \tfrac{1}{2}A(2I + 1)\sqrt{1 + 4x^2}$$

and the correction due to the magnetic field is

$$\begin{aligned}
&\approx A(2I + 1)x^2 \\
&= \frac{k^2}{A}(2I + 1) \\
&= \frac{\mu_B^2 B_z^2(g_J - g_I')^2}{A(2I + 1)}
\end{aligned} \tag{6-30}$$

where we have included the nuclear g–factor, g_I. Using (6-28) we obtain the change $\Delta\omega$ in angular frequency ω for the clock transitions as

$$\Delta\omega = \frac{\mu_B^2 B_z^2(g_J - g_I')^2}{2\hbar \Delta E_{hfs}} \tag{6-31}$$

6.5 ELECTRIC QUADRUPOLE HYPERFINE STRUCTURE

Nuclei with $I >= 1$ exhibit deformation from spherical symmetry. The interaction between the charged nucleons and electrons is in general defined by the electrostatic interaction between the appropriate charge densities,

$$H_{EM} = -e^2 \int_{\tau_e} \int_{\tau_n} \frac{\rho_e(r_e)\rho_n(r_n)d\tau_e d\tau_n}{|\mathbf{r}_e - \mathbf{r}_n|} \tag{6-32}$$

where $e\rho_e(r_e)$ and $e\rho_n(r_n)$ are the electron and nucleon charge densities and r_e and r_n are measured relative to the nuclear centre. Expanding the denominator of (6-32) in spherical harmonics (as in (2-29)) the interaction has the form of the multipole expansion

$$\frac{1}{|\mathbf{r}_e - \mathbf{r}_n|} = \sum_k \frac{r_n^k}{r_e^{k+1}}\left(\mathbf{C}_e^{(k)} \cdot \mathbf{C}_n^{(k)}\right) \tag{6-33}$$

where in general the rank of the multipoles localized on both interacting centers k is not limited. In practice however, the parity considerations for the non-vanishing corrections to the energy (in particular defined within the single configuration approximation) allow only the even rank multipoles. Thus all the terms for $k = odd$ in (6-33) (in particular dipole interactions) do not contribute to the energy. However when the scheme is extended by the inter-shell interactions and the interaction between various excited configurations are taken into account, the odd terms also play a contributing role. This aspect of the interactions via the electric multipole hyperfine operator is discussed separately in chapter 20 which is devoted to the hyperfine induced transitions in the lanthanide ions.

Note that in practice the multipole expansion is limited to a few terms by the triangular condition for the non-vanishing reduced matrix elements of the spherical tensors that define the angular part of H_{EM}. Limiting the expansion in (6-33) to the values of $k = even$, among all terms, the most important role is played by the quadrupole interaction for $k = 2$. The $k = 0$ term corresponds to a monopole (or single charge) that is represented by a tensor operator, which is a scalar (zero order), and therefore does not introduce any modification to the energy pattern and may be dropped. At the same time, the higher electric multipole terms are not considered here.

Thus for the electric-quadrupole interaction the additional term of the Hamiltonian has the form

$$H_{EQ} = -e^2 \int_{\tau_e} \int_{\tau_n} \rho_e(r_e)\rho_n(r_n)\frac{r_n^2}{r_e^3} \left(\mathbf{C}_e^{(2)} \cdot \mathbf{C}_n^{(2)}\right) d\tau_e d\tau_n \qquad (6\text{-}34)$$

The matrix elements of H_{EQ} may be evaluated to give for the diagonal elements

$$\langle \alpha JIF|\mathbf{H}_{EQ}|\alpha' JIF\rangle = -e^2(-1)^{J+I+F} \begin{Bmatrix} J & J & 2 \\ I & I & F \end{Bmatrix} \langle \alpha J \|r_e^{-3}C_e^{(2)}\|\alpha' J\rangle$$

$$\langle I\|r_n^2 C_n^{(2)}\|I\rangle \qquad (6\text{-}35)$$

The nuclear quadrupole moment Q is commonly defined as the matrix element over the space of the nuclear coordinates evaluated when $\mathbf{I}$ has its largest component in the $z-$direction, that is,

$$Q = \langle II|r_n^2(3cos^2\theta - 1)|II\rangle_{av} = 2\langle II|r_n^2 C_{n0}^{(2)}|II\rangle$$

$$= 2\begin{pmatrix} I & 2 & I \\ -I & 0 & I \end{pmatrix}\langle I\|r_n^2 C_n^{(2)}\|I\rangle \qquad (6\text{-}36)$$

Evaluating the $3j-$symbol explicitly we obtain

$$Q = 2\sqrt{\frac{I(2I-1)}{(I+1)(2I+1)(2I+3)}}\langle I\|r_n^2 C_n^{(2)}\|I\rangle \qquad (6\text{-}37)$$

Inserting (6-37) into (6-35) and evaluating the $6j-$symbol explicitly leads to the first order correction to the energy evaluated as a matrix element of the perturbation H_{EM}

(if the interactions are weak) between the zero order functions,

$$\langle \alpha J I F | H_{EQ} | \alpha' J I F \rangle = \frac{-e^2 Q < r^{-3} >}{I(2I-1)} \langle \alpha J \| C_e^{(2)} \| \alpha' J \rangle$$

$$\times \frac{\left[\frac{3}{4}K(K+1) - I(I+1)J(J+1)\right]}{\sqrt{(2J-1)J(2J+1)(J+1)(2J+3)}}$$

$$= b_\ell X_J \left[\frac{\frac{3}{4}K(K+1) - I(I+1)J(J+1)}{I(2I-1)J(2J-1)}\right] \quad (6\text{-}38)$$

where

$$b_\ell = e^2 Q < r^{-3} > \quad (6\text{-}39)$$

The *electric quadrupole hyperfine constant B* is usually defined as

$$B = b_\ell X_J \quad (6\text{-}40)$$

and K is defined as in (5-35c). X_J is different for different electronic states while b_ℓ is a constant over the states of a given configuration. Specifically,

$$X_J = -\sqrt{\frac{J(2J-1)}{(J+1)(2J+1)(2J+3)}} \langle \alpha J \| C_e^{(2)} \| \alpha' J \rangle \quad (6\text{-}41)$$

Thus, for nuclei with $I \geq 1$ we have two corrections to the energy that originate from the hyperfine interactions,

$$H_{hfs} = \frac{1}{2}AK + B \left[\frac{\frac{3}{4}K(K+1) - I(I+1)J(J+1)}{I(2I-1)J(2J-1)}\right] \quad (6\text{-}42)$$

where the first contribution is due to the magnetic hyperfine interactions, and the second results from the electric quadrupole hyperfine interactions. In the $J M_J I M_I$ scheme, when the electronic and nuclear angular momenta are uncoupled, the matrix elements of H_{EQ} have the form

$$\langle \alpha J M_J I M_I | H_{EQ} | \alpha' J M_J \pm q I M_I \mp q \rangle$$

$$= (-1)^{J-M_J \mp q} \begin{pmatrix} J & 2 & J \\ -M_J & \mp q & M_J \pm q \end{pmatrix} (-1)^{I-M_I} \begin{pmatrix} I & 2 & I \\ -M_I & \pm q & M_I \mp q \end{pmatrix}$$

$$\times \frac{B}{2} \sqrt{\frac{(2I+1)(I+1)(2I+3)(2J+1)(J+1)(2J+3)}{I(2I-1)J(2J-1)}} \quad (6\text{-}43)$$

where q is limited to the values of 0, 1, 2. Note that $M_F = M_J + M_I$ remains a good quantum number.

The matrix elements that are diagonal in M_J and M_I are found by explicit evaluation of the two $3j-$ symbols to give

$$\langle \alpha J M_J I M_I | H_{EQ} | \alpha' J M_J I M_I \rangle = \frac{B}{2} \left[\frac{3M_I^2 - I(I+1)}{IJ(2I-1)(2J-1)}\right]$$

$$\times [3M_J^2 - J(J+1)] \quad (6\text{-}44)$$

Finally, assuming that the off-diagonal matrix elements may be ignored, the corrections to the energy are determined by the following expression derived for the JM_JIM_I coupling scheme

$$\langle \alpha JM_JIM_I|H_{EQ}|\alpha'JM_JIM_I\rangle = M_JM_IA$$
$$+\frac{B}{2}\left[\frac{3M_I^2-I(I+1)}{IJ(2I-1)(2J-1)}\right][3M_J^2-J(J+1)]$$

$$(6\text{-}45)$$

which is the analogue of energy corrections in (6-42) evaluated with the functions expressed in the scheme of coupled electronic and nuclear angular momenta.

6.6   EXERCISES

6-1 Show that

$$\langle \alpha SLJ\|L^{(1)}+g_sS^{(1)}\|\alpha SLJ\rangle = g(SLJ)\sqrt{J(J+1)(2J+1)} \quad (6\text{-}46)$$

6-2 Show that

$$\langle \alpha SLJ\|L^{(1)}+g_sS^{(1)}\|\alpha SLJ+1\rangle$$
$$\frac{1-g_s}{2}\sqrt{\frac{(J+L+S+2)(J+L-S+1)(J-L+S+1)(S+L-J)}{J+1}}$$

$$(6\text{-}47)$$

Note that in the above two exercises the matrix elements are diagonal in α, S and L. Note also the special 6-j symbols

$$\left\{\begin{matrix} a & b & c \\ 1 & c & b \end{matrix}\right\} = (-1)^s\frac{2[a(a+1)-b(b+1)-c(c+1)]}{\sqrt{2b(2b+1)(2b+2)2c(2c+1)(2c+2)}} \quad (6\text{-}48)$$

and

$$\left\{\begin{matrix} a & b & c \\ 1 & c-1 & b \end{matrix}\right\} = (-1)^s\sqrt{\frac{2(s+1)(s-2a)(s-2b)(s-2c+1)}{2b(2b+1)(2b+2)(2c-1)2c(2c+1)}} \quad (6\text{-}49)$$

where $s = a+b+c$ and in (12-16) $g(SLJ)$ is the Lande g-factor defined in (2-54).

6-3 Show that

$$\langle \alpha JIFM_F|H_{mag}|\alpha JIF+1M_F\rangle$$
$$= \mu_BB_z(g_J-g_I)\sqrt{(F+1)^2-M_F^2}$$
$$\times\sqrt{\frac{(F+J+I+2)(F+J-I+1)(F-J+I+1)(I+J-F)}{4(F+1)^2(2F+1)(2F+3)}}$$

$$(6\text{-}50)$$

6-4 Show that

$$\langle \alpha SLJIFM_F|H_{mag}|\alpha SLJ + 1IFM_F\rangle$$

$$= \mu_B B_z \frac{M_F(1 - g_s)}{4F(F + 1)(J + 1)}$$

$$\times \sqrt{\frac{(F + I + J + 2)(F + J - I + 1)(J + I - F + 1)(I + F - J - 1)}{(2J + 1)(2J + 3)}}$$

$$\times \sqrt{(J + L + S + 2)(J + L - S + 1)(J - L + S + 1)(S + L - J)}$$

$$(6\text{-}51)$$

6-5 Show that

$$\langle \alpha SLJIFM_F|H_{mag}|\alpha SLJ + 1IF + 1M_F\rangle$$

$$= \mu_B B_z \frac{(g_s - 1)}{4(F + 1)(J + 1)} \sqrt{\frac{(F + 1)^2 - M_F^2}{(2F + 1)(2F + 3)(2J + 1)(2J + 3)}}$$

$$\times \sqrt{(F + J + I + 2)(F + J + I + 3)(F + J - I + 1)(F + J - I + 2)}$$

$$\times \sqrt{(J + L + S + 2)(J + L - S + 1)(J - L + S + 1)(S + L - J)}$$

$$(6\text{-}52)$$

In this exercise you need to note that

$$\begin{Bmatrix} a & b & c \\ 1 & c - 1 & b - 1 \end{Bmatrix} = (-1)^s \sqrt{\frac{s(s + 1)(s - 2a - 1)(s - 2a)}{(2b - 1)2b(2b + 1)(2c - 1)2c(2c + 1)}}$$

$$(6\text{-}53)$$

where s = a + b + c.

6-6 Show that

$$\langle \alpha SLJIFM_F|H_{mag}|\alpha SLJ + 1IF - 1M_F\rangle$$

$$= \mu_B B_z \frac{(g_s - 1)\sqrt{F^2 - M_F^2}}{4F(J + 1)}$$

$$\times \sqrt{(I + F - J - 1)(I + F - J)(I + J - F + 1)(I + J - F + 2)}$$

$$\times \frac{\sqrt{(J + L + S + 2)(J + L - S + 1)(J - L + S + 1)(S + L - J)}}{(2J + 1)(2J + 3)(2F - 1)(2F + 1)}$$

$$(6\text{-}54)$$

REFERENCES

1. Steck, D A, (2001) http://george.ph.utexas.edu/~dsteck/ alkalidata.
2. Steck, D, (2002) *Rubidium 87 D Line Data* http://george.ph.utexas.edu/~dsteck/ alkalidata/rubidium.
3. Arimondo E, Inguscio M and Violino P, (1977) Experimental Determinations of the Hyperfine Structure in the Alkali Atoms *Rev. Mod. Phys.*, **49** 31.
4. Breit G and Rabi I I, (1931) Measurement of Nuclear Spin *Phys. Rev.,* **38** 2082.

7 Intensities of Electronic Transitions

When the great innovation appears, it will almost certainly be in a muddled, incomplete and confusing form. To the discoverer himself it will be only half-understood; to everybody else it will be a mystery. For any speculation which does not at first glance look crazy, there is no hope. The reason why new concepts in any branch of science are hard to grasp is always the same; contemporary scientists try to picture the new concepts in terms of ideas which existed before.

Freeman J Dyson

In this chapter we look at questions relating to the intensities of electronic transitions, first in the absence of hyperfine structure and then with it. Our emphasis is on relative intensities rather than absolute intensities. The latter are difficult to reliably compute and to measure. Two distinct types of transitions arise. Parity *allowed* transitions that require a change of parity of the one-electron orbitals, and parity *forbidden* transitions, which occur when there is no change in the parity of the one-electron states. Recall that the parity, $\mathcal{P}$, of the states of an electron configuration is

$$\mathcal{P} = (-1)^{\sum \ell} \tag{7-1}$$

where the summation is over all one-electron orbitals of the configuration. The parity is said to be *even* or *odd* as $\mathcal{P}$ is *even* or *odd*. Closed shells are necessarily of even parity and may be omitted in the summation.

Electric dipole transitions occur between states of *opposite* parity and are commonly referred to as *allowed* transitions. Magnetic dipole and electric quadrupole transitions only occur between states of the *same* parity and are referred to as *forbidden* transitions in the sense that they are forbidden as electric dipole transitions. In general the allowed transitions are orders of magnitude more intense than the forbidden transitions.

Thus in atomic spectroscopy all transitions that violate the rigorous selection rules for electric dipole radiation in free atoms are termed *forbidden transitions*[1]. In order to see the difference between the conditions that have to be satisfied in the case of each kind of transition, the appropriate selection rules are presented in Table 7-1 below.

The earliest observation of electric quadrupole transitions was of the $^2D \rightarrow {}^2S$ transitions in potassium by Datta[2] in 1922, followed by Lord Rayleigh's observations on Hg spectra[3] in 1927. Babcock[4] observed in the Aurora Borealis a green line at 5577.3Å, which was later attributed to the forbidden OI line resulting from the transition $(2p^4)^1 S_0^e \rightarrow (2p^4)^1 D_2^e$. Bowen[5] found in 1928 that many of the strong spectral lines found in gaseous nebulae originate from the forbidden transitions in OII, $OIII$ and NII. In a sense gaseous nebulae form an extraterrestial atomic spectroscopy laboratory, because they constitute physical conditions appropriate to the observation

79

TABLE 7-1

Selection Rules for Atomic Spectra

Electric Dipole	Magnetic Dipole	Electric Quadrupole
$\Delta J = 0, \pm 1$	$\Delta J = 0, \pm 1$	$\Delta J = 0, \pm 1, \pm 2$
$0 \nleftrightarrow 0$	$0 \nleftrightarrow 0$	$0 \nleftrightarrow 0, 1$ or $\frac{1}{2} \nleftrightarrow \frac{1}{2}$
$\Delta M = 0, \pm 1$	$\Delta M = 0, \pm 1$	$\Delta M = 0, \pm 1, \pm 2$
Parity change	No parity change	No parity change
$\Delta \ell = \pm 1$	$\Delta \ell = 0$	$\Delta \ell = 0, \pm 2$
	$\Delta n = 0$	
$\Delta S = 0$	$\Delta S = 0$	$\Delta S = 0$
$\Delta L = 0, \pm 1$	$\Delta L = 0$	$\Delta L = 0, \pm 1, \pm 2$
$0 \nleftrightarrow 0$	$0 \nleftrightarrow 0$	$0 \nleftrightarrow 0, 1$

Source: From Garstang, R.H., in *Atomic and Molecular Processes* (ed. D.R. Bates), Academic Press, New York, 1962.

of strongly forbidden transitions that would be impossible to realize on earth. The very low densities but very large volume of gaseous nebulae reduce the possibilities of collisional de-excitation. The calculation of the transition probabilities of magnetic dipole and electric quadrupole transitions has been reviewed by Garstang[1] who was largely responsible for introducing the Racah methods to the subject (see references therein).

7.1   ELECTRIC DIPOLE TRANSITIONS IN ATOMS

In the absence of perturbing fields each energy level characterized by a total angular momentum J is $2J + 1$−fold degenerate, each of the different states being identified by a different value of M_J. A *spectral line* is defined as the radiation associated with all possible transitions between the states belonging to two levels. The radiation resulting from a transition between a particular pair of states is called a *component* of the line.

Electric dipole transitions are induced by the electric dipole operator

$$P = -e \sum_i \mathbf{r}_i = -e \sum_i \sum_{q=-1}^{q=1} r_i C_q^{(1)}{}_i \qquad (7\text{-}2)$$

where $q = -1, 0, 1$. The intensities of a given transition are proportional to the absolute square of the matrix element of P coupling the two states. Applying the Wigner-Eckart theorem (2-32) we have

$$\langle \alpha J M_J | P | \alpha' J' M_J' \rangle = (-1)^{J-M_J} \begin{pmatrix} J & 1 & J' \\ -M_J & q & M_J' \end{pmatrix} \langle \alpha J \| P \| \alpha' J' \rangle \qquad (7\text{-}3)$$

The triangular condition on the top row of the $3jm$−symbol gives the selection rule

$$\Delta J = 0, \pm 1 \qquad (7\text{-}4a)$$

while the requirement that the arguments of the bottom row sum to zero gives the selection rule

$$\Delta M_J = 0, \pm 1 \quad \text{or} \quad M_J \to M_J - q \tag{7-4b}$$

If the light, absorbing or emitting, is polarized in the z direction, then only $q = 0$ is active. If the light is right circularly polarized, σ_+, $q = +1$, while if it is left circularly polarized, σ_-, $q = -1$. This, as we shall see in chapter 8, assumes great importance in laser cooling down to nanoKelvin temperatures. Using laser beams left or right circularly polarized one can selectively induce transitions between different hyperfine levels, which themselves have been split in a magnetic field. This is an important feature of magneto-optical traps and in Bose-Einstein-Condensation.

If the light is unpolarized, and we observe the complete set of transitions from an initial state with M_J to the levels with M'_J, then the intensity involves

$$\sum_{M'_J, q} |\langle \alpha J M_J | P | \alpha' J' M'_J \rangle|^2 = |\langle \alpha J \| P \| \alpha' J' \rangle|^2 / (2J' + 1) \tag{7-5}$$

which follows from the orthogonality relation of $3jm$−symbols. The quantity

$$S(\alpha J; \alpha' J') = |\langle \alpha J \| P \| \alpha' J' \rangle|^2 \tag{7-6}$$

is commonly referred to as the *line strength* of the transition. Often one is interested in first computing the *square root* of the line strength, $S^{1/2}(\alpha J; \alpha' J')$, the so-called transition amplitude. This is a relevant term to compute, if one wishes to make corrections for intermediate coupling, for example, where preservation of phase information must be retained. After making the corrections the resultant transition amplitude is squared to give the line strength.

7.2 RATIO OF THE LINE STRENGTHS FOR THE *D* LINES OF ALKALI ATOMS

Expanding the state description for an alkali atom we have, noting (2-46),

$$\langle s\ell j \| P \| s\ell' j' \rangle$$
$$= -e < r >_{\ell\ell'} \langle s\ell j \| C^{(1)} \| s\ell' j' \rangle$$
$$= -e < r >_{\ell\ell'} (-1)^{s+\ell'+j+1} \sqrt{(2j+1)(2j'+1)} \begin{Bmatrix} j & 1 & j' \\ \ell' & s & \ell \end{Bmatrix} \langle \ell \| C^{(1)} \| \ell' \rangle \tag{7-7}$$

where the parity of ℓ is opposite to the parity of ℓ' (see the definition of the reduced matrix element of the spherical tensor). Using this result in (7-6) we have for the ratio of the line strengths of the $D_2 : D_1$ transitions

$$\frac{S(ns\,^2S_{1/2}; np\,^2P_{3/2})}{S(ns\,^2S_{1/2}; np\,^2P_{1/2})} = 2 \tag{7-8}$$

which may be compared with the experimental value[6] of 1.9809(9) found for $^{133}_{55}Cs$.

7.3   LINE STRENGTHS FOR MANY-ELECTRON ATOMS

In $LS-$coupling we can enlarge our state description and obtain the $J-$dependence of the line strengths as

$$S^{\frac{1}{2}}(\alpha SLJ;\alpha'S'L'J') = (-1)^{J+L'+S+1}\sqrt{(2J+1)(2J'+1)}\begin{Bmatrix} J & 1 & J' \\ L' & S & L \end{Bmatrix}$$

$$\times - e\langle \alpha SL\| \sum_i r_i C_i^{(1)} \|\alpha'S'L'\rangle\delta(S,S') \qquad (7\text{-}9)$$

The triangular conditions on the arguments of the $6j-$symbol lead to the selection rules

$$\Delta J = 0, \pm 1 \ (0 \nrightarrow 0) \qquad\qquad (7\text{-}10a)$$
$$\Delta L = 0, \pm 1 (0 \nrightarrow 0) \qquad\qquad (7\text{-}10b)$$
$$\Delta S = 0 \qquad\qquad (7\text{-}10c)$$

The J selection rule is, in the absence of hyperfine interactions and external fields, independent of the coupling scheme. Recall that the spin-orbit interaction may mix states of different S and L leading to a breakdown of (7-10b,c). In that case intermediate coupling must be considered.

To obtain the intermediate coupling corrected line strength we may write

$$S^{\frac{1}{2}}(aJ;bJ') = \sum_{\alpha,\beta}(aJ|\alpha J)S^{\frac{1}{2}}(\alpha J;\beta J')(\beta J'|bJ') \qquad (7\text{-}11)$$

where $(aJ|\alpha J)$ represents the transformation matrix that transforms the states $|\alpha J\rangle$ into the states $|aJ\rangle$ of the actual coupling (and analogously for $(\beta J'|bJ')$).

7.4   RELATIVE LINE STRENGTHS IN $LS-$COUPLING

It is sometimes useful to have explicit expressions for the relative line strengths in $LS-$coupling. These can be obtained from the square of (7-9) with explicit evaluation of the $6j-$symbol to give,

$S(\alpha SL, J;\alpha'SL, J)$
$$= \frac{(2J+1)[(S(S+1)-J(J+1)-L(L+1)]^2}{4J(J+1)}\mathcal{P} \qquad (7\text{-}12a)$$

$S(\alpha SL, J;\alpha'SL, J+1)$
$$= \frac{(S+L+J+2)(L+J-S+1)(S-L+J+1)(S+L-J)}{4(J+1)}\mathcal{P} \qquad (7\text{-}12b)$$

$S(\alpha SL, J;\alpha'SL, J-1)$
$$= \frac{(S+L+J+1)(L+J-S)(S-L+J)(S+L-J+1)}{4J}\mathcal{P} \qquad (7\text{-}12c)$$

$S(\alpha SL, J;\alpha'SL+1, J)$

$$= \frac{(2J+1)(S+L+J+2)(L+J-S+1)(S-J+L+1)(S+J-L)}{4J(J+1)}\mathcal{P}'$$

(7-12d)

$$S(\alpha SL, J; \alpha' SL - 1, J)$$
$$= \frac{(2J+1)(S+L+J+1)(L+J-S)(S-J+L)(S+J-L+1)}{4J(J+1)}\mathcal{P}''$$

(7-12e)

$$S(\alpha SL, J; \alpha' SL + 1, J + 1)$$
$$= \frac{(S+L+J+2)(S+L+J+3)(L+J-S+1)(L+J-S+2)}{4(J+1)}\mathcal{P}'$$

(7-12f)

$$S(\alpha SL, J; \alpha' SL - 1, J + 1)$$
$$= \frac{(S+L-J-1)(S+L-J)(S-L+J+1)(S-L+J+2)}{4(J+1)}\mathcal{P}'' \quad (7\text{-}12\text{g})$$

$$S(\alpha SL, J; \alpha' SL + 1, J - 1)$$
$$= \frac{(S-L+J-1)(S-L+J)(S+L-J+1)(S+L-J+2)}{4J}\mathcal{P}' \quad (7\text{-}12\text{h})$$

where

$$\mathcal{P} = \frac{|\langle \alpha SL \| P \| \alpha' SL \rangle|^2}{L(L+1)(2L+1)},$$

(7-13a)

$$\mathcal{P}' = \frac{|\langle \alpha SL \| P \| \alpha' SL + 1 \rangle|^2}{(L+1)(2L+1)(2L+3)}$$

(7-13b)

and

$$\mathcal{P}'' = \frac{|\langle \alpha SL \| P \| \alpha' SL - 1 \rangle|^2}{L(2L-1)(2L+1)}$$

(7-13c)

Thus we have from (7-12d,f) the line strength ratio

$$\frac{S(\alpha SL, J; \alpha' SL + 1, J + 1)}{S(\alpha SL, J; \alpha' SL + 1, J)} = \frac{J(S+L+J+3)(L+J-S+2)}{(2J+1)(S-J+L+1)(S+J-L)}$$

(7-14)

which is, of course, consistent with (7-8).

7.5 RELATIVE LINE STRENGTHS FOR HYPERFINE LEVELS

The treatment of the line strengths for electric dipole transitions between hyperfine levels, in the absence of external fields, parallels our earlier work on atomic transitions. Thus the relative line strengths for the transitions between the hyperfine levels can be derived from (7-12) to (7-14) by simply making the replacements

$$S \to I, \quad L \to J, \quad J \to F$$

(7-15)

and hence

$$S(\alpha JI, F; \alpha' JI, F)$$
$$= \frac{(2F+1)[I(I+1) - F(F+1) - J(J+1)]^2}{4F(F+1)} \mathcal{P} \qquad (7\text{-}16a)$$

$$S(\alpha JI, F; \alpha' JI, F+1)$$
$$= \frac{(I+J+F+2)(J+F-I+1)(I-J+F+1)(I+J-F)}{4(F+1)} \mathcal{P} \qquad (7\text{-}16b)$$

$$S(\alpha JI, F; \alpha' JI, F-1)$$
$$= \frac{(I+J+F+1)(J+F-I)(I-J+F)(I+J-F+1)}{4F} \mathcal{P} \qquad (7\text{-}16c)$$

$$S(\alpha JI, F; \alpha' J+1I, F)$$
$$= \frac{(2F+1)(I+J+F+2)(J+F-I+1)(I-F+J+1)(I+F-J)}{4F(F+1)} \mathcal{P}'$$
$$\qquad (7\text{-}16d)$$

$$S(\alpha JI, F; \alpha' J-1I, F)$$
$$= \frac{(2F+1)(I+J+F+1)(J+F-I)(I-F+J)(I+F-J+1)}{4F(F+1)} \mathcal{P}''$$
$$\qquad (7\text{-}16e)$$

$$S(\alpha JI, F; \alpha' J+1I, F+1)$$
$$= \frac{(I+J+F+2)(I+J+F+3)(J+F-I+1)(J+F-I+2)}{4(F+1)} \mathcal{P}'$$
$$\qquad (7\text{-}16f)$$

$$S(\alpha JI, F; \alpha' J-1I, F+1)$$
$$= \frac{(I+J-F-1)(I+J-F)(I-J+F+1)(I-J+F+2)}{4(F+1)} \mathcal{P}'' \qquad (7\text{-}16g)$$

$$S(\alpha JI, F; \alpha' J+1I, F-1)$$
$$= \frac{(I-J+F-1)(I-J+F)(I+J-F+1)(I+J-F+2)}{4F} \mathcal{P}' \qquad (7\text{-}16h)$$

where

$$\mathcal{P} = \frac{|\langle \alpha JI \| P \| \alpha' JI \rangle|^2}{J(J+1)(2J+1)}, \qquad (7\text{-}17a)$$

$$\mathcal{P}' = \frac{|\langle \alpha JI \| P \| \alpha' J+1I \rangle|^2}{(J+1)(2J+1)(2J+3)} \qquad (7\text{-}17b)$$

and

$$\mathcal{P}'' = \frac{|\langle \alpha JI \| P \| \alpha' J-1I \rangle|^2}{J(2J-1)(2J+1)} \qquad (7\text{-}17c)$$

Thus we have from (7-16d,f) the line strength ratio

$$\frac{S(\alpha JI, F; \alpha' J+1I, F+1)}{S(\alpha JI, F; \alpha' J+1I, F)} = \frac{F(I+J+F+3)(J+F-I+2)}{(2F+1)(I-F+J+1)(I+F-J)} \qquad (7\text{-}18)$$

7.6 RELATIVE LINE STRENGTHS FOR THE *D*2 TRANSITIONS OF $^{87}_{37}Rb$

Recall that for $^{87}_{37}Rb$ the nuclear spin is $I = \frac{3}{2}$, and the ground state is $5s\ ^2S_{1/2}$. The spin-orbit interaction splits the first excited state into the two sublevels $5p\ ^2P_{1/2}$ and $5p\ ^2P_{3/2}$ with the latter sublevel being highest. The D2 transitions are associated with transitions $5s\ ^2S_{1/2} \rightarrow 5p\ ^2P_{3/2}$. The magnetic hyperfine interaction results in the ground state splitting into two sublevels with $F = 1$ and $F = 2$, respectively, while for excited level $5p\ ^2P_{3/2}$ we obtain four sublevels with $F' = 0, 1, 2, 3$. Here we are interested in the transition array $5s\ ^2S_{1/2}F \rightarrow 5p\ ^2P_{3/2}F'$ and their relative line strengths.

The relative line strengths may be calculated directly from (7-17b) together with (7-16d,f,h), and normalized to give the results below,

$$
\begin{array}{c}
\begin{array}{cccc} 0 & 1 & 2 & 3 \end{array} \\
\begin{array}{c} 1 \\ 2 \end{array}
\left(
\begin{array}{cccc}
\frac{1}{6} & \frac{5}{12} & \frac{5}{12} & 0 \\
0 & \frac{1}{20} & \frac{1}{4} & \frac{7}{10}
\end{array}
\right)
\end{array}
$$

7.7 EFFECTIVE OPERATORS AND PERTURBATION THEORY

The Stark effect involves the interaction of an atomic electric dipole moment, **p**, with an external electric field, **E**, which may be represented by the operator[7]

$$H_E = -\mathbf{E} \cdot \mathbf{p} \tag{7-19}$$

For a H−atom there is a first-order Stark effect due to the degeneracy of states of opposite parity for a given principal quantum number n. For alkali atoms the degeneracy is lifted by the Coulomb field, there is no first-order energy shift of the states, and one must go to second-order to obtain a Stark effect. Before proceeding to the quadratic Stark effect in atoms a few remarks about effective operators[8,9] and perturbation theory are made.

Let us assume that the exact Hamiltonian H of the system can be split into two Hermitian operators, H_0 and V,

$$H = H_0 + \lambda V \tag{7-20}$$

where λ is the perturbing parameter. The spectral resolution of H_0 is assumed to be known explicitly and the eigenvalues of H_0 are normally degenerate. Following the standard procedure of perturbation approach the eigenvalues and the wave functions are expressed in the terms of power series in λ,

$$\mathcal{E} = \sum_j \lambda^j E^j \qquad \Psi = \sum_j \lambda^j \Psi^j \tag{7-21}$$

where E^j and Ψ^j are the corrections to the energy and the wave function.

For a particular configuration A we have the eigenvalue equation

$$H_0|A\alpha_i\rangle = E_{A\alpha}|A\alpha_i\rangle \tag{7-22}$$

where i labels the $g-$degenerate eigenstates, and $E_{A\alpha} \equiv E^0$.

The *first-order* correction to the eigenvalue $E_{A\alpha}$, which is due to the effect of the perturbation operator V, is found by diagonalizing the $g \times g$ energy matrix formed by the matrix elements $\langle A\alpha|V|A\alpha'\rangle$. The diagonalization of this matrix usually results in a lowering of the initial degeneracy. Furthermore, the obtained corrections to the energy lower its value if the partitioning of the Hamiltonian in (7-20) is properly performed. In addition, the partitioning of the Hamiltonian leads to the separation of the sub-space P, spanned by the wave functions of the unperturbed Hamiltonian Ψ^0, from its orthogonal complement Q. If H_0 describes the system within the single configuration approximation, then P projects onto the space associated with a particular configuration A, while Q contains all the components built of the states of excited configurations $B, C, \ldots$. Therefore, the perturbing operator V in (7-20) denotes in fact the inter- and intra-shell interactions,

$$V \equiv (P + Q)V(P + Q) \tag{7-23}$$

where $P + Q$ spans the whole space. In this case V normally also couples the states of the configuration A to those of $B, C, \ldots$. If one or more of these configurations is approximately degenerate with A, then it is desirable to diagonalize the energy matrix for that set of configurations. Only if the perturbing configuration is well separated from A may their effect on the energy levels be studied by perturbation theory.

In the standard Rayleigh-Schrödinger perturbation theory the *first-order* correction to the eigenfunction $|A\alpha'\rangle$, due to the perturbation by the states $|B\beta\rangle$ of the configuration B taken into account via V, is given by

$$|A\alpha'\rangle_1 = \sum_\beta \frac{|B\beta\rangle\langle B\beta|V|A\alpha'\rangle}{E_{A\alpha'} - E_{B\beta}} \tag{7-24}$$

where the summation is over such states (in general, including those of A) for which the energy denominators do not vanish.

The *second-order* correction to the energy matrix is given by the matrix elements

$$\langle A\alpha|V|A\alpha'\rangle_1 = \sum_\beta \frac{\langle A\alpha|V|B\beta\rangle\langle B\beta|V|A\alpha'\rangle}{E_{A\alpha'} - E_{B\beta}} \tag{7-25}$$

where the summation is over all the states $|B\beta\rangle$ of the configuration B. In order to include the perturbing influence of all the configurations other than A (all the excited configurations), the summation over B has to be performed in (7-25). This impact taken into account breaks the limitations of the single configuration approximation of the zeroth order eigenvalue problem. At the same time, summation in (7-25) covers also the components that contain the states of configuration A that are other than those present in the energy denominator.

Let us for the moment consider the object

$$\sum_\beta \langle A\alpha|V|B\beta\rangle\langle B\beta|V|A\alpha'\rangle \tag{7-26}$$

Recall the *closure theorem* that for a *complete* set of states gives

$$\sum_\beta |B\beta\rangle\langle B\beta| = 1 \tag{7-27}$$

since these are the solutions of an eigenvalue problem of a hermitian operator H_0. Clearly we are not entitled to use (7-27) in (7-26), since some of the intermediate states include states of A, and furthermore in (7-25) there are energy denominators that prevent performing the closure. The first problem is solved when the configuration interaction is regarded in a perturbative way, and the perturbing potential V is limited to its inter-shell part $QVP + PVQ$. This means that since $PQ = 0$, the intermediate states have to belong to any configuration other than A, otherwise a vanishing result is obtained. The selection of excited configurations contributing to the second-order correction depends on the nature, and most of all on the parity, of the perturbing operator. Introducing additional approximations (for details see chapter 17) it is possible to use the closure theorem to obtain the result

$$\sum_\beta \langle A\alpha|V|B\beta\rangle\langle B\beta|V|A\alpha'\rangle = \langle A\alpha|V_{AB}V_{BA}|A\alpha'\rangle \tag{7-28}$$

The product operator $V_{AB}V_{BA}$ now plays the role of an *effective operator* acting within the zero-order states $|A\alpha\rangle$ of the configuration of interest.

The problem is then to define suitable operators, $V_{AB}V_{BA}$, that have the desired properties and that reproduce the original interactions via the intermediate states acting only within the particular configuration A. One obvious way would be to introduce annihilation and creation operators[10], and to perform their contraction. Alternatively it is possible to use tensor operators that are essentially coupled products of annihilation and creation operators. We shall exploit the latter approach. To that end let us introduce tensor operators whose single particle operators are defined by their reduced matrix element

$$\langle a\|v^{(k)}(c,d)\|b\rangle = [k]^{\frac{1}{2}}\delta(a,c)\delta(b,d) \tag{7-29}$$

where $|a\rangle$, $|b\rangle$, $|c\rangle$, $|d\rangle$ designate single particle states, and $[k] = 2k + 1$. These operators are the inter-shell objects, and therefore they are a natural extension of unit tensor operators that act within a particular shell. For orbital operators we have the basic commutator relation

$$[v^{(k_1)}(n\ell, n'\ell')v^{(k_2)}(n''\ell'', n'''\ell''')]_q^{(k)} - (-1)^{k_1+k_2-k}[v^{(k_2)}(n''\ell'', n'''\ell''')v^{(k_1)}(n\ell, n'\ell')]_q^{(k)}$$

$$= [k_1, k_2]^{\frac{1}{2}}[\delta(n'\ell', n''\ell'')(-1)^{\ell+\ell''+k}\begin{Bmatrix} k_1 & k_2 & k \\ \ell''' & \ell & \ell' \end{Bmatrix}v_q^{(k)}(n\ell, n'''\ell''')$$

$$+ \delta(n\ell, n'''\ell''')(-1)^{\ell'+\ell''+k_1+k_2}\begin{Bmatrix} k_1 & k_2 & k \\ \ell'' & \ell' & \ell \end{Bmatrix}v_q^{(k)}(n''\ell'', n'\ell')] \tag{7-30}$$

Consider the case where in (7-25) the configuration A contains just one open shell $n\ell^N$ of N equivalent particles, and B differs by a single excitation of a particle from the $n\ell^N$ into an empty shell $n'\ell'$. Then in general for one-particle spin-independent interactions represented by an operator $V_{q_1}^{(k_1)}$, a typical term in (7-25) for the perturbation $V_{q_2}^{(k_2)}$ is (to within a multiplicative factor that includes radial integrals, energy denominator and one particle reduced matrix elements of spherical tensor)

$$\sum_\beta \langle A\alpha | V_{q_1}^{(k_1)}(\ell\ell') | B\beta \rangle \langle B\beta | V_{q_2}^{(k_2)}(\ell'\ell) | A\alpha' \rangle \tag{7-31}$$

The summation over the states of B is accomplished by simply removing the intermediate states $|B\beta\rangle\langle B\beta|$ to give

$$\langle A\alpha | V_{q_1}^{(k_1)}(\ell\ell') V_{q_2}^{(k_2)}(\ell'\ell) | A\alpha' \rangle \tag{7-32}$$

The two operators may then be coupled together to yield

$$\sum_{k,q} \langle k_1 q_1 k_2 q_2 | k_1 k_2 kq \rangle \langle A\alpha | [V^{(k_1)}(\ell\ell') V^{(k_2)}(\ell'\ell)]_q^{(k)} | A\alpha' \rangle \tag{7-33}$$

The matrix element may now be simplified using the commutation relation (7-30) and remembering that since $\langle A\alpha|$ and $|A\alpha'\rangle$ do not contain any particles in ℓ' orbitals the matrix elements of the operators $[V^{(k_2)}(\ell'\ell)V^{(k_1)}(\ell\ell')]_q^{(k)}$ and $V_q^{(k)}(\ell'\ell')$ (that are in fact coupled products of the annihilation and creation operators) vanish to give the final result

$$\sum_\beta \langle A\alpha | V_{q_1}^{(k_1)}(\ell\ell') | B\beta \rangle \langle B\beta | V_{q_2}^{(k_2)}(\ell'\ell) | A\alpha' \rangle$$
$$= \sum_{k,q} \langle k_1 q_1 k_2 q_2 | k_1 k_2 kq \rangle (-1)^{2\ell+k} [k_1,k_2]^{\frac{1}{2}} \begin{Bmatrix} k_1 & k_2 & k \\ \ell & \ell & \ell' \end{Bmatrix}$$
$$\times \langle A\alpha | V_q^{(k)} | A\alpha' \rangle \tag{7-34}$$

where the effective operator $V_q^{(k)}$, acting only within the configuration $A \equiv n\ell^N$ reproduces the inter-shell interactions between A and singly excited configurations $B \equiv n\ell^{N-1}n'\ell'$. Similarly as before, the parity of the excited configurations B and the limitations for ℓ' are determined by the triangular conditions for the non-vanishing angular momenta coupling coefficients.

7.8   THE QUADRATIC STARK EFFECT IN ATOMS

With the above outline of the relevant perturbation theory we can consider the quadratic Stark effect for atoms. For simplicity let us assume that the applied electric field is in the z−direction and that the electric dipole moment is as given earlier in (7-2). The perturbation term V_E is then

$$V_E = eE_z \sum_i r_i C_0^{(1)}{}_i \tag{7-35}$$

Let us assume that we are interested in a configuration $A = n\ell^N$ that is perturbed by states from a configuration $B = \ell^{N-1}n'\ell'$ as considered above. For convenience, we introduce a factor, $\mathcal{A}$, which is defined as

$$\mathcal{A} = \frac{1}{3}\frac{e^2 E_z^2}{E_A - E_B}\langle n\ell|r|n'\ell'\rangle\langle n'\ell'|r|n\ell\rangle\langle\ell\|C^{(1)}\|\ell'\rangle\langle\ell'\|C^{(1)}\|\ell\rangle \qquad (7\text{-}36)$$

where in the energy denominator the average energy of a configuration is used. This energy denominator is independent of the summation index in (7-25) and therefore it is included in factor A (it has the same value for all states of a given configuration). We note the parity selection rule $\Delta\ell = \pm 1$ is inherent in (7-36). Given the above definition, the perturbation sum in (7-25) may be immediately found using (7-34) to give

$$\mathcal{A}\sum_\beta \langle n\ell^N\alpha|V_0^{(1)}(\ell\ell')|n\ell^{N-1}n'\ell'\beta\rangle\langle n\ell^{N-1}n'\ell'\beta|V_0^{(1)}(\ell'\ell)|n\ell^N\alpha'\rangle$$

$$= 3\mathcal{A}\sum_k\langle 10,10|(1,1)k0\rangle\langle n\ell^N\alpha|V_0^{(k)}(\ell\ell)|n\ell^N\alpha'\rangle(-1)^k\begin{Bmatrix}1 & 1 & k\\ \ell & \ell & \ell'\end{Bmatrix}$$

$$= 3\mathcal{A}\sum_k[k]^{\frac{1}{2}}\begin{pmatrix}1 & 1 & k\\ 0 & 0 & 0\end{pmatrix}\langle n\ell^N\alpha|V_0^{(k)}(\ell\ell)|n\ell^N\alpha'\rangle(-1)^k\begin{Bmatrix}1 & 1 & k\\ \ell & \ell & \ell'\end{Bmatrix} \qquad (7\text{-}37)$$

The $3jm$—symbol vanishes except for $k = 0,\ 2$ and has the following values

$$\begin{pmatrix}1 & 1 & 0\\ 0 & 0 & 0\end{pmatrix} = -\frac{\sqrt{3}}{3} \quad \text{and} \quad \begin{pmatrix}1 & 1 & 2\\ 0 & 0 & 0\end{pmatrix} = \frac{\sqrt{30}}{15} \qquad (7\text{-}38)$$

The term with $k = 0$ is a scalar and has the effect of giving, for $\alpha = \alpha'$, a uniform shift to all levels of the configuration A.

In the case of the $k = 2$ contribution let us put

$$\gamma_{\ell\ell'} = \mathcal{A}\begin{Bmatrix}\ell & \ell' & 1\\ 1 & 2 & \ell\end{Bmatrix} \qquad (7\text{-}39)$$

so that the $k = 2$ contribution becomes, after expanding the state description and suppressing the factor $\gamma_{\ell\ell'}$

$$\sqrt{6}\langle\alpha SLJIFM_F|V_0^{(2)}(\ell\ell)|\alpha'SL'J'IF'M_F\rangle$$

$$= \sqrt{6}(-1)^{F-M_F}\begin{pmatrix}F & 2 & F'\\ -M_F & 0 & M_F\end{pmatrix}\langle\alpha SLJIF\|V^{(2)}(\ell\ell)\|\alpha'SL'J'IF'\rangle$$

$$= \sqrt{6}(-1)^{F-M_F}\begin{pmatrix}F & 2 & F'\\ -M_F & 0 & M_F\end{pmatrix}(-1)^{J'+I+F}[F,F']^{\frac{1}{2}}$$

$$\times \begin{Bmatrix}F & 2 & F'\\ J' & I & J\end{Bmatrix}\langle\alpha SLJ\|V^{(2)}(\ell\ell)\|\alpha'SL'J'\rangle \qquad (7\text{-}40)$$

For $N \geq 2$ the final reduced matrix element can be evaluated using Judd[11] (Eq. (7-52)), or the tables of Nielson and Koster[12]. For a single electron outside of closed shells we have

$$\langle s\ell j \| V^{(2)}(\ell\ell) \| s\ell j' \rangle = (-1)^{s+\ell+j} \sqrt{5}[j, j']^{\frac{1}{2}} \left\{ \begin{array}{ccc} 2 & j & j' \\ \frac{1}{2} & \ell & \ell \end{array} \right\} \tag{7-41}$$

In particular

$$\langle ^2P_{3/2} \| V^{(2)}(p, p) \| ^2P_{3/2} \rangle = \frac{\sqrt{30}}{3} \tag{7-42}$$

The $6j$-symbol vanishes unless $\ell \geq 1$ and $j+j' \geq 2$. If the fine structure splittings are much greater than the hyperfine splittings, then only the matrix elements diagonal in j need be considered. Note that the effect of an external electric field on the hyperfine levels is to, unlike the case of magnetic fields, only partially lift the degeneracy. Indeed the states with $\pm M_F$ remain degenerate and hence, apart from states with $M_F = 0$, each sublevel is still two-fold degenerate.

7.9   EXAMPLE OF $^{133}_{55}Cs$

In the case of $^{133}_{55}Cs$ we confine our attention to the $6p(^2P_{3/2})$ hyperfine multiplet where, since the nuclear spin is $I = \frac{7}{2}$, the hyperfine levels have $F = 2, 3, 4, 5$. It follows from (7-39) that $\ell' = 0, 2$. Since the dependence on ℓ' is entirely contained in (7-39), let us consider just the matrix elements

$$\sqrt{6} \langle ^2P_{3/2} \tfrac{7}{2} F M_F | V_0^{(2)}(p, p) |^2 P_{3/2} \tfrac{7}{2} F' M_F \rangle$$
$$= (-1)^{M_F+1} 2\sqrt{5}[F, F']^{\frac{1}{2}} \left(\begin{array}{ccc} F & 2 & F' \\ -M_F & 0 & M_F \end{array} \right) \left\{ \begin{array}{ccc} F & 2 & F' \\ \frac{3}{2} & \frac{7}{2} & \frac{3}{2} \end{array} \right\} \tag{7-43}$$

where the value of the reduced matrix element (7-42) is inserted. It is a simple matter to write a short MAPLE programme to evaluate the relevant matrix elements and to construct a matrix for each value of M_F. These matrices are given below. The states are designated by just $|F M_F\rangle$, since all other quantum numbers are fixed.

$$\begin{array}{cc} & M_F = \pm 5 \quad |5 \pm 5\rangle \\ \langle 5 \pm 5| & (\quad 1 \quad) \end{array}$$

$$\begin{array}{c|cc} M_F = \pm 4 & |5 \pm 4\rangle & |4 \pm 4\rangle \\ \langle 5 \pm 4| & \frac{2}{5} & \mp\frac{\sqrt{21}}{5} \\ \langle 4 \pm 4| & \mp\frac{\sqrt{21}}{5} & -\frac{2}{5} \end{array}$$

$$\begin{array}{c|ccc} M_F = \pm 3 & |5 \pm 3\rangle & |4 \pm 3\rangle & |3 \pm 3\rangle \\ \langle 5 \pm 3| & -\frac{1}{15} & \mp\frac{\sqrt{21}}{5} & \frac{\sqrt{35}}{15} \\ \langle 4 \pm 3| & \mp\frac{\sqrt{21}}{5} & -\frac{1}{10} & \mp\frac{\sqrt{15}}{10} \\ \langle 3 \pm 3| & \frac{\sqrt{35}}{15} & \mp\frac{\sqrt{15}}{10} & -\frac{5}{6} \end{array}$$

$M_F = \pm 2$	$\|5\pm2\rangle$	$\|4\pm2\rangle$	$\|3\pm2\rangle$	$\|2\pm2\rangle$
$\langle5\pm2\|$	$-\frac{2}{5}$	$\mp\frac{7}{10}$	$\frac{\sqrt{35}}{10}$	0
$\langle4\pm2\|$	$\mp\frac{7}{10}$	$\frac{4}{35}$	$\mp\frac{2\sqrt{35}}{35}$	$\frac{5\sqrt3}{14}$
$\langle3\pm2\|$	$\frac{\sqrt{35}}{10}$	$\mp\frac{2\sqrt{35}}{35}$	0	$\pm\frac{\sqrt{105}}{14}$
$\langle2\pm2\|$	0	$\frac{5\sqrt3}{14}$	$\pm\frac{\sqrt{105}}{14}$	$\frac{2}{7}$

$M_F = \pm 1$	$\|5\pm1\rangle$	$\|4\pm1\rangle$	$\|3\pm1\rangle$	$\|2\pm1\rangle$
$\langle5\pm1\|$	$-\frac{3}{5}$	$\mp\frac{\sqrt{14}}{10}$	$\frac{\sqrt2}{2}$	0
$\langle4\pm1\|$	$\mp\frac{\sqrt{14}}{10}$	$\frac{17}{70}$	$\mp\frac{\sqrt7}{14}$	$\frac{5\sqrt6}{14}$
$\langle3\pm1\|$	$\frac{\sqrt2}{2}$	$\mp\frac{\sqrt7}{14}$	$\frac{1}{2}$	$\pm\frac{\sqrt{42}}{14}$
$\langle2\pm1\|$	0	$\frac{5\sqrt6}{14}$	$\pm\frac{\sqrt{42}}{14}$	$-\frac{1}{7}$

$M_F = 0$	$\|50\rangle$	$\|40\rangle$	$\|30\rangle$	$\|20\rangle$
$\langle50\|$	$-\frac{2}{3}$	0	$\frac{\sqrt5}{3}$	0
$\langle40\|$	0	$\frac{2}{7}$	0	$\frac{3\sqrt5}{7}$
$\langle30\|$	$\frac{\sqrt5}{3}$	0	$\frac{2}{3}$	0
$\langle20\|$	0	$\frac{3\sqrt5}{7}$	0	$-\frac{2}{7}$

Diagonalization of the above matrices yields just two distinct eigenvalues $+1$ and -1 each with a degeneracy of 16. This is exactly what we would expect, if we let the hyperfine structure constant approach zero. The Stark effect would lead to two levels, one for $m_j = \pm\frac{3}{2}$ and one for $m_j = \pm\frac{1}{2}$. The degeneracies associated with each $|m_j|$ would be $2(2I + 1)$. Indeed we have

$$\sqrt6\langle{}^2P_{3/2}m_j|V_0^{(2)}(p,p)|{}^2P_{3/2}m_j\rangle = (-1)^{|m_j|+\frac{1}{2}} \tag{7-44}$$

REFERENCES

1. Garstang R H, (1962) in *Atomic and Molecular Processes* ed. D R Bates (Academic Press) p1.
2. Datta S, (1922) The Absorption Spectrum of Potassium Vapour *Proc. R. Soc.*, **A101** 539.
3. Rayleigh, Lord, (1927) The Line Spectrum of Mercury in Absorption. Occurrence of the "Forbidden" Line λ 2270 $1^1S_0 - 1^3P_2$ *Proc. R. Soc.*, **A117** 294.
4. Babcock H D, (1923) A Study of the Green Auroral Line by the Interference Method *Astrophys. J.* **57** 209.
5. Bowen, I S, (1960) Wave Lengths of Forbidden Nebular Lines. II *Astrophys. J.* **132** 1.
6. Rafac, R J and Tanner, C E, (1998) Measurement of the Ratio of the Cesium D−line Transition Strengths, *Phys. Rev.,* **A58** 1087.
7. Schmieder, R W, (1972) Matrix Elements of the Quadratic Stark Effect on Atoms with Hyperfine Structure, *Am. J. Phys.* **40** 297.

8. Wybourne, B G, (1967) *Coupled Products of Annihilation and Creation Operators in Crystal Energy Problems*, pp35-52 in *Optical Properties of Ions in Crystals*, Wiley - Interscience Publication, John Wiley and Sons Inc., New York.

9. Wybourne, B G, (1968) Effective Operators and Spectroscopic Properties, *J. Chem. Phys.,* **48** 2596.

10. Judd, B R, (1967) *Second Quantization and Atomic Spectroscopy*, Baltimore, Md: Johns Hopkins Univ. Press.

11. Judd B R, (1998) *Operator Techniques in Atomic Spectroscopy* Princeton: Princeton University Press.

12. Nielson, C W and Koster, G F, (1963) *Spectroscopic Coefficients for the p^n, d^n, and f^n Configurations*, Cambridge: MIT Press.

8 Hyperfine Interactions and Laser Cooling

Here a largely qualitative picture of the practical application of hyperfine and magnetic interactions to laser cooling is given. This process is essential in areas of physics that involve the cooling of materials to nanokelvin temperatures such as is required in Bose-Einstein Condensation (BEC). For an excellent interactive website on laser cooling and BEC it is strongly recommended to visit:
http://www.colorado.edu/physics/PhysicsInitiative/Physics2000.03.99/bec/index.html
Other sites of interest can be found at
http://www.physicscentral.com/action/action-00-4.html
http://www.colorado.edu/physics/PhysicsInitiative/Physics2000/index.pl

8.1 MOTION AND TEMPERATURE

Recall Boltzman's equipartition of energy and the property, that molecules in thermal equilibrium have the same average energy, $\frac{1}{2}kT$, associated with each independent degree of freedom of their motion. Thus the average kinetic energy associated with three degrees of translation freedom is

$$\frac{1}{2}mv^2 = \frac{3}{2}kT \qquad (8\text{-}1)$$

and hence can be taken as a definition of *kinetic temperature* with

$$T = \frac{mv^2}{3k} \qquad (8\text{-}2)$$

Or for a given temperature T the average speed of a particle of mass m is

$$v = \sqrt{\frac{3kT}{m}} \qquad (8\text{-}3)$$

This suggests that if we can reduce the average speed of the particles we are equivalently lowering their temperature.

8.2   SOME BASIC QUANTUM RESULTS

Let us recall some basic properties of photons and particles. For a photon we have the standard energy and momentum relation

$$E = hf = \frac{hc}{\lambda} \tag{8-4}$$

$$p = \frac{h}{\lambda} \tag{8-5}$$

and for a non-relativistic particle of mass m traveling with a speed v we have for the de Broglie wavelength, λ_{dB},

$$\lambda_{dB} = \frac{h}{mv} \tag{8-6}$$

From (8-3) and (8-6) we can relate the de Broglie wavelength for an atom of mass m to temperature by writing

$$T = \frac{h^2}{3mk\lambda_{dB}^2} \tag{8-7}$$

In thermal physics it is common to define the *thermal de Broglie wavelength* as

$$\lambda_{th} = \frac{h}{\sqrt{2\pi mkT}} \tag{8-8}$$

Note that λ_{th} decreases as the square root of the temperature T. BEC arises when atoms are cooled to the temperature where λ_{th} is comparable with the interatomic spacing of the atoms. In this situation the atomic wave packets overlap to make a gas of indistinguishable particles with the result that a cloud of atoms all occupy the same quantum state. Note that for alkali atoms the BEC involves the alkali atoms as a very dilute vapor, *not* a solid or liquid. Indeed typical densities are of the order of 10^{-5} of that of air. The vapor must consist of weakly interacting atoms, otherwise liquidification or solidification would occur before BEC can happen. Typical BEC's have dimensions $\sim 100\mu m$ which may be compared with the Bohr radius of a $H-$atom of $1a_0 = 0.0529nm$. For one atomic mass unit (amu)

$$1amu = 1.6605402 \times 10^{-27}kg,$$

and specifically, it is found that in amu

$$^{23}Na : 22.989769$$
$$^{85}Rb : 84.911789$$
$$^{87}Rb : 86.909180$$
$$^{133}Cs : 132.905451$$

Furthermore,

$$h = 6.6260693 \times 10^{-34} Js \qquad k = 1.380651 \times 10^{-23} J K^{-1}$$

To obtain BEC we clearly have to be able to cool the alkali vapor to orders of nanokelvin temperatures. This means that sufficiently low temperatures for BEC to occur requires unconventional cooling techniques. Here we only sketch some of the broad features of these techniques.

It has been known since the mid-1920's that for BEC to occur the phase space density ρ, defined as the number of particles per cubic thermal de Broglie wavelength λ_{th}, must be of the order of unity. Indeed it was shown early on that if the atoms are bosons then BEC occurs when ρ reaches the critical value of 2.612. For an early account of BEC see Ref. 1. In most experiments BEC requires temperatures between $500nK$ and $2\mu K$ with densities between 10^{14} and 10^{15} atoms per cm^3.

8.3   ABSORPTION AND EMISSION OF PHOTONS

The absorption or emission of a photon can possibly occur between two states of energy E_1 and E_2, if the energy of the photon matches the energy difference $\Delta E = |E_1 - E_2|$,

$$\Delta E = \hbar\omega \tag{8-9}$$

where the wavelength λ is related to the angular frequency ω_0 in vacuum and to the wavenumber k_L by

$$\lambda = \frac{2\pi c}{\omega_0} \quad \text{and} \quad k_L = \frac{2\pi}{\lambda} \tag{8-10}$$

Note we say *possibly* as satisfaction of the energy criterion is not sufficient, since in most cases certain selection rules must also be satisfied.

If a resonant photon is absorbed or emitted by an atom, the atom velocity is changed by a recoil velocity, v_r, given by

$$v_r = \frac{\hbar k_L}{m} \tag{8-11}$$

The recoil energy is related to the photon energy such that

$$\hbar\omega_r = \frac{\hbar^2 k_L^2}{2m} \tag{8-12}$$

This gives the idea of how to use a laser to lower the velocity of a cloud of *Rb* vapor by using a laser tuned to a particular energy separation. Atoms approaching the laser beam have their velocity reduced by absorption of photons, while those traveling away from the beam have their velocity increased. Of course in reality things are more complicated. The photons of the incident light are Doppler shifted due to the motion of the incident atoms. The Doppler shift of the incident light of frequency ω_0 due to the motion of the atom is

$$\Delta\omega_d = \frac{v_r}{c}\omega_0 \tag{8-13}$$

which, upon noting (8-12) and (8-10), becomes simply

$$\Delta\omega_d = 2\omega_r \tag{8-14}$$

8.4   LASER COOLING

To observe BEC and related phenomena it is necessary to cool the extremely low pressure vapor, commonly an alkali metal vapor such as $^{87}_{37}Rb$, to nanokelvin temperatures. This is accomplished using lasers, magnetic fields and sometimes electric fields. Usually a combination of methods is applied, such as laser cooling, Doppler cooling, polarized gradient cooling and evaporative cooling. In laser cooling one usually chooses a specific atomic transition and uses three intersecting orthogonal pairs (mutually perpendicular) of counter-propagating laser beams of equal intensity and frequency tuned to just below the atomic transition frequency with the vapor trapped at the point of intersection of the beams. The absorption of photons slows down the atoms by transferring their momentum to the atoms against the direction of motion of the atom; spontaneous emission by the atom occurs in randomly oriented directions. Eventually, repetition of this process results in a slowing down of the atoms and hence a dramatic lowering of the temperature.

The Doppler effect plays a key role in laser cooling. Consider an atom moving along the axis of a particular laser beam that has been tuned just below the resonance frequency of the atomic transition. If the atom is traveling towards the laser beam it is Doppler shifted into resonance, whereas if the atom is traveling in the opposite direction, it is Doppler shifted further away from the resonance frequency. As noted earlier the faster the atom is moving the greater the Doppler shift. If the atom is moving too fast towards the laser beam, it shifts the atom beyond the resonance frequency. We now see the rationale for using counter-propagating laser beams of identical frequency and intensity. If the two beams are collinear then an atom moving in any direction along the beam axis has a greater absorption of photons traveling in the opposite direction. As a result the atom experiences, on average, a loss of momentum, and therefore it slows down, which gives its effective cooling.

Thus it is as if the atom is being forced, in three dimensions, through a dense molasses like liquid, and hence the picturesque terminology *optical molasses*. In this case the viscous medium is actually light. So, where do hyperfine structure and the Zeeman effect enter?

8.5   MAGNETO-OPTICAL TRAPS

Two practical problems arise, containment of the vapor and tuning the laser frequency. The walls of any container are "hot" so the vapor must be kept away from the walls. As to tuning the laser there are two possibilities. One is to adjust the frequency of the photons or change the frequency of the atomic transition. An alkali atom normally exhibits hyperfine structure. The $D2$ transitions, in the absence of hyperfine structure, correspond to the optical transitions $^2S_{1/2} \rightarrow {}^2P_{3/2}$ (see section 7.6) As a result of hyperfine interactions, in the case of $^{87}_{37}Rb$, with $F = 0, 1, 2, 3$, the $^2S_{1/2}$ ground state splits into two hyperfine levels with $F = I \pm \frac{1}{2}$, and the $^2P_{3/2}$ level into up to four hyperfine levels. If we apply a weak external magnetic field, we split each hyperfine level into $2F + 1$ Zeeman sublevels, each characterized by a magnetic quantum number, M_F, where

$$M_F = F, F - 1, \ldots, -F \qquad (8\text{-}15)$$

The magnetic field shifts the energies of the hyperfine levels in proportion to the quantum number M_F and the strength of the magnetic field. Thus we can use a weak magnetic field to change the frequency of the atomic transitions rather than needing to change the frequency of the lasers. Furthermore, if we use circularly polarized laser beams, we can have σ_+ (right-circularly polarized) in one direction and σ_- (left-circularly polarized) in the opposite direction. The σ_+ light is only absorbed if the value of M_F is *increased* by one unit, while the σ_- light is only absorbed if M_F is *decreased* by one unit.

As a concrete example let us assume that an atom is in its ground state $^2S_{1/2}$ with say $F = 1$, and the lasers are tuned to just below the resonance frequency of the $F = 2$ hyperfine level of the excited $^2P_{3/2}$ level. The application of a weak magnetic field results in both levels splitting. The ground state splits into three Zeeman sublevels with $M_F = \pm 1, 0$, with the $M_F = -1$ lowest in energy. The $F = 2$ excited level splits into five sublevels with $M_F = \pm 2, \pm 1, 0$. Assuming circularly polarized light, there are two allowed transitions from the $M_F = -1$ ground state, $M_F = -1 \rightarrow M_F = -2$ for σ_- absorption, and $M_F = -1 \rightarrow M_F = 0$ for σ_+ absorption. Let us now assume that the atom is moving along the z-axis away from $z = 0$, and that the $M_F = -2$ level is closer to the resonance frequency than is the $M_F = 0$ level. As a consequence there is more absorption of the σ_- light than of the σ_+ light, and hence the effect of the net momentum imparted by the photons is to push the atom back towards $z = 0$. Adding a magnetic field that is zero at $z = 0$, the point of intersection of the polarized laser beams, and linearly increasing with distance z, results in any atom not at the center ($z = 0$) having its Zeeman resonance frequency shifted towards the region of zero magnetic field. Such an arrangement confines the cooled atoms to the region of zero magnetic field and hence constitutes a *magneto-optical trap*. Such a trap can reduce the temperature of $^{87}_{37}Rb$ atoms to $\sim 240 \mu K$. In practice one can cool to much lower temperatures using so-called polarized gradient cooling to get to $\sim 10 \mu K$. To reach still lower temperatures requires additional technology such as evaporative cooling where "hot" atoms are boiled off leaving "cold" atoms behind much as in the traditional cooling of a hot cup of coffee. Such techniques allow one to explore temperatures in the nanokelvin range.

REFERENCES

1. Mayer J E and M G Mayer, (1940) *Statistical Mechanics*, New York: J Wiley & Sons p 416.

9 Ions in Crystals

He who can, does; he who cannot, teaches.

George Bernard Shaw, *Man & Superman* (1903)

Those who can, do; those who can't, attend conferences.

Daily Telegraph (6th August, 1979)

So far we have discussed isolated atoms and ions. Now we consider the additional effects that arise when an ion is confined in a crystalline environment. Attention is restricted here to the particular cases of lanthanides and actinides, and hence principally to ions that contain electrons in the $4f-$ or $5f-$shell. For these systems the spin-orbit interaction cannot be ignored and crystal field splittings are usually two orders of magnitude larger than hyperfine splittings; this means working in a $|\alpha SLJM_J IM_I\rangle$ basis rather than a $|\alpha SLJIFM_F\rangle$ basis. Furthermore, a crystalline environment usually means describing the crystal fields surrounding the ions of interest in terms of finite point symmetry groups that are subgroups of the rotation group SO_3, commonly extended to its covering group, SU_2. In some cases the crystal field is dominated by the symmetry of a point group G with a small departure from that symmetry to that described by a subgroup, $H \in G$. By way of examples the discussion is primarily concentrated on the ions Pr^{3+} and Ho^{3+}, which involve the electron configurations $4f^2$ and $4f^{10}$, respectively, and whose hyperfine structure has been studied experimentally in some detail[1-9].

9.1 CRYSTAL FIELD SPLITTINGS

An ion in free space has spherical symmetry and each energy level may be characterized by its total angular momentum **J** (neglecting, at this moment, nuclear angular momentum) and each level is $2J + 1-$fold degenerate. This degeneracy is at least partially lifted, if the ion is placed in a crystal. The degree of degeneracy lifting depends on the value of **J** and the symmetry surrounding the ion. The number of components into which a state of a given angular momentum **J** splits in a symmetry field characterized by a site point group G may be determined from a knowledge of the *branching rules* for the decomposition of the irreducible representation $[J]$ under the group-subgroup reduction $SU_2 \rightarrow G$. This problem was first considered by Bethe[10] with later corrections by Opechowski[11]. Detailed tables have been given by Koster *et al*[12]. In order to discuss the Pr^{3+} ion in a $CsCdBr_3$ crystalline environment and the Ho^{3+} ion in $LiYF_4$ crystals, relevant data are provided below.

9.2   DATA ON THE FINITE GROUPS $O \sim S_4$ AND $C_{3v} \sim S_3$

The finite groups $O \sim S_4$ and $C_{3v} \sim S_3$ play a key role in understanding the Pr^{3+} ion in a $CsCdBr_3$ crystalline environment. The crystal field is predominantly octahedral with a small trigonal component (C_{3v}). Here some data on the finite groups $O \sim S_4$ and $C_{3v} \sim S_3$ are collected. The natural stable isotope of Pr has a nuclear spin $I = \frac{5}{2}$, while the electronic angular momentum has integer values. As a result we need information on the double groups. Below is a collection of character tables (Tables 9-1 to 9-4) and branching rules that are required for further analysis of chosen example. Note that the irreps Γ_4, Γ_5 form a complex pair (or doublet), sometimes designated as Y, with Γ_6 designated as X.

TABLE 9-1

The Character Table of the Double Octahedral Group O

O	E	$\bar{E}$	$8C_3$	$8\bar{C}_3$	$3C_2, 3\bar{C}_2$	$6C_2', 6\bar{C}_2'$	$6C_4$	$6\bar{C}_4$
Γ_1	1	1	1	1	1	1	1	1
Γ_2	1	1	1	1	1	−1	−1	−1
Γ_3	2	2	−1	−1	2	0	0	0
Γ_4	3	3	0	0	−1	−1	1	1
Γ_5	3	3	0	0	−1	1	−1	−1
Γ_6	2	−2	1	−1	0	0	$\sqrt{2}$	$-\sqrt{2}$
Γ_7	2	−2	1	−1	0	0	$-\sqrt{2}$	$\sqrt{2}$
Γ_8	4	−4	−1	1	0	0	0	0

In Tables 9-5 and 9-6 the terms involved in the *symmetric* part of the Kronecker squares are enclosed in curly brackets {, } and the antisymmetric terms in square brackets [,].

TABLE 9-2

The Character Table of the Double Group C_{3v}

C_{3v}	E	$\bar{E}$	$2C_3$	$2\bar{C}_3$	$3\sigma_v$	$3\bar{\sigma}_v$
Γ_1	1	1	1	1	1	1
Γ_2	1	1	1	1	−1	−1
Γ_3	2	2	−1	−1	0	0
Γ_4	1	−1	−1	1	i	$-i$
Γ_5	1	−1	−1	1	$-i$	i
Γ_6	2	−2	1	−1	0	0

TABLE 9-3

$SO(3) \rightarrow O$ **Branching Rules**

$\mathcal{D}_J$	J	O
1	0	Γ_1
3	1	Γ_4
5	2	$\Gamma_3 + \Gamma_5$
7	3	$\Gamma_2 + \Gamma_4 + \Gamma_5$
9	4	$\Gamma_1 + \Gamma_3 + \Gamma_4 + \Gamma_5$
11	5	$\Gamma_3 + 2\Gamma_4 + \Gamma_5$
13	6	$\Gamma_1 + \Gamma_2 + \Gamma_3 + \Gamma_4 + 2\Gamma_5$
2	1/2	Γ_6
4	3/2	Γ_8
6	5/2	$\Gamma_7 + \Gamma_8$
8	7/2	$\Gamma_6 + \Gamma_7 + \Gamma_8$
10	9/2	$\Gamma_6 + 2\Gamma_8$
12	11/2	$\Gamma_6 + \Gamma_7 + 2\Gamma_8$

TABLE 9-4

$O \rightarrow C_{3v}$ **Branching Rules**

$\mathcal{D}_\Gamma$	O	C_{3v}
1	Γ_1	Γ_1
1	Γ_2	Γ_2
2	Γ_3	Γ_3
3	Γ_4	$\Gamma_2 + \Gamma_3$
3	Γ_5	$\Gamma_1 + \Gamma_3$
2	Γ_6	Γ_6

9.3 DATA ON THE FINITE GROUPS FOR Ho^{3+} IONS IN $LiYF_4$ CRYSTALS

Ho^{3+} substitutes for the Y^{3+} in $LiYF_4$ at sites of tetragonal symmetry described by the point group S_4 (not to be confused with the symmetric group on four objects!). Since the ionic radii of Ho^{3+} and Y^{3+} are almost the same, there is little, if any, lattice distortion.

The group S_4 is a cyclic group isomorphic to C_4, consisting of the identity, E, the rotation-reflection $S_4 = IC_4^{-1}$, a two-fold rotation C_2, and the inverse operator $S_4^{-1} = IC_4$. All rotations are taken about the z−axis. The character table is given in Table 9-7. The Kronecker products for S_4 may be easily established from this character table to yield the results given in Table 9-8.

TABLE 9-5
Kronecker Products for the Octahedral Group O

	Γ_1	Γ_2	Γ_3	Γ_4	Γ_5	Γ_6	Γ_7	Γ_8
Γ_1	Γ_1	Γ_2	Γ_3	Γ_4	Γ_5	Γ_6	Γ_7	Γ_8
Γ_2		Γ_1	Γ_3	Γ_5	Γ_4	Γ_7	Γ_6	Γ_8
Γ_3			$\{\Gamma_1+\Gamma_3\}+[\Gamma_2]$	$\Gamma_4+\Gamma_5$	$\Gamma_4+\Gamma_5$	Γ_8	Γ_8	$\Gamma_6+\Gamma_7+\Gamma_8$
Γ_4				$\{\Gamma_1+\Gamma_3+\Gamma_4\}+[\Gamma_5]$	$\Gamma_2+\Gamma_3+\Gamma_4+\Gamma_5$	$\Gamma_6+\Gamma_8$	$\Gamma_7+\Gamma_8$	$\Gamma_6+\Gamma_7+2\Gamma_8$
Γ_5					$\{\Gamma_1+\Gamma_3+\Gamma_4\}+[\Gamma_5]$	$\Gamma_7+\Gamma_8$	$\Gamma_6+\Gamma_8$	$\Gamma_6+\Gamma_7+2\Gamma_8$
Γ_6						$\{\Gamma_4\}+[\Gamma_1]$	$\Gamma_2+\Gamma_5$	$\Gamma_3+\Gamma_4+\Gamma_5$
Γ_7							$\{\Gamma_4\}+[\Gamma_1]$	$\Gamma_3+\Gamma_4+\Gamma_5$
Γ_8								$\{\Gamma_2+2\Gamma_4+\Gamma_5\}+[\Gamma_1+\Gamma_3+\Gamma_5]$

TABLE 9-6
Kronecker Products for the Trigonal Group C_{3v}

	Γ_1	Γ_2	Γ_3	Γ_4	Γ_5	Γ_6
Γ_1	Γ_1	Γ_2	Γ_3	Γ_4	Γ_5	Γ_6
Γ_2	Γ_2	Γ_1	Γ_3	Γ_5	Γ_4	Γ_6
Γ_3	Γ_3	Γ_3	$\{\Gamma_1+\Gamma_3\}+[\Gamma_2]$	Γ_6	Γ_6	$\Gamma_4+\Gamma_5+\Gamma_6$
Γ_4	Γ_4	Γ_5	Γ_6	Γ_2	Γ_1	Γ_3
Γ_5	Γ_5	Γ_4	Γ_6	Γ_1	Γ_2	Γ_3
Γ_6	Γ_6	Γ_6	$\Gamma_4+\Gamma_5+\Gamma_6$	Γ_3	Γ_3	$\{\Gamma_2+\Gamma_3\}+[\Gamma_1]$

TABLE 9-7
Character Table for the Double Group S_4

	E	$\bar{E}$	S_4^{-1}	$\bar{S}_4^{-1}$	C_2	$\bar{C}_2$	S_4	$\bar{S}_4$
Γ_1	1	1	1	1	1	1	1	1
Γ_2	1	1	-1	-1	1	1	-1	-1
Γ_3	1	1	i	i	-1	-1	$-i$	$-i$
Γ_4	1	1	$-i$	$-i$	-1	-1	i	i
Γ_5	1	-1	ω	$-\omega$	i	$-i$	$-\omega^3$	ω
Γ_6	1	-1	$-\omega^3$	ω^3	$-i$	i	ω	$-\omega^3$
Γ_7	1	-1	$-\omega$	ω	i	$-i$	ω^3	$-\omega$
Γ_8	1	-1	ω^3	$-\omega^3$	$-i$	i	$-\omega$	ω^3

TABLE 9-8
Kronecker Products for the Point Group S_4

	Γ_1	Γ_2	Γ_3	Γ_4	Γ_5	Γ_6	Γ_7	Γ_8
Γ_1	Γ_1	Γ_2	Γ_3	Γ_4	Γ_5	Γ_6	Γ_7	Γ_8
Γ_2	Γ_2	Γ_1	Γ_4	Γ_3	Γ_7	Γ_8	Γ_5	Γ_6
Γ_3	Γ_3	Γ_4	Γ_2	Γ_1	Γ_8	Γ_5	Γ_6	Γ_7
Γ_4	Γ_4	Γ_3	Γ_1	Γ_2	Γ_6	Γ_7	Γ_8	Γ_5
Γ_5	Γ_5	Γ_7	Γ_8	Γ_6	Γ_3	Γ_1	Γ_4	Γ_2
Γ_6	Γ_6	Γ_8	Γ_5	Γ_7	Γ_1	Γ_4	Γ_2	Γ_3
Γ_7	Γ_7	Γ_5	Γ_6	Γ_8	Γ_4	Γ_2	Γ_3	Γ_1
Γ_8	Γ_8	Γ_6	Γ_7	Γ_5	Γ_2	Γ_3	Γ_1	Γ_4

TABLE 9-9

Branching Rules for $O_3 \Rightarrow S_4$

D_J^+	S_4		D_J^-	S_4
D_0^+	Γ_1		D_0^-	Γ_2
D_1^+	$\Gamma_1 + \Gamma_3 + \Gamma_4$		D_1^-	$\Gamma_2 + \Gamma_3 + \Gamma_4$
D_2^+	$\Gamma_1 + 2\Gamma_2 + \Gamma_3 + \Gamma_4$		D_2^-	$2\Gamma_1 + \Gamma_2 + \Gamma_3 + \Gamma_4$
D_3^+	$\Gamma_1 + 2\Gamma_2 + 2\Gamma_3 + 2\Gamma_4$		D_3^-	$2\Gamma_1 + \Gamma_2 + 2\Gamma_3 + 2\Gamma_4$
D_4^+	$3\Gamma_1 + 2\Gamma_2 + 2\Gamma_3 + 2\Gamma_4$		D_4^-	$2\Gamma_1 + 3\Gamma_2 + 2\Gamma_3 + 2\Gamma_4$
D_5^+	$3\Gamma_1 + 2\Gamma_2 + 3\Gamma_3 + 3\Gamma_4$		D_5^-	$2\Gamma_1 + 3\Gamma_2 + 3\Gamma_3 + 3\Gamma_4$
D_6^+	$3\Gamma_1 + 4\Gamma_2 + 3\Gamma_3 + 3\Gamma_4$		D_6^-	$4\Gamma_1 + 3\Gamma_2 + 3\Gamma_3 + 3\Gamma_4$
D_7^+	$3\Gamma_1 + 4\Gamma_2 + 4\Gamma_3 + 4\Gamma_4$		D_7^-	$4\Gamma_1 + 3\Gamma_2 + 4\Gamma_3 + 4\Gamma_4$
D_8^+	$5\Gamma_1 + 4\Gamma_2 + 4\Gamma_3 + 4\Gamma_4$		D_8^-	$4\Gamma_1 + 5\Gamma_2 + 4\Gamma_3 + 4\Gamma_4$
$D_{1/2}^+$	$\Gamma_5 + \Gamma_6$		$D_{1/2}^-$	$\Gamma_7 + \Gamma_8$
$D_{3/2}^+$	$\Gamma_5 + \Gamma_6 + \Gamma_7 + \Gamma_8$		$D_{3/2}^-$	$\Gamma_5 + \Gamma_6 + \Gamma_7 + \Gamma_8$
$D_{5/2}^+$	$\Gamma_5 + \Gamma_6 + 2\Gamma_7 + 2\Gamma_8$		$D_{5/2}^-$	$2\Gamma_5 + 2\Gamma_6 + \Gamma_7 + \Gamma_8$
$D_{7/2}^+$	$2\Gamma_5 + 2\Gamma_6 + 2\Gamma_7 + 2\Gamma_8$		$D_{7/2}^-$	$2\Gamma_5 + 2\Gamma_6 + 2\Gamma_7 + 2\Gamma_8$
$D_{9/2}^+$	$3\Gamma_5 + 3\Gamma_6 + 2\Gamma_7 + 2\Gamma_8$		$D_{9/2}^-$	$2\Gamma_5 + 2\Gamma_6 + 3\Gamma_7 + 3\Gamma_8$
$D_{11/2}^+$	$3\Gamma_5 + 3\Gamma_6 + 3\Gamma_7 + 3\Gamma_8$		$D_{11/2}^-$	$3\Gamma_5 + 3\Gamma_6 + 3\Gamma_7 + 3\Gamma_8$
$D_{13/2}^+$	$3\Gamma_5 + 3\Gamma_6 + 4\Gamma_7 + 4\Gamma_8$		$D_{13/2}^-$	$4\Gamma_5 + 4\Gamma_6 + 3\Gamma_7 + 3\Gamma_8$
$D_{15/2}^+$	$4\Gamma_5 + 4\Gamma_6 + 4\Gamma_7 + 4\Gamma_8$		$D_{15/2}^-$	$4\Gamma_5 + 4\Gamma_6 + 4\Gamma_7 + 4\Gamma_8$
$D_{17/2}^+$	$5\Gamma_5 + 5\Gamma_6 + 4\Gamma_7 + 4\Gamma_8$		$D_{17/2}^-$	$4\Gamma_5 + 4\Gamma_6 + 5\Gamma_7 + 5\Gamma_8$
$D_{19/2}^+$	$5\Gamma_5 + 5\Gamma_6 + 5\Gamma_7 + 5\Gamma_8$		$D_{19/2}^-$	$5\Gamma_5 + 5\Gamma_6 + 5\Gamma_7 + 5\Gamma_8$
$D_{21/2}^+$	$5\Gamma_5 + 5\Gamma_6 + 6\Gamma_7 + 6\Gamma_8$		$D_{21/2}^-$	$6\Gamma_5 + 6\Gamma_6 + 5\Gamma_7 + 5\Gamma_8$
$D_{23/2}^+$	$6\Gamma_5 + 6\Gamma_6 + 6\Gamma_7 + 6\Gamma_8$		$D_{23/2}^-$	$6\Gamma_5 + 6\Gamma_6 + 6\Gamma_7 + 6\Gamma_8$

The degeneracies of the states of a given J in a crystal field of S_4 symmetry are determined by the $O_3 \Rightarrow S_4$ branching rules presented in Table 9-9, where O_3 is the *full* orthogonal group (since the point group S_4 includes reflections and hence improper rotations). The irreducible representations of O_3 are labeled with a $+$ or $-$ superscript to distinguish those irreducible representations that are *even* under inversion ($+$) from those that are *odd* ($-$). The results are given in the table below for integer and half-integer values of J. The decompositions of the D_J^- irreducible representations of O_3 may be obtained from those of D_J^+ by multiplication by Γ_2. Note that since the spin irreducible representations of S_4 are all two-dimensional, then for half-integer angular momentum the levels in a crystal with point group symmetry S_4 must necessarily remain two-fold degenerate. An external magnetic field is required to lift this residual Kramer's degeneracy.

In the case of Ho^{3+} in $LiYF_4$ the electronic angular momentum J is an integer, and the Stark electric field degeneracies follow from the appropriate $O_3 \Rightarrow S_4$ branching rules. Adding the half-integer angular momentum of the Ho nucleus results in states of total angular momentum **F** which is half-integer, and hence the degeneracies are always two-fold. The hyperfine interaction also changes selection rules, as discussed in the following chapters.

The point group D_{2d} contains S_4 as a subgroup, and therefore is regarded as an *approximate* symmetry for describing Ho^{3+} in $LiYF_4$ crystals. D_{2d} is isomorphic to the group D_4 and consists of the operations of D_2. In addition it has the operations S_4 and S_4^{-1} about one of the two-fold axes of rotation about the z—axis, as well as two reflections σ_d through perpendicular planes containing the axis of S_4, which bisect the angles between the two rotations of D_2 about the axes x and y, C_2'. The character table, Kronecker products, and $O_3 \Rightarrow D_{2d}$ decompositions are given in Koster *et al* [12].

9.4   THE CRYSTAL FIELD EXPANSION

While group theory provides information about the *number* of levels into which a free ion level of angular momentum **J** splits when the symmetry of the ion is reduced to that of some point group $\mathcal{G}$, it tells us nothing about the *size* of the splittings. To that end it is useful to consider a Hamiltonian H that consists of a free ion part, H_F, and a perturbing part, V_{cryst}, such that

$$H = H_F + V_{cryst} \tag{9-1}$$

The "free ion" part, H_F, includes such terms as the Coulomb and spin-orbit interactions and possibly additional effective interactions that are necessary for a proper description of a many electron system, as for example configuration interaction. The perturbing part, V_{cryst}, represents terms that attempt to explicitly take into account the perturbation produced by the crystal. In fact only in the case of the so-called weak crystal field (as in the particular case of the lanthanides) it is appropriate to treat V_{cryst} as a perturbation, otherwise the matrix of the whole Hamiltonian in (9-1) has to be diagonalized to evaluate the energy of a system. Taking into account an electrostatic interaction between the ion and its environment that is expressed in the terms of the multipole expansion, the crystal field potential is defined as a linear combination of spherical tensors $C_q^{(k)}$ of various ranks,

$$V_{cryst} = \sum_{k,q} B_q^k \sum_i r_i^k \left(C_q^{(k)} \right)_i \tag{9-2}$$

where B_q^k, are the structural factors, commonly called the *crystal field parameters* that represent a symmetry of the environment. The second summation is over all of the electrons of the ion of interest. Here we limit our attention to single particle type operators. The values of k and q are restricted by the symmetry of the point group and by the type of electron orbitals being considered. If the bra and ket states are of the same parity, then k is necessarily *even*, while for bra and ket states of opposite parity, k is necessarily *odd*. Furthermore, if the electron orbitals are ℓ and ℓ' then, taking into account the triangular condition for the non-vanishing $3j$-symbol of reduced matrix element of spherical tensor (2-37c),

$$\ell + \ell' \geq k \geq |\ell - \ell'| \tag{9-3}$$

Thus within a f^N configuration we have $\ell = \ell' = 3$, and

$$k = 0, \ 2, \ 4, \ 6; \tag{9-4}$$

whereas the states of the f^N configuration are coupled to the states of the opposite parity configurations, such as $f^{N-1}d$ or $f^{N-1}g$, by the spherical tensors with *odd* values of k

$$k = \begin{cases} 1, \ 3, \ 5 & \ell = 3, \ \ell' = 2 \\ 1, \ 3, \ 5, \ 7 & \ell = 3, \ \ell' = 4 \end{cases} \tag{9-5}$$

These odd terms play a key role in the Judd-Ofelt theory[13,14] of intensities for lanthanide and actinide ions in crystals.

The term with $k = q = 0$ is spherically symmetric, so it leads to a uniform shift of all the levels of the configuration of interest. At the same time, while the B_q^k are real functions of the radial distance, the non-cylindrical terms (i.e. terms with $|q| > 0$) are not necessarily real functions of the angular coordinates.

For f^N configurations the matrix elements of the crystal field potential V_{cryst} have the form

$$\langle f^N \alpha SLJM_J | V_{cryst} | f^N \alpha' SL'J'M'_J \rangle$$
$$= \sum_{k,q} B_q^k \langle nf|r^k|nf \rangle \langle f^N \alpha SLJM_J | U_q^{(k)} | f^N \alpha' SL'J'M'_J \rangle \langle f||C^{(k)}||f \rangle \tag{9-6}$$

When the semi-empirical procedure is applied to evaluate the energy of lanthanide ion in a crystal, then instead of the structural parameters B_q^k, the crystal field parameters A_q^k, that contain the radial integrals are determined,

$$A_q^k = B_q^k \langle nf|r^k|nf \rangle \tag{9-7}$$

In this way, the problems with the evaluation of radial integrals in (9-6) are avoided. If H_F in (9-1) is defined within the single configuration approximation (as in the case of the Hartree-Fock model), only the intra-shell interactions via crystal field potential defined in (9-6) contribute to the energy. Thus, inspection of the reduced matrix element of spherical tensor indicates that k has to be even, and therefore the energy is determined by even rank crystal field parameters only. This means that through the fitting procedure it is possible to estimate only even rank structural parameters. However, in the case of the *ab initio*-type calculations the radial basis set has to be generated and all the radial integrals, as well as the structural factors, have to be evaluated directly.

From the Wigner-Eckart theorem, the matrix element of the unit tensor operator in (9-6) has the form

$$\langle f^N \alpha SLJM_J | U_q^{(k)} | f^N \alpha' SL'J'M'_J \rangle$$
$$= (-1)^{J-M_J} \begin{pmatrix} J & k & J' \\ -M_J & q & M'_J \end{pmatrix} \langle f^N \alpha SLJ||U^{(k)}||f^N \alpha' SL'J' \rangle \tag{9-8}$$

where

$$\langle f^N \alpha SLJ||U^{(k)}||f^N \alpha' SL'J' \rangle$$
$$= (-1)^{S+L+J'+k}[J, J']^{\frac{1}{2}} \begin{Bmatrix} J & J' & k \\ L' & L & S \end{Bmatrix} \langle f^N \alpha SL||U^{(k)}||f^N \alpha' SL' \rangle \tag{9-9}$$

The doubly reduced matrix elements represent the genealogy of both states of f^N configuration and they may be taken from the tables of Nielson and Koster[15] or from various computer programmes[16]. Nielson and Koster list the matrix elements for the even values of k for the configurations ℓ^N ($\ell = p, d, f$) for $N <= 2\ell + 1$. The corresponding matrix elements for $N > 2\ell + 1$ are obtained by multiplication by -1 of those for $\ell^{4\ell+2-N}$. Note that the matrix elements are diagonal in the spin S. Furthermore, the axial terms, $U_0^{(k)}$, lead to a splitting of terms with different M_J, and the non-axial terms, $U_q^{(k)}$, mix states with $M_J - M_J' = q$. As a result J and M_J cease to be good quantum numbers.

9.5 POINT GROUP SYMMETRY RESTRICTIONS

The possible values of q are restricted by two requirements. The first being that $k >= |q|$, and the second that the crystal field potential, V_{cryst}, be invariant with respect to all the symmetry operations of the relevant point group. This invariance amounts to the requirement that the potential transforms as the identity irreducible representation Γ_1 of the point group $\mathcal{G}$. The number of independent expansion coefficients A_q^k for a given value of k is just the number of times Γ_1 occurs in the decomposition $O_3 \Rightarrow \mathcal{G}$ of the O_3 irreducible representation D_k^+. In the case of the group S_4, we find from Table 9-9 that the identity irreducible representation Γ_1 occurs once for $k = 2$ and three times for $k = 4$ and $k = 6$. This may be compared with the higher symmetry group D_{2d} where Γ_1 occurs once for $k = 2$ and twice for each of $k = 4$ and $k = 6$. Thus in D_{2d} the crystal field expansion for the states of f^N configurations is

$$D_{2d} : V = A_0^2 C_0^{(2)} + A_0^4 C_0^{(4)} + A_4^4(C_{-4}^{(4)} + C_4^{(4)}) + A_0^6 C_0^{(6)} + A_4^6(C_{-4}^{(6)} + C_4^{(6)}) \quad (9\text{-}10)$$

The potential is Hermitian with the expansion coefficients A_q^k (defined by (9-7)) all *real*.

The lower symmetry of the point group S_4 manifests itself in the need for an extra expansion coefficient for each of the non-axial terms. This can be realized by taking the non-axial terms as *complex* rather than real. Thus for S_4 the crystal field potential becomes

$$S_4 : V = A_0^2 C_0^{(2)} + A_0^4 C_0^{(4)} + A_{\pm 4}^4 C_{\pm 4}^{(4)} + A_0^6 C_0^{(6)} + A_{\pm 4}^6 C_{\pm 4}^{(6)} \quad (9\text{-}11)$$

where

$$A_{\pm q}^k = A_q^k \pm i a_q^k \quad (9\text{-}12)$$

with A_q^k and a_q^k both real. Thus the S_4 crystal field is associated with seven independent crystal field parameters whereas D_{2d} has five independent parameters. This crystal field expansion is relevant for Ho^{3+} doped $LiYF_4$ crystals.

9.6   AN OCTAHEDRAL CRYSTAL FIELD

In the case of Pr^{3+} doped $CsCdBr_3$ crystals, to a first approximation, the Pr^{3+} ion sees a predominantly octahedral (O) crystal field which may be written as

$$V_{cryst} = A_4 \left[C_0^{(4)} - \frac{\sqrt{70}}{7} \left(C_3^{(4)} - C_{-3}^{(4)} \right) \right]$$

$$+ A_6 \left[C_0^{(6)} + \frac{\sqrt{210}}{24} \left(C_3^{(6)} - C_{-3}^{(6)} \right) + \frac{\sqrt{231}}{24} \left(C_6^{(6)} + C_{-6}^{(6)} \right) \right] \quad (9\text{-}13)$$

being the fourth and sixth order invariants associated with the integrity basis[17,18] for $O \rightarrow T \rightarrow C_3$. Note that an *integrity basis* is the minimal set of independent invariants associated with a group $\mathcal{G}$ such that all other invariants are polynomials of those of the minimal set. The success of parameterized crystal field calculations has much to do about getting the integrity basis right for the appropriate symmetry group.

Let us consider in some detail the behavior of the $4f^2\ {}^3F_3$ level of Pr^{3+} in a pure octahedral field. This is the only $J = 3$ term for the $4f^2$ configuration so there is no intermediate coupling, and therefore it can be treated as a pure LS−coupled term.

We first note that for the ${}^3F_3$ term of f^2 we have

$$\langle f^2\ {}^3F_3 || C^{(2)} || f^2\ {}^3F_3 \rangle = \frac{\sqrt{105}}{30} \quad (9\text{-}14a)$$

$$\langle f^2\ {}^3F_3 || C^{(4)} || f^2\ {}^3F_3 \rangle = -\frac{\sqrt{154}}{198} \quad (9\text{-}14b)$$

$$\langle f^2\ {}^3F_3 || C^{(6)} || f^2\ {}^3F_3 \rangle = -\frac{5\sqrt{3003}}{858} \quad (9\text{-}14c)$$

Furthermore,

$$\langle f^2\ {}^3F_{3M} | C_q^{(k)} | f^2\ {}^3F_{3M'} \rangle$$

$$= (-1)^{3-M} \begin{pmatrix} 3 & k & 3 \\ -M & q & M' \end{pmatrix} \langle f^2\ {}^3F_3 || C^{(k)} || f^2\ {}^3F_3 \rangle \quad (9\text{-}15)$$

Let us construct the matrices for the operator

$$X_4 = 594 \left[C_0^{(4)} - \frac{\sqrt{70}}{7} \left(C_3^{(4)} - C_{-3}^{(4)} \right) \right] \quad (9\text{-}16)$$

where we have chosen the number 594 to yield simple matrix elements and consequently integer eigenvalues. Using (9-15) and (9-16) we obtain the matrices

$$
\begin{array}{cc}
 & \begin{array}{cc} |{}^3F_{3,1}\rangle & |{}^3F_{3,-2}\rangle \end{array} \\
\begin{array}{c} \langle {}^3F_{3,1}| \\ \langle {}^3F_{3,-2}| \end{array} &
\begin{pmatrix} -1 & 2\sqrt{5} \\ 2\sqrt{5} & 7 \end{pmatrix}
\end{array}
\quad (9\text{-}17a)
$$

$$
\begin{array}{cc}
 & \begin{array}{cc} |{}^3F_{3,-1}\rangle & |{}^3F_{3,2}\rangle \end{array} \\
\begin{array}{c} \langle {}^3F_{3,-1}| \\ \langle {}^3F_{3,2}| \end{array} &
\begin{pmatrix} -1 & -2\sqrt{5} \\ -2\sqrt{5} & 7 \end{pmatrix}
\end{array}
\quad (9\text{-}17b)
$$

$$
\begin{array}{c}
\begin{array}{ccc} |^3F_{3,3}\rangle & |^3F_{3,0}\rangle & |^3F_{3,-3}\rangle \end{array} \\
\begin{array}{c} \langle^3F_{3,3}| \\ \langle^3F_{3,0}| \\ \langle^3F_{3,-3}| \end{array}
\begin{pmatrix} -3 & -3\sqrt{10} & 0 \\ -3\sqrt{10} & -6 & 3\sqrt{10} \\ 0 & 3\sqrt{10} & -3 \end{pmatrix}
\end{array}
\tag{9-17c}
$$

Diagonalizing the matrices yields three distinct eigenvalues

$$
-18(1) \qquad -3(3) \qquad 9(3) \tag{9-18}
$$

with degeneracies encased in parentheses (,). The degeneracies are as expected for octahedral symmetry. The associated eigenvectors are,

$$
|(-18)\rangle = \frac{1}{3}\left[-\sqrt{2}|^3F_{3,3}\rangle - \sqrt{5}|^3F_{3,0}\rangle + \sqrt{2}|^3F_{3,-3}\rangle\right] \tag{9-19a}
$$

$$
|(-3)a\rangle = \frac{1}{\sqrt{2}}\left[|^3F_{3,3}\rangle + |^3F_{3,-3}\rangle\right] \tag{9-19b}
$$

$$
|(-3)b\rangle = \frac{1}{\sqrt{6}}\left[-\sqrt{5}|^3F_{3,1}\rangle + |^3F_{3,-2}\rangle\right] \tag{9-19c}
$$

$$
|(-3)c\rangle = \frac{1}{\sqrt{6}}\left[\sqrt{5}|^3F_{3,-1}\rangle + |^3F_{3,2}\rangle\right] \tag{9-19d}
$$

$$
|(9)a\rangle = \frac{1}{6}\left[-\sqrt{10}|^3F_{3,3}\rangle + 4|^3F_{3,0}\rangle + \sqrt{10}|^3F_{3,-3}\rangle\right] \tag{9-19e}
$$

$$
|(9)b\rangle = \frac{1}{\sqrt{6}}\left[|^3F_{3,1}\rangle + \sqrt{5}|^3F_{3,-2}\rangle\right] \tag{9-19f}
$$

$$
|(9)c\rangle = \frac{1}{\sqrt{6}}\left[|^3F_{3,-1}\rangle - \sqrt{5}|^3F_{3,2}\rangle\right] \tag{9-19g}
$$

Let us repeat the above for the sixth-order operator

$$
X_6 = 6864\left[C_0^{(6)} + \frac{\sqrt{210}}{24}\left(C_3^{(6)} - C_{-3}^{(6)}\right) + \frac{\sqrt{231}}{24}\left(C_6^{(6)} + C_{-6}^{(6)}\right)\right] \tag{9-20}
$$

This time we obtain the matrices

$$
\begin{array}{c}
\begin{array}{cc} |^3F_{3,1}\rangle & |^3F_{3,-2}\rangle \end{array} \\
\begin{array}{c} \langle^3F_{3,1}| \\ \langle^3F_{3,-2}| \end{array}
\begin{pmatrix} -300 & 105\sqrt{5} \\ 105\sqrt{5} & 120 \end{pmatrix}
\end{array}
\tag{9-21a}
$$

$$
\begin{array}{c}
\begin{array}{cc} |^3F_{3,-1}\rangle & |^3F_{3,2}\rangle \end{array} \\
\begin{array}{c} \langle^3F_{3,-1}| \\ \langle^3F_{3,2}| \end{array}
\begin{pmatrix} -300 & -105\sqrt{5} \\ -105\sqrt{5} & 120 \end{pmatrix}
\end{array}
\tag{9-21b}
$$

$$
\begin{array}{c}
\begin{array}{ccc} |^3F_{3,3}\rangle & |^3F_{3,0}\rangle & |^3F_{3,-3}\rangle \end{array} \\
\begin{array}{c} \langle^3F_{3,3}| \\ \langle^3F_{3,0}| \\ \langle^3F_{3,-3}| \end{array}
\begin{pmatrix} -20 & 35\sqrt{10} & -385 \\ 35\sqrt{10} & 400 & -35\sqrt{10} \\ -385 & -35\sqrt{10} & -20 \end{pmatrix}
\end{array}
\tag{9-21c}
$$

and the eigenvalues are,

$$540(1), \qquad 225(3), \qquad -405(3) \qquad (9\text{-}22)$$

The eigenvectors are the same as found in (9-19a-g), to within an overall phase, with the correlations $(540 \rightarrow -18)$, $(225 \rightarrow 9)$ and $(-405 \rightarrow -3)$. We now give the matrices of the complete octahedral crystal field $V_{Oh} = X_4 A^4 + X_6 A^6$,

$$
\begin{array}{cc}
 & |^3F_{3,1}\rangle \qquad\qquad |^3F_{3,-2}\rangle \\
\begin{array}{c} \langle ^3F_{3,1}| \\ \langle ^3F_{3,-2}| \end{array}
\begin{pmatrix}
-A^4 - 300A^6 & \sqrt{5}(2A^4 + 105A^6) \\
\sqrt{5}(2A^4 + 105A^6) & 7A^4 + 120A^6
\end{pmatrix}
\end{array}
\qquad (9\text{-}23a)
$$

$$
\begin{array}{cc}
 & |^3F_{3,-1}\rangle \qquad\qquad |^3F_{3,2}\rangle \\
\begin{array}{c} \langle ^3F_{3,-1}| \\ \langle ^3F_{3,2}| \end{array}
\begin{pmatrix}
-A^4 - 300A^6 & -\sqrt{5}(2A^4 + 105A^6) \\
-\sqrt{5}(2A^4 + 105A^6) & 7A^4 + 120A^6
\end{pmatrix}
\end{array}
\qquad (9\text{-}23b)
$$

$$
\begin{array}{c}
\qquad |^3F_{3,3}\rangle \qquad\qquad\qquad |^3F_{3,0}\rangle \qquad\qquad\qquad |^3F_{3,-3}\rangle \\
\begin{array}{c} \langle ^3F_{3,3}| \\ \langle ^3F_{3,0}| \\ \langle ^3F_{3,-3}| \end{array}
\begin{pmatrix}
-3A^4 - 20A^6 & \sqrt{10}(-3A^4 + 35A^6) & -385A^6 \\
\sqrt{10}(-3A^4 + 35A^6) & -6A^4 + 400A^6 & \sqrt{10}(3A^4 - 35A^6) \\
-385A^6 & \sqrt{10}(3A^4 - 35A^6) & -3A^4 - 20A^6
\end{pmatrix}
\end{array}
$$
$$(9\text{-}23c)$$

The above matrices may be rewritten in terms of the octahedral states given in (9-19a-g) to give

$$
\begin{array}{c}
\qquad |(-3)a\rangle \qquad\qquad |(-3)b\rangle \qquad\qquad |(-3)c\rangle \\
\begin{array}{c} \langle(-3)a| \\ \langle(-3)b| \\ \langle(-3)c| \end{array}
\begin{pmatrix}
-3A^4 - 405A^6 & 0 & 0 \\
0 & -3A^4 - 405A^6 & 0 \\
0 & 0 & -3A^4 - 405A^6
\end{pmatrix}
\end{array}
\qquad (9\text{-}24a)
$$

$$
\begin{array}{c}
\qquad |(9)a\rangle \qquad\qquad |(9)b\rangle \qquad\qquad |(9)c\rangle \\
\begin{array}{c} \langle(9)a| \\ \langle(9)b| \\ \langle(9)c| \end{array}
\begin{pmatrix}
9A^4 + 225A^6 & 0 & 0 \\
0 & 9A^4 + 225A^6 & 0 \\
0 & 0 & 9A^4 + 225A^6
\end{pmatrix}
\end{array}
\qquad (9\text{-}24b)
$$

$$
\begin{array}{c}
|(-18)\rangle \\
\langle(-18)| \begin{pmatrix} -18A^4 + 540A^6 \end{pmatrix}
\end{array}
\qquad (9\text{-}24c)
$$

Not surprisingly, the matrices are diagonal in the octahedral basis.

9.7 IDENTIFICATION OF THE OCTAHEDRAL STATES FOR $^3F_3$

Inspection of Table 9-3 shows that in an octahedral field a state with $J = 3$ splits into three sublevels that belong to the Γ_2, Γ_4, Γ_5 irreducible representations of the octahedral group. The Γ_2 irreducible representation is one-dimensional, while the Γ_4 and Γ_5 irreducible representations are both three-dimensional. This is consistent with our finding of three distinct eigenvalues, one non-degenerate and two that were three-fold degenerate. Thus the eigenvector $|(-18)\rangle$ given in (9-19a) must be associated with the Γ_2 irreducible representation of the octahedral group. The correspondences for the other six eigenvectors must now be determined. This could be achieved by examining their behavior under the symmetry operations of the octahedral group. It is instructive to adopt an alternative approach. Note that the angular momentum operator $\mathbf{J}$ is a rank one tensor operator and transforms like a $J = 1$ state. From Table 9-3 it is seen that its three components must span the Γ_4 irreducible representation. Furthermore, Table 9-5 gives the Kronecker product

$$\Gamma_2 \times \Gamma_4 = \Gamma_5 \tag{9-25}$$

This implies that the matrix elements of $\langle \Gamma_2 | J_z | \Gamma_4 \alpha \rangle$ must necessarily vanish. It is readily seen that this is indeed the case, if Γ_2 is identified with the ket vector $|(-18)\rangle$ given by (9-19a), and the three ket vectors $|(-3)\alpha\rangle$ given by (9-19b,c,d) are taken as belonging to Γ_4. Conversely, (9-25) is satisfied, if the three ket vectors $|(9)\alpha\rangle$ are identified as three components of the Γ_5 irreducible representation.

9.8   INFLUENCE OF THE TRIGONAL C_{3V} CRYSTAL FIELD

The crystal field is predominantly octahedral with a smaller trigonal component. The basic effect can be seen by introducing the operator

$$T = 120C_0^{(2)} \tag{9-26}$$

and considering the matrix elements

$$\langle (-18)\Gamma_2 | T | (-18)\Gamma_2 \rangle = 0$$
$$\langle (9)a\Gamma_4 | T | (9)a\Gamma_4 \rangle \quad = 2$$
$$\langle (9)b\Gamma_4 | T | (9)b\Gamma_4 \rangle \quad = -1 = \langle (9)c\Gamma_4 | T | (9)c\Gamma_4 \rangle$$
$$\langle (-3)a\Gamma_5 | T | (-3)a\Gamma_5 \rangle = 10$$
$$\langle (-3)b\Gamma_5 | T | (-3)b\Gamma_5 \rangle = -5 = \langle (-3)c\Gamma_5 | T | (-3)c\Gamma_5 \rangle \tag{9-27}$$

The above results are consistent with the $O \rightarrow C_{3v}$ branching rules

$$\Gamma_2 \rightarrow \Gamma_2$$
$$\Gamma_4 \rightarrow \Gamma_2 + \Gamma_3$$
$$\Gamma_5 \rightarrow \Gamma_1 + \Gamma_3 \tag{9-28}$$

This means that the octahedral Γ_2 irrep remains non-degenerate, while the Γ_4 and Γ_5 split into a singlet and a doublet. The trigonal splitting for the octahedral Γ_5 irrep is five times larger than for the Γ_4 irrep. Experimentally[2] one finds a ratio of the splittings to be of the order of four. The above result shows that one can often gain insight into the physics of a problem by relatively simple calculations.

REFERENCES

1. Chukalina E P, Popova M N, Antic-Fidancev E and Chaminade J P, (1999) Hyperfine Structure in Optical Spectra of CsCdBr$_3$: Pr^{3+} *Phys. Lett.,* **A 258** 375–8.

2. Popova M N, Chukalina E P, Malkin B Z, Iskhakova A I, Antic-Fidancev E, Porcher P and Chaminade J P, (2001) High-resolution Infrared Absorption Spectra, Crystal-field Levels, and Relaxation Processes in CsCdBr$_3$: Pr^{3+} *Phys. Rev.,* **B 63** 075103-1–9.

3. Ramaz F, Macfarlane R M, Vial J C, Chaminade J P and Madeore F, (1993) Laser and Zeeman Spectroscopy of Pr^{3+} : CsCdBr$_3$: a Simplified Crystal Field Model *J. Lumin.* **55** 173.

4. Wells J-P R, Jones Glynn D and Reeves Roger J, (1999) Zeeman and Hyperfine Infrared Spectra of Pr^{3+} Centers in Alkaline Earth Fluoride Crystals *Phys. Rev.,* **B 60** 851.

5. Agladze N I and Popova M N, (1985) Hyperfine Structure in Optical Spectra of LiYF$_4$: Ho *Sol. St. Comm.,* **55** 1097.

6. Agladze N I, Vinogradov E A and Popova M N, (1986) Manifestation of Quadrupole Hyperfine Interaction and of Interlevel Interaction in the Optical Spectrum of the LiYF$_4$: Ho Crystal *JETP* **64** 716.

7. Agladze N I, Popova M N, Zhizhin G N, Egorov V J and Petrova M A, (1991) Isotope Structure in Optical Spectra of LiYF$_4$: Ho^{3+} *Phys. Rev. Lett.,* **66** 477.

8. Agladze N I, Popova M N, Zhizhin G N and Becucci M, (1993) Study of Isotope Composition in Crystals by High Resolution Spectroscopy of Monoisotope Impurity *J. Exper. Theor. Phys.* **76** 1110.

9. Agladze N I, Popova M N, Koreiba M A and Malkin B Z, (1993) Isotope Effects in the Lattice Structure and Vibrational and Optical Spectra of ^{6}Li$_x$^{7}Li$_{1-x}$YF$_4$: Ho Crystals *J. Exper. Theor. Phys.* **77** 1021.

10. Bethe H, (1929) Termaufspaltung in Kristallen *Ann. Physik* **3** 133.

11. Opechowski W, (1940) Sur les Groupes Cristallographiques "Doubles" *Physica* **7** 552.

12. Koster G F, Dimock J O, Wheeler R G, and Statz H, (1963) *Properties of the Thirty-two Point Groups* Cambridge, Mass.: MIT Press.

13. Judd B R, (1962) Optical Absorption Intensities of Rare-Earth Ions *Phys. Rev.,* **127** 750.

14. Ofelt G S, (1962) Intensities of Crystal Spectra of Rare-Earth Ions *J. Chem. Phys.,* **37** 511.

15. Nielson, C W and Koster, G F (1963) *Spectroscopic Coefficients for the p^n, d^n, and f^n Configurations*, Cambridge: MIT Press.

16. Newman D J and Ng B, (2000) *Crystal Field Handbook* Cambridge: Cambridge University Press.

17. Bickerstaff R P and Wybourne B G, (1976) Integrity Bases, Invariant Operators and the State Labeling Problem for Finite Subgroups of SO$_3$ *J. Phys. A* **9** 1051.

18. Weyl H, (1946) *The Classical Groups* Princeton: Princeton University Press.

10 Some Aspects of Crystal Field Theory

Both liberty and equality are among the primary goals pursued by human beings through many centuries; but total liberty for wolves is death to lambs, total liberty of the powerful, the gifted, is not compatible with the rights to a decent existence of the weak and less gifted.

Isaiah Berlin, *On the Pursuit of the Ideal* (March 17, 1988)

Before returning to the subject of hyperfine interactions in crystals some further remarks on crystal field theory and on the Judd-Ofelt[1,2] theory of intensities are required. Since the classical paper of Bethe[3], considerable attention has been given to the interpretation of the spectra of $f-$ electron ions substituted into various crystal lattices. The simple theory[4-8] based, initially, on the assumption that the substituted ion "sees" a purely electrostatic field with a point symmetry of the lattice site it occupies, met with considerable success. It allowed meaningful assignments to be made of most of the observed crystal field levels and, for the rare earths, the crystal field parameters usually reproduce the observed spectra to a standard deviation of $\sim 10 cm^{-1}$.

In this chapter first the selection rules that follow from the point group symmetry of the immersed ion are discussed for a particular point group S_4 introduced in section 9.3. The relevant selection rules for electric dipole and magnetic dipole transitions in crystals are derived with no reference to specific physical mechanisms. However it must be noted that hyperfine interactions can lead to a violation of these selection rules.

10.1 SELECTION RULES FOR TRANSITIONS IN IONS IN A CRYSTAL FIELD OF S_4 POINT SYMMETRY

Electric dipole (E.d) transitions involve the matrix elements of z for polarization parallel to the $z-$axis ($\pi-$polarization) and matrix elements of $x \pm iy$ for polarization perpendicular to the $z-$axis ($\sigma-$polarization). In the case of S_4 point symmetry z transforms as the Γ_2 representation and $x \pm iy$ as the (Γ_3, Γ_4) complex pair of representations. The E.d selection rules follow from inspection of the Kronecker product presented in Table 9-8 of section 9.3 to give for an *even* number of electrons

$$
\begin{array}{c}
E.d \quad \begin{matrix} \Gamma_1 & \Gamma_2 & \Gamma_3 & \Gamma_4 \end{matrix} \\
\begin{matrix} \Gamma_1 \\ \Gamma_2 \\ \Gamma_3 \\ \Gamma_4 \end{matrix}
\begin{pmatrix}
- & \pi & \sigma & \sigma \\
\pi & - & \sigma & \sigma \\
\sigma & \sigma & - & \pi \\
\sigma & \sigma & \pi & -
\end{pmatrix}
\end{array}
\tag{10-1}
$$

and for an *odd* number of electrons

$$
\begin{array}{c|cccc}
E.d & \Gamma_5 & \Gamma_6 & \Gamma_7 & \Gamma_8 \\
\hline
\Gamma_5 & - & \sigma & \pi & \sigma \\
\Gamma_6 & \sigma & - & \sigma & \pi \\
\Gamma_7 & \pi & \sigma & - & \sigma \\
\Gamma_8 & \sigma & \pi & \sigma & -
\end{array}
\tag{10-2}
$$

For magnetic dipole transitions $(M.d)$ we need the matrix elements of J_z for σ−polarization and $J_x \pm i J_y$ for π−polarization. For S_4 symmetry J_z transforms as Γ_1 and $J_x \pm i J_y$ as the (Γ_3, Γ_4) complex pair of representations. This leads to the selection rules for an *even* number of electrons,

$$
\begin{array}{c|cccc}
M.d & \Gamma_1 & \Gamma_2 & \Gamma_3 & \Gamma_4 \\
\hline
\Gamma_1 & \sigma & - & \pi & \pi \\
\Gamma_2 & - & \sigma & \pi & \pi \\
\Gamma_3 & \pi & \pi & \sigma & - \\
\Gamma_4 & \pi & \pi & - & \sigma
\end{array}
\tag{10-3}
$$

and for an *odd* number of electrons

$$
\begin{array}{c|cccc}
M.d & \Gamma_5 & \Gamma_6 & \Gamma_7 & \Gamma_8 \\
\hline
\Gamma_5 & \sigma & \pi & - & \pi \\
\Gamma_6 & \pi & \sigma & \pi & - \\
\Gamma_7 & - & \pi & \sigma & \pi \\
\Gamma_8 & \pi & - & \pi & \sigma
\end{array}
\tag{10-4}
$$

Note that the crystal field potential can mix states of different J and L, lifting the $\Delta J, \Delta L = 0, \pm 1$ restrictions of the free ion, while the spin-orbit interaction can lead to a breakdown of the spin selection rule $\Delta S = 0$.

Magnetic dipole transitions are allowed between states of the same parity. In the free ion in pure LS−coupling we have the magnetic dipole selection rules

$$
\Delta S, \Delta L = 0, \quad \Delta J = 0, \pm 1
\tag{10-5}
$$

Again these selection rules can be broken by spin-orbit and crystal field interactions. Nevertheless the group-theoretical selection rules (10-1)–(10-4) are rigorous. An exception can arise, if hyperfine interactions exist, and they mix close by crystal field levels.

Consider the trivalent ion, Ho^{3+}, having the electron configuration $4f^{10}$ (an *even* number of electrons), and thus with integer electronic angular momentum. However, the Ho nucleus is half-integer, and hence the net angular momentum is half-integer. As a result the crystal field levels, in the presence of the hyperfine interaction, involve the spin irreducible representations of S_4, and hence the selection rules are given by (10-3) and (10-4) rather than (10-1) and (10-2). Taking into account the degeneracy of the pairs Γ_{56} and Γ_{78} we see that some of the transitions occur in pure $\pi-$ or $\sigma-$ polarization with the remaining ones with $\sigma\pi$−polarization, as shown below

$$
\begin{array}{c|cc}
E.d & \Gamma_{56} & \Gamma_{78} \\
\hline
\Gamma_{56} & \sigma & \sigma\pi \\
\Gamma_{78} & \sigma\pi & \sigma
\end{array}
\tag{10-6}
$$

and

$$
\begin{array}{cc}
M.d & \Gamma_{56} \quad \Gamma_{78} \\
\begin{array}{c} \Gamma_{56} \\ \Gamma_{78} \end{array} & \begin{pmatrix} \sigma\pi & \pi \\ \pi & \sigma\pi \end{pmatrix}
\end{array}
\tag{10-7}
$$

which allows one to distinguish the different symmetries by polarization measurements.

10.2   CRYSTAL FIELD QUANTUM NUMBERS

The crystal field matrix elements must satisfy the selection rule (see (9-8)),

$$
\Delta M_J = q \tag{10-8}
$$

to be non-vanishing. Following Hellwege[9] it is convenient to introduce a set of crystal quantum numbers μ such that

$$
M_J = \mu (mod\ q) \tag{10-9}
$$

Thus for S_4 we have $q = 0, \pm 4$ leading to the crystal quantum numbers

$$
\mu = \begin{cases} 0, \pm 1, 2 & J \text{ integer} \\ \pm\frac{1}{2}, \pm\frac{3}{2} & J \text{ half-integer} \end{cases} \tag{10-10}
$$

In the case of $J = 8$, the following $|J M_J\rangle$ basis states are coupled by the S_4 crystal field

$$
\mu = 0 \ |80\rangle, \ |8\pm 4\rangle, \ |8\mp 8\rangle \tag{10-11a}
$$

$$
\mu = \pm 1 \ |8\pm 1\rangle, \ |8\mp 3\rangle, \ |8\pm 5\rangle, \ |8\mp 7\rangle \tag{10-11b}
$$

$$
\mu = 2 \ |82\rangle, \ |8-2\rangle, \ |86\rangle, \ |8-6\rangle \tag{10-11c}
$$

while for a hyperfine level with $F = \frac{15}{2}$, or an angular momentum level for an odd number of electrons with $J = \frac{15}{2}$, we would have the following basis states

$$
\mu = \pm\tfrac{1}{2} \ |\tfrac{15}{2} \pm \tfrac{1}{2}\rangle, \ |\tfrac{15}{2} \mp \tfrac{7}{2}\rangle, \ |\tfrac{15}{2} \pm \tfrac{9}{2}\rangle, \ |\tfrac{15}{2} \mp \tfrac{15}{2}\rangle \tag{10-12a}
$$

$$
\mu = \pm\tfrac{3}{2} \ |\tfrac{15}{2} \pm \tfrac{3}{2}\rangle, \ |\tfrac{15}{2} \mp \tfrac{5}{2}\rangle, \ |\tfrac{15}{2} \pm \tfrac{11}{2}\rangle, \ |\tfrac{15}{2} \mp \tfrac{13}{2}\rangle \tag{10-12b}
$$

The correlation between the crystal quantum numbers and the group representations of S_4 is readily seen to be

$$
\mu = \begin{cases} 0 : \Gamma_1 \\ \pm 1 : \Gamma_{34} \\ 2 : \Gamma_2 \end{cases} \tag{10-13a}
$$

$$
\mu = \begin{cases} \pm\frac{1}{2} : \Gamma_{56} \\ \pm\frac{3}{2} : \Gamma_{78} \end{cases} \tag{10-13b}
$$

where we have combined the complex pairs of irreducible representations of S_4, emphasizing that they form doubly degenerate states in the crystal field. Note that matrix elements of the crystal field potential vanish between states that belong to different crystal quantum numbers. Furthermore, it is useful to introduce, for the symmetric and antisymmetric linear combinations relevant to the Γ_1 and Γ_2 irreducible representations

$$|JM_J\rangle_\pm = \frac{1}{\sqrt{2}}(|JM_J\rangle \pm |J - M_J\rangle) \tag{10-14}$$

We then have

$$\langle JM_J|V_{cryst}|JM'_J\rangle_{++} = \langle JM_J|V_{cryst}|JM'_J\rangle_{--}, \; real \tag{10-15a}$$

$$\langle JM_J|V_{cryst}|JM'_J\rangle_{+-} = \langle JM_J|V_{cryst}|JM'_J\rangle^*_{-+}, \; imaginary \tag{10-15b}$$

10.3 INTENSITIES OF TRANSITIONS AND EFFECTIVE OPERATORS FOR IONS IN CRYSTALS

In the preceding two sections we have developed the selection rules for electric dipole transitions for ions in crystals. These selection rules were developed independently of any specific physical mechanism. Thus in the case of the S_4 point group symmetry we were considering the selection rules that arise if we had constructed an effective operator with three components transforming as the irreducible representations, Γ_2, for the π−polarization, and the (Γ_3, Γ_4) complex pair of representations, for the two σ−polarization components. This may be compared with the construction of an effective crystal field operator, V_{cryst}, expressed as a linear combination of spherical tensor operators $C_q^{(k)}$ that transform themselves as the identity irreducible representation Γ_1. In the latter case one was simply constructing an integrity basis, and then endowing it with an interpretation based upon the parameterization of the crystal field.

As pointed out by McLellan[10−12], the concept of an integrity basis can be generalized to polynomial bases for specific representations of a group $\mathcal{G}$. Thus one could construct an effective operator to represent the intensities of transitions by defining, once again, linear combinations of the operators $C_q^{(k)}$. This time however it should be required that the linear combinations transform as the Γ_2, Γ_3 and Γ_4 irreducible representations of S_4. These linear combinations would be associated with parameters depending upon the allowed values of (k, q). If the operators are expressed in terms of the $C_q^{(k)}$ operators that act within the f^N electron configuration, then, as discussed in chapter 9, the rank k must be *even*. It would, however, be perfectly possible to construct operators with the appropriate transformation properties having *both* even and odd ranks, if the unit tensor operators $U_q^{(k)}$ were used. Exclusion of the *odd* rank operators from the effective operators that represent the intensity of a transition requires a physical argument such as that applied in the Judd-Ofelt theory[1,2] of $f \longleftrightarrow f$ transitions. In this approach based on second-order perturbation theory the odd rank operators approximately cancel, and the degree of cancelation depends upon the choice of the zero-order eigenvalue problem, and most of all, on the physical reality of investigated system. At the third-order analysis however[13,14] in which electron correlation effects are taken into account, new objects appear and they are the two-particle effective operators with even and also with odd ranks (see chapter 17).

10.4   A SIMPLIFIED CRYSTAL FIELD CALCULATION

The Pr^{3+} ion has $4f^2$ as its lowest energy electron configuration having the spectroscopic terms $^3PFH\ ^1SDGI$. Hund's rules give the ground term as $^3H_4$. Placed in a $LaCl_3$ single crystal the Pr^{3+} ion sees a nearest neighbor point symmetry environment of D_{3h}. (The 'exact' point group symmetry is C_{3h}). The crystal field levels may be labeled by the irreducible representations of the point group D_{3h} whose character table is given below (Table 10.1).

TABLE 10-1
Character Table for the Ordinary Irreducible
Representations of D_{3h}

	E	$2C_3$	$3C_2$	σ_h	$2S_3$	$3\sigma_v$
Γ_1	1	1	1	1	1	1
Γ_2	1	1	−1	1	1	−1
Γ_3	1	1	−1	−1	−1	1
Γ_4	1	1	1	−1	−1	−1
Γ_5	2	−1	0	2	−1	0
Γ_6	2	−1	0	−2	1	0

Under $SO_3 \rightarrow D_{3h}$ we have the branching rules defined in Table 10-2.

The irreducible representations of D_{3h} form a wider class of labels than do the crystal quantum numbers. Levels with $\mu = 0$ may be divided according to whether they transform as Γ_1 or Γ_2, and those with $\mu = 3$ divided according to the irreducible representations Γ_3 or Γ_4. The $\mu = \pm 1$ and $\mu = \pm 2$ levels span the two-dimensional Γ_5 and Γ_6 irreducible representations respectively. In the absence of magnetic fields the Γ_5 and Γ_6 levels remain two-fold degenerate.

For a crystal field of D_{3h} symmetry the potential may be written as

$$V_{cryst} = A_0^2 C_0^{(2)} + A_0^4 C_0^{(4)} + A_0^{(6)} C_0^{(6)} + A_6^6 \left(C_6^{(6)} + C_{-6}^{(6)} \right) \qquad (10\text{-}16)$$

where the crystal field parameters defined in (9-7) are used.

It is easy to evaluate the matrix elements of V_{cryst} for states of the $4f^2$ configuration following the standard tensor operator algebra using for example MAPLE. The test calculations presented here are limited to just the triplets ^3PFH for which any departures from pure LS−coupling (explored earlier in chapter 3) and J−mixing are ignored. There is no splitting for the $^3P_0$ level, which is a non-degenerate Γ_1 level. For the $^3P_1$ term, since it has three components in a pure LS−coupling, two levels are expected, a non-degenerate Γ_2 and a two-fold degenerate Γ_6 level. These levels

TABLE 10-2

$SO_3 \rightarrow D_{3h}$ Branching Rules

J	D_{3h} Irreducible Representations
0	Γ_1
1	$\Gamma_2 + \Gamma_6$
2	$\Gamma_1 + \Gamma_5 + \Gamma_6$
3	$\Gamma_2 + \Gamma_3 + \Gamma_4 + \Gamma_5 + \Gamma_6$
4	$\Gamma_1 + \Gamma_3 + \Gamma_4 + 2\Gamma_5 + \Gamma_6$
5	$\Gamma_2 + \Gamma_3 + \Gamma_4 + 2\Gamma_5 + 2\Gamma_6$
6	$2\Gamma_1 + \Gamma_2 + \Gamma_3 + \Gamma_4 + 2\Gamma_5 + 2\Gamma_6$

posses crystal field energies of (we suppress the SL labels, just giving the $|JM_J\rangle$ numbers),

$$\langle 1,0|V_{cryst}|1,0\rangle = +\frac{A_0^2}{5} \quad \text{and} \quad \langle 1,\pm 1|V_{cryst}|1,\pm 1\rangle = -\frac{A_0^2}{10} \qquad (10\text{-}17)$$

and thus a crystal field energy separation of

$$E(\Gamma_6) - E(\Gamma_2) = \frac{3A_0^2}{10} \qquad (10\text{-}18)$$

Experimentally[15] the separation is found to be $30.3 cm^{-1}$ suggesting from (10-18) a value for A_0^2 of $\sim 103 cm^{-1}$. Indeed on the basis of a least squares fit with the inclusion of intermediate coupling and $J-$mixing, Margolis[16], obtained the parameter set

$$A_0^2 = 95 cm^{-1}, \quad A_0^4 = -325 cm^{-1}, \quad A_0^6 = -634 cm^{-1}, \quad A_6^6 = 427 cm^{-1} \qquad (10\text{-}19)$$

For the $^3P_2$ level we expect from Table 10-1 a non-degenerate Γ_1 and two two-fold degenerate Γ_5, Γ_6 levels. Indeed we find the matrix elements

$$\langle 2,0|V_{cryst}|2,0\rangle = -\frac{A_0^2}{5}, \quad \langle 2,\pm 1|V_{cryst}|2,\pm 1\rangle = -\frac{A_0^2}{10}, \quad \langle 2,\pm 2|V_{cryst}|2,\pm 2\rangle = \frac{A_0^2}{5} \qquad (10\text{-}20)$$

Again the matrix elements are determined by just one crystal field parameter, A_0^2, whereas the matrix elements for the $^3F_2$ term involve A_0^4 as well as A_0^2, though the number of levels remains the same. Indeed the energies for the three levels are

$$\langle 2,0|V_{cryst}|2,0\rangle = -\frac{8A_0^2}{105} - \frac{2A_0^4}{63} \qquad (10\text{-}21a)$$

$$\langle 2,\pm 1|V_{cryst}|2,\pm 1\rangle = -\frac{4A_0^2}{105} + \frac{4A_0^4}{189} \qquad (10\text{-}21b)$$

$$\langle 2,\pm 2|V_{cryst}|2,\pm 2\rangle = \frac{8A_0^2}{105} - \frac{A_0^4}{189} \qquad (10\text{-}21c)$$

In calculating the matrix elements for the $^3F_3$ term we take advantage of the symmetric and antisymmetric linear combinations $|3,3\rangle_\pm$ defined in (10-14) with the symmetric combination belonging to Γ_4, and the antisymmetric to Γ_3.

$$\langle 3, 0|V_{cryst}|3, 0\rangle \quad = -\frac{A_0^2}{15} - \frac{A_0^4}{99} + \frac{25A_0^6}{429} \tag{10-22a}$$

$$\langle 3, \pm1|V_{cryst}|3, \pm1\rangle = -\frac{A_0^2}{20} - \frac{A_0^4}{594} - \frac{25A_0^6}{572} \tag{10-22b}$$

$$\langle 3, \pm2|V_{cryst}|3, \pm2\rangle = \frac{7A_0^4}{594} + \frac{5A_0^6}{286} \tag{10-22c}$$

$$\pm\langle 3, 3|V_{cryst}|3, 3\rangle_\pm \quad = \frac{A_0^2}{12} - \frac{A_0^4}{198} - \frac{5A_0^6}{1716} \mp \frac{5\sqrt{231}A_0^6}{858} \tag{10-22d}$$

$$\pm\langle 3, 3|V_{cryst}|3, 3\rangle_\mp \quad = \mp\langle 3, 3|V_{cryst}|3, 3\rangle_\pm = 0 \tag{10-22e}$$

Here we see that by applying the symmetrized states, the rank 2 matrix formed by the $|3, \pm3\rangle$ states has effectively been put in diagonal form.

As a final example we consider $^3H_4$ ground term.

$$\langle 4, 0|V_{cryst}|4, 0\rangle = \frac{104A_0^2}{495} - \frac{12A_0^4}{121} - \frac{1360A_0^6}{14157} \tag{10-23a}$$

$$\langle 4, \pm1|V_{cryst}|4, \pm1\rangle = \frac{442A_0^2}{2475} - \frac{6A_0^4}{121} + \frac{68A_0^6}{14157} \tag{10-23b}$$

$$
\begin{array}{cc}
\mu = \pm2 & |4, 2\rangle_\pm \qquad\qquad\qquad |4, 4\rangle_\pm \\
\pm\langle 4, 2| & \left(\begin{array}{cc} \frac{208A_0^2}{2475} + \frac{2A_0^4}{33} + \frac{136A_0^6}{1287} & \pm\frac{272\sqrt{33}A_0^6}{14157} \\ \pm\langle 4, 4| & \pm\frac{272\sqrt{33}A_0^6}{14157} & -\frac{728A_0^2}{2475} - \frac{28A_0^4}{363} + \frac{272A_0^6}{14157} \end{array} \right)
\end{array} \tag{10-23c}
$$

$$\pm\langle 4, 3|V_{cryst}|4, 3\rangle_\pm = -\frac{182A_0^2}{2475} + \frac{14A_0^4}{121} - \frac{1156A_0^6}{14157} \pm \frac{136\sqrt{231}A_0^6}{14157} \tag{10-23d}$$

Notice that (10-23c) corresponds to a pair of rank 2 matrices which have identical pairs of eigenvalues though different eigenfunctions.

10.5   THE MAPLE PROGRAMME

The preceding calculations were done by the simple MAPLE programme given below. Note that the crystal field parameters A_q^k are declared globally. If you do not specify them, then you get the general expressions given earlier. If you declare the values of the parameters, the output is numerical.

```
print('This routine calculates the crystal field matrix elements for');
print('the triplet states of the f^2 configuration with D3h symmetry');
read"njsym";
fck:=proc(L,J,M1,M2,k,q)
local result;
result:=simplify(combine((((-1)^(J-M1))*threej(J,k,J,-M1,q,M2)*
(-2)*(2*L+1)*sixj(L,k,L,3,3,3)*ck(3,3,k)*((-1)^(1+L+J+k))*(2*J+1)*
sixj(J,k,J,L,1,L))));
end:
V:=proc(L,J,M1,M2)
local me;
global A20,A40,A60,A66;
me:=A20*fck(L,J,M1,M2,2,0)+A40*fck(L,J,M1,M2,4,0)
+A60*fck(L,J,M1,M2,6,0)
+A66*(fck(L,J,M1,M2,6,6)+fck(L,J,M1,M2,6,-6));
end:
print('Example to calculate the matrix element ⟨3H43|V|3H43⟩');
print('Enter V(5,4,3,3);');
V(5,4,3,3);
```

Using the Margolis parameter set (10-19) in (10-23a–d) leads to the numerical results (in cm^{-1})

$$\langle 4, 0|V_{cryst}|4, 0\rangle = 113 \tag{10-24a}$$

$$\langle 4, \pm 1|V_{cryst}|4, \pm 1\rangle = 30 \tag{10-24b}$$

$$
\mu = \pm 2 \quad
\begin{array}{cc}
|4, 2\rangle_{\pm} & |4, 4\rangle_{\pm} \\
\end{array}
$$
$$
\begin{array}{c}
{}_{\pm}\langle 4, 2| \\
{}_{\pm}\langle 4, 4|
\end{array}
\left(
\begin{array}{cc}
-79 & \pm 47 \\
\pm 47 & -15
\end{array}
\right) \tag{10-24c}
$$

$$
{}_{\pm}\langle 4, 3|V_{cryst}|4, 3\rangle_{\pm} = 7 \pm 62 \tag{10-24d}
$$

Diagonalization of the matrix in (10-24c) yields the eigenvalues -55 and 69. Adjusting the eigenvalues found in (10-24a-d) so that the lowest eigenvalue matches the lowest crystal field level for the ground state, we have the comparison with the experimental data[17] given below,

D_{3h}	μ	E_{expt}	E_{calc}	
Γ_5	2	0	0	
Γ_3	3	49	33	
Γ_5	2	114	96	(10-25)
Γ_6	1	134	130	
Γ_4	3	173	137	
Γ_1	0	217	199	

The agreement between the experimental and calculated levels is reasonable considering the simplicity of our calculation, where in fact we have been more interested in illustrating principles rather than precision; however, the order of the levels is correctly reproduced.

Crystal field calculations that involve parameterized fits of symmetrized sets of spherical harmonics are to a large extent model independent with a "good fit" being indicative of the correct symmetry. As noted earlier such calculations are what one expects if all the invariants of the appropriate integrity basis are used. In many respects the parameterized Judd-Ofelt theory of intensities can be regarded as essentially a generalized integrity basis. Thus for the point group D_{3h} one knows that the electric dipole selection rules arise from the knowledge that z belongs to the Γ_4 irrep and $x \pm iy$ to Γ_6. We may attempt to construct an effective operator from the even rank unit tensor operators $U_q^{(k)}$ that mimic the selection rules. Two such operators would be

$$O(\Gamma_4) = T_3^4(U_3^{(4)} + U_{-3}^{(4)}) + T_3^6(U_3^{(6)} + U_{-3}^{(6)}) \tag{10-26a}$$

$$O(\Gamma_6) = T_2^2(U_2^{(2)} + U_{-2}^{(2)}) + T_2^4(U_2^{(4)} + U_{-2}^{(4)}) + T_4^4(U_4^{(4)} + U_{-4}^{(4)})$$
$$+ T_2^6(U_2^{(6)} + U_{-2}^{(6)}) + T_4^6(U_4^{(6)} + U_{-4}^{(6)}) \tag{10-26b}$$

The operator $O(\Gamma_4)$ with two parameters determines the π−polarized transitions, and $O(\Gamma_6)$ that involve five parameters the σ−polarized transitions.

REFERENCES

1. Judd B R, (1962) Optical Absorption Intensities of Rare-Earth Ions *Phys. Rev.*, **127** 750.
2. Ofelt G S, (1962) Intensities of Crystal Spectra of Rare-Earth Ions *J. Chem. Phys.*, **37** 511.
3. Bethe H, (1929) Termaufspaltung in Kristallen *Ann. Physik* **3** 133.
4. Bleaney B and Stevens K W H, (1953) Paramagnetic Resonance *Rep. Prog. Phys.*, **16** 108.
5. Elliott J P, Judd B R and Runciman W A, (1957) Energy Levels in Rare-earth Ions *Proc. R. Soc.*, **A 240** 509.
6. Griffiths J S, (1961) *The Theory of Transition-Metal Ions* Cambridge: Cambridge University Press.
7. Ballhausen C J, (1962) *Introduction to Ligand Field Theory* New York: McGraw-Hill.
8. Hutchings M T, (1964) Energy Levels of Ions in Crystal Fields *Solid State Physics* **16** 227.
9. Hellwege K H, (1949) Electronenterme und Strahlung von Atomen in Kristallen *Ann. Physik* **4** 95.
10. McLellan A G, (1974) Invariant Functions and Homogeneous Bases of Irreducible Representations of the Crystal Point Groups, with Applications to Thermodynamic Properties of Crystals under Strain *J. Phys. C: Solid State Phys.*, **7**, 3326.
11. McLellan A G, (1979) The Derivation of Generalized Integrity Bases for Finite Orthogonal Groups and Counting Rules for Numerator Invariants and Covariants *J. Phys. C: Solid State Phys.*, **12** 753.
12. McLellan A G, (1980) *The Classical Thermodynamics of Deformable Materials* Cambridge: Cambridge University Press.

13. Wybourne, B G, (1967) Coupled Products of Annihilation and Creation Operators in Crystal Energy Problems, Ed. H. M. Crosswhite and H. W. Moos, pp 35-52 in *Optical Properties of Ions in Crystals*, Wiley-Interscience Publication, John Wiley and Sons Inc., New York.
14. Wybourne, B G, (1968) Effective Operators and Spectroscopic Properties, *J. Chem. Phys.*, **48** 2596.
15. Dieke G H and Sarup R, (1958) Fluorescence Spectrum of $PrCl_3$ and the Levels of the Pr^{+++} Ion *J. Chem. Phys.*, **29** 741.
16. Margolis Jack S, (1961) Energy Levels of $PrCl_3$ *J. Chem. Phys.*, **35** 1367.
17. Rana R S and Kaseta F W, (1983) Laser Excited Fluorescence and Infrared Absorption Spectra of Pr^{3+} : $LaCl_3$ *J. Chem. Phys.*, **79** 5280.

11 Hyperfine Interactions in Crystals: Pr^{3+} in Octahedral Field

Example isn't another way to teach, it is the only way to teach.

Albert Einstein

The earliest studies on hyperfine interactions came from electron spin resonance. Later work by the groups of Hellwege in Darmstadt and Dieke at Johns Hopkins in Baltimore reported observations of hyperfine structure in optical absorption spectra of Pr^{3+} and Ho^{3+} ions in crystals. The substantive development came with the very high resolution studies by M Popova and her associates[1-7] at Troitsk, near Moscow. In this chapter the primary aim is to calculate the magnetic dipole hyperfine structure in the $^3F_3$ term of the $4f^2$ configuration of the Pr^{3+} ion in a $CsCdBr_3$ crystalline environment with the assumption that the Pr^{3+} ion is at a site of $SO(3) \rightarrow O_h \rightarrow C_{3v}$ symmetry. The objective is to explain, using an example and specific calculations, the principal features of the observed hyperfine structure, since the high precision of the discussion is not the aim of the discussion but rather a relatively simple description of the broad features of the observed hyperfine structure.

In order to evaluate the hyperfine structure resulting from the magnetic dipole interactions first of all the dominant, physical mechanisms and their major contributions to the energy have to be taken into account. These are the parts of many-electron Hamiltonian that represent Coulomb and spin-orbit interactions.

The f^2 configuration involves just seven ^{2S+1}L terms whose Coulomb energies in terms of the Slater F_k integrals are

$$E(^1S) = F_0 + 60F_2 + 198F_4 + 1716F_6$$
$$E(^3P) = F_0 + 45F_2 + 33F_4 - 1287F_6$$
$$E(^1D) = F_0 + 19F_2 - 99F_4 + 715F_6$$
$$E(^3F) = F_0 - 10F_2 - 33F_4 - 286F_6$$
$$E(^1G) = F_0 - 30F_2 + 97F_4 + 78F_6$$
$$E(^3H) = F_0 - 25F_2 - 51F_4 - 13F_6$$
$$E(^1I) = F_0 + 25F_2 + 9F_4 + F_6 \tag{11-1}$$

The complete spin-orbit interaction matrices for all the $^{2S+1}L_J$ levels of the f^2 electron configuration are displayed below. Notice that there is no spin-orbit coupling between the $^3F_3$ level and any other levels, and hence there are no intermediate coupling corrections to be included in this case. In this sense $^3F_3$ is a pure state, similarly as the excited states $^3P_1$ and $^3H_5$.

$$
\begin{array}{c} \\ ^3P_0 \\ ^1S_0 \end{array}
\begin{array}{cc} ^3P_0 & ^1S_0 \\ \begin{pmatrix} -1 & -2\sqrt{3} \\ -2\sqrt{3} & 0 \end{pmatrix} \end{array}
\qquad
\begin{array}{c} \\ ^3P_1 \end{array}
\begin{array}{c} ^3P_1 \\ \left(-\tfrac{1}{2}\right) \end{array}
\qquad
\begin{array}{c} \\ ^3P_2 \\ ^1D_2 \\ ^3F_2 \end{array}
\begin{array}{ccc} ^3P_2 & ^1D_2 & ^3F_2 \\ \begin{pmatrix} \tfrac{1}{2} & \tfrac{3}{2}\sqrt{2} & 0 \\ \tfrac{3}{2}\sqrt{2} & 0 & -\sqrt{6} \\ 0 & -\sqrt{6} & -2 \end{pmatrix} \end{array}
$$

$$
\begin{array}{c} \\ ^3F_3 \end{array}
\begin{array}{c} ^3F_3 \\ \left(-\tfrac{1}{2}\right) \end{array}
\qquad
\begin{array}{c} \\ ^3F_4 \\ ^1G_4 \\ ^3H_4 \end{array}
\begin{array}{ccc} ^3F_4 & ^1G_4 & ^3H_4 \\ \begin{pmatrix} \tfrac{3}{2} & \tfrac{\sqrt{33}}{3} & 0 \\ \tfrac{\sqrt{33}}{3} & 0 & -\tfrac{\sqrt{30}}{3} \\ 0 & -\tfrac{\sqrt{30}}{3} & -3 \end{pmatrix} \end{array}
\qquad
\begin{array}{c} \\ ^3H_5 \end{array}
\begin{array}{c} ^3H_5 \\ \left(-\tfrac{1}{2}\right) \end{array}
\qquad
\begin{array}{c} \\ ^3H_6 \\ ^1I_6 \end{array}
\begin{array}{cc} ^3H_6 & ^1I_6 \\ \begin{pmatrix} \tfrac{5}{2} & \tfrac{\sqrt{6}}{2} \\ \tfrac{\sqrt{6}}{2} & 0 \end{pmatrix} \end{array}
$$

$$(11\text{-}2)$$

The individual matrix elements of (11-2) are to be multiplied by the spin-orbit coupling constant ζ, which in the case of central field potential is related to the radial integral of r^{-3} with the one-electron states of $4f$ symmetry.

11.1   MATRIX ELEMENTS OF MAGNETIC DIPOLE HYPERFINE INTERACTIONS

The electronic part of the magnetic dipole hyperfine interactions are represented by the operator (defined in (5-30)),

$$
\mathbf{N}_i^{(1)} = \mathbf{l}_i^{(1)} - \sqrt{10}(\mathbf{s}^{(1)}\mathbf{C}^{(2)})_i^{(1)}
$$
$$
= [\mathbf{l}_i - \sqrt{10}\mathbf{X}_i^{(1)}] \tag{11-3}
$$

where $X^{(1)}$ is a tensorial product of $\mathbf{s}^{(1)}$ and $\mathbf{C}^{(2)}$, and it is associated with the double tensor operator $V^{(12)1}$ or with the unit double tensor operator $W^{(12)1}$. For the $N-$ electron system

$$
\mathbf{N}^{(1)} = \sum_{i=1}^{N} N_i^{(1)} \tag{11-4}
$$

where the sum is over a group of equivalent electrons in the configuration ℓ^N.

The interaction of a nuclear magnetic moment with the orbital and spin moments of N electrons in the above tensor operator notation has a form of a scalar product of tensor operators of rank 1,

$$
\mathbf{H}_{hfs} = a_\ell(\mathbf{N}^{(1)} \cdot \mathbf{I}^{(1)}) \tag{11-5}
$$

where the constant factor is defined as follows

$$
a_\ell = 2\mu_B^2(m_e/m_p)g_I\langle r^{-3}\rangle \tag{11-6}
$$

with μ_B the Bohr magneton, g_I the nuclear g factor and $\langle r^{-3}\rangle$ the average inverse-cube radius of the electron orbital ℓ. In the case of the coupled electron and nuclear angular momenta scheme, $JIFM_F$, the matrix element of H_{hfs} is as follows

$$
\langle \alpha JIFM|a_\ell(\mathbf{N}^{(1)} \cdot \mathbf{I}^{(1)})|\alpha'J'IF'M_F'\rangle
$$
$$
= a_\ell(-1)^{J'+I+F}\delta_{F,F'}\delta_{M_F,M_F'}
\begin{Bmatrix} J' & I & F \\ I & J & 1 \end{Bmatrix}
\langle \alpha J||N^{(1)}||\alpha'J'\rangle\langle I||I^{(1)}||I\rangle \tag{11-7}
$$

For matrix elements diagonal in J (11-7) simplifies to

$$\langle \alpha J I F M_F | a_\ell (\mathbf{N}^{(1)} \cdot \mathbf{I}^{(1)}) | \alpha' J I F M_F \rangle$$
$$= a_\ell \frac{K}{2\sqrt{J(J+1)(2J+1)}} \langle \alpha J || N^{(1)} || \alpha' J \rangle \qquad (11\text{-}8)$$

where

$$K = F(F+1) - J(J+1) - I(I+1) \qquad (11\text{-}9)$$

While there is no difficulty in calculating matrix elements with J−mixing, we leave that extension as an exercise. The calculation of the electronic part of the magnetic dipole hyperfine interaction involves two parts, an orbital part ($\mathcal{L}$) and a spin part ($\mathcal{S}$).

For the orbital part we have

$$\mathcal{L} = \frac{\langle \alpha S L J || L^{(1)} || \alpha' S' L' J \rangle}{\sqrt{J(J+1)(2J+1)}} = \delta_{\alpha,\alpha'} \delta_{S,S'} \delta_{L,L'} (2 - g) \qquad (11\text{-}10)$$

where g is the usual Lande g−factor

$$2 - g = \frac{J(J+1) + L(L+1) - S(S+1)}{2J(J+1)} \qquad (11\text{-}11)$$

This part may be corrected for intermediate coupling by simply replacing g by its intermediate coupling value, as discussed in chapter 5. However, for pure states, as $^3F_3$, as mentioned above, there is no improvement resulting from the intermediate coupling (in fact this is the reason that this state is presented here as an instructive example).

For the spin part we have

$$\mathcal{S} = -\frac{\sqrt{10} \langle \alpha S L J || \sum_{i=1}^{N} (\mathbf{s}^{(1)} \mathbf{C}^{(2)})_i^{(1)} || \alpha' S' L' J \rangle}{\sqrt{J(J+1)(2J+1)}} \qquad (11\text{-}12)$$

which is determined by

$$\mathcal{S} = (-1)^{\ell+1} (2\ell+1) \begin{pmatrix} \ell & \ell & 2 \\ 0 & 0 & 0 \end{pmatrix} \begin{Bmatrix} S & S' & 1 \\ L & L' & 2 \\ J & J & 1 \end{Bmatrix}$$
$$\times \sqrt{\frac{30(2J+1)}{J(J+1)}} \langle \alpha SL || V^{(12)} || \alpha' S' L' \rangle \qquad (11\text{-}13)$$

In the matrix element at the right-hand-side there is a double tensor operator $\mathbf{V}^{(12)}$ that acts in the spin and orbital spaces. This operator is defined by its one-electron reduced matrix element (which is not equal to 1, as in the case of the operator w)

$$\langle \ell || v^{(12)} || \ell \rangle = \sqrt{\frac{3}{2}} \qquad (11\text{-}14)$$

The matrix element of the double tensor operator at the right hand side of (11-13) contains the information about the genealogy of the states of the $4f^N$ configuration.

The particular case of the two electron configuration of Pr^{3+} ion, $4f^2$, is the simplest one. For this configuration the fractional parentage coefficients are equal to 1, since there is only one parent state and the coefficients are normalized. Therefore the matrix element is essentially determined by the 6−j symbols,

$$\langle \alpha SL||V^{(12)}||\alpha'S'L'\rangle = \sqrt{6(2S+1)(2S'+1)(2L+1)(2L'+1)}$$
$$\times \left\{ \begin{matrix} S & 1 & S' \\ \frac{1}{2} & \frac{1}{2} & \frac{1}{2} \end{matrix} \right\} \left\{ \begin{matrix} L & 2 & L' \\ \ell & \ell & \ell \end{matrix} \right\} \tag{11-15}$$

From (11-10) and (11-12) we find for $f^2(^3F_3)$ the values

$$\mathcal{L} = \frac{11}{12} \quad \text{and} \quad \mathcal{S} = -\frac{1}{36} \tag{11-16}$$

The magnetic hyperfine structure constant, A, as normally defined, is given by

$$A = a_\ell[\mathcal{L} + \mathcal{S}] \tag{11-17}$$

and hence

$$A(^3F_3) = a_\ell \left[\frac{8}{9} \right] \tag{11-18}$$

since, as mentioned above, for the $^3F_3$ state there is no intermediate coupling correction.

The result obtained in (11-18) was evaluated for the free ionic system that possesses a spherical symmetry, and therefore the energy states are identified by the angular and spin momenta that are keeping their status of good quantum numbers. The situation is different when the spherical symmetry is perturbed by the crystal field, and the energy states have to be described by their transformation properties defined in the terms of irreducible representations of a certain group.

Continuing the discussion on the free ion case and evaluation of the magnetic dipole hyperfine matrix elements in the scheme of the coupled electron and nuclear angular momenta, $|JIFM_F\rangle$, we apply the general results of chapter 6 to the particular case of our example. For the appropriate values of various momenta, and for a particular electronic state $^3F_3$, the matrix elements contributing to the hyperfine structure of the $4f^2$ configuration have the form

$$\begin{array}{cc} & M_F = \frac{11}{2} \quad |33\frac{5}{2}\frac{5}{2}\rangle \\ \langle 33\frac{5}{2}\frac{5}{2}| & \left(\begin{array}{c} \frac{15}{2} \end{array} \right) \end{array} \tag{11-19a}$$

$$\begin{array}{ccc} & M_F = \frac{9}{2} \quad |33\frac{5}{2}\frac{3}{2}\rangle \quad |32\frac{5}{2}\frac{5}{2}\rangle \\ \langle 33\frac{5}{2}\frac{3}{2}| & \left(\begin{array}{cc} \frac{9}{2} & \frac{\sqrt{30}}{2} \\ \frac{\sqrt{30}}{2} & 5 \end{array} \right) \\ \langle 32, \frac{5}{2}\frac{5}{2}| & \end{array} \tag{11-19b}$$

$$M_F = \tfrac{7}{2} \quad |33\tfrac{5}{2}\tfrac{1}{2}\rangle \quad |32\tfrac{5}{2}\tfrac{3}{2}\rangle \quad |31\tfrac{5}{2}\tfrac{5}{2}\rangle$$

$$\begin{array}{c} \langle 33\tfrac{5}{2}\tfrac{1}{2}| \\ \langle 32\tfrac{5}{2}\tfrac{3}{2}| \\ \langle 31\tfrac{5}{2}\tfrac{5}{2}| \end{array} \left(\begin{array}{ccc} \tfrac{3}{2} & 2\sqrt{3} & 0 \\ 2\sqrt{3} & 3 & \tfrac{5}{2}\sqrt{2} \\ 0 & \tfrac{5}{2}\sqrt{2} & \tfrac{5}{2} \end{array} \right) \qquad \text{(11-19c)}$$

$$M_F = \tfrac{5}{2} \quad |33\tfrac{5}{2}-\tfrac{1}{2}\rangle \quad |32\tfrac{5}{2}\tfrac{1}{2}\rangle \quad |31\tfrac{5}{2}\tfrac{3}{2}\rangle \quad |30\tfrac{5}{2}\tfrac{5}{2}\rangle$$

$$\begin{array}{c} \langle 33\tfrac{5}{2}-\tfrac{1}{2}| \\ \langle 32\tfrac{5}{2}\tfrac{1}{2}| \\ \langle 31\tfrac{5}{2}\tfrac{3}{2}| \\ \langle 30,\tfrac{5}{2}\tfrac{5}{2}| \end{array} \left(\begin{array}{cccc} -\tfrac{3}{2} & \tfrac{3}{2}\sqrt{6} & 0 & 0 \\ \tfrac{3}{2}\sqrt{6} & 1 & 2\sqrt{5} & 0 \\ 0 & 2\sqrt{5} & \tfrac{3}{2} & \sqrt{15} \\ 0 & 0 & \sqrt{15} & 0 \end{array} \right) \qquad \text{(11-19d)}$$

$$M_F = \tfrac{3}{2} \quad |33\tfrac{5}{2}-\tfrac{3}{2}\rangle \quad |32\tfrac{5}{2}-\tfrac{1}{2}\rangle \quad |31\tfrac{5}{2}\tfrac{1}{2}\rangle \quad |30\tfrac{5}{2}\tfrac{3}{2}\rangle \quad |3-1\tfrac{5}{2}\tfrac{5}{2}\rangle$$

$$\begin{array}{c} \langle 33\tfrac{5}{2}-\tfrac{3}{2}| \\ \langle 32\tfrac{5}{2}-\tfrac{1}{2}| \\ \langle 31\tfrac{5}{2}\tfrac{1}{2}| \\ \langle 30\tfrac{5}{2}\tfrac{3}{2}| \\ \langle 3-1\tfrac{5}{2}\tfrac{5}{2}| \end{array} \left(\begin{array}{ccccc} -\tfrac{9}{2} & 2\sqrt{3} & 0 & 0 & 0 \\ 2\sqrt{3} & -1 & \tfrac{3}{2}\sqrt{10} & 0 & 0 \\ 0 & \tfrac{3}{2}\sqrt{10} & \tfrac{1}{2} & 2\sqrt{6} & 0 \\ 0 & 0 & 2\sqrt{6} & 0 & \sqrt{15} \\ 0 & 0 & 0 & \sqrt{15} & -\tfrac{5}{2} \end{array} \right) \qquad \text{(11-19e)}$$

$$M_F = \tfrac{1}{2} \quad |33\tfrac{5}{2}-\tfrac{5}{2}\rangle \quad |32\tfrac{5}{2}-\tfrac{3}{2}\rangle \quad |31\tfrac{5}{2}-\tfrac{1}{2}\rangle \quad |30\tfrac{5}{2}\tfrac{1}{2}\rangle \quad |3-1\tfrac{5}{2}\tfrac{3}{2}\rangle \quad |3-2\tfrac{5}{2}\tfrac{5}{2}\rangle$$

$$\begin{array}{c} \langle 33\tfrac{5}{2}-\tfrac{5}{2}| \\ \langle 32\tfrac{5}{2}-\tfrac{3}{2}| \\ \langle 31\tfrac{5}{2}-\tfrac{1}{2}| \\ \langle 30\tfrac{5}{2}\tfrac{1}{2}| \\ \langle 3-1\tfrac{5}{2}\tfrac{3}{2}| \\ \langle 3-2\tfrac{5}{2}\tfrac{5}{2}| \end{array} \left(\begin{array}{cccccc} -\tfrac{15}{2} & \tfrac{1}{2}\sqrt{30} & 0 & 0 & 0 & 0 \\ \tfrac{1}{2}\sqrt{30} & -3 & 2\sqrt{5} & 0 & 0 & 0 \\ 0 & 2\sqrt{5} & -\tfrac{1}{2} & 3\sqrt{3} & 0 & 0 \\ 0 & 0 & 3\sqrt{3} & 0 & 2\sqrt{6} & 0 \\ 0 & 0 & 0 & 2\sqrt{6} & -\tfrac{3}{2} & \tfrac{5}{2}\sqrt{2} \\ 0 & 0 & 0 & 0 & \tfrac{5}{2}\sqrt{2} & -5 \end{array} \right)$$

$$\text{(11-19f)}$$

$$M_F = -\tfrac{1}{2} \quad |32\tfrac{5}{2}-\tfrac{5}{2}\rangle \quad |31\tfrac{5}{2}-\tfrac{3}{2}\rangle \quad |30\tfrac{5}{2}-\tfrac{1}{2}\rangle \quad |3-1\tfrac{5}{2}\tfrac{1}{2}\rangle \quad |3-2\tfrac{5}{2}\tfrac{3}{2}\rangle \quad |3-3\tfrac{5}{2}\tfrac{5}{2}\rangle$$

$$\begin{array}{c} \langle 32\tfrac{5}{2}-\tfrac{5}{2}| \\ \langle 31\tfrac{5}{2}-\tfrac{3}{2}| \\ \langle 30\tfrac{5}{2}-\tfrac{1}{2}| \\ \langle 3-1\tfrac{5}{2}\tfrac{1}{2}| \\ \langle 3-2\tfrac{5}{2}\tfrac{3}{2}| \\ \langle 3-3\tfrac{5}{2}\tfrac{5}{2}| \end{array} \left(\begin{array}{cccccc} -5 & \tfrac{5}{2}\sqrt{2} & 0 & 0 & 0 & 0 \\ \tfrac{5}{2}\sqrt{2} & -\tfrac{3}{2} & 2\sqrt{6} & 0 & 0 & 0 \\ 0 & 2\sqrt{6} & 0 & 3\sqrt{3} & 0 & 0 \\ 0 & 0 & 3\sqrt{3} & -\tfrac{1}{2} & 2\sqrt{5} & 0 \\ 0 & 0 & 0 & 2\sqrt{5} & -3 & \tfrac{1}{2}\sqrt{30} \\ 0 & 0 & 0 & 0 & \tfrac{1}{2}\sqrt{30} & -\tfrac{15}{2} \end{array} \right)$$

$$\text{(11-19g)}$$

$$M_F = -\tfrac{3}{2} \qquad
\begin{array}{c}
\\
\langle 31\tfrac{5}{2}-\tfrac{5}{2}| \\
\langle 30\tfrac{5}{2}-\tfrac{3}{2}| \\
\langle 3-1\tfrac{5}{2}-\tfrac{1}{2}| \\
\langle 3-2\tfrac{5}{2}\tfrac{1}{2}| \\
\langle 3-3\tfrac{5}{2}\tfrac{3}{2}|
\end{array}
\begin{array}{ccccc}
|31\tfrac{5}{2}-\tfrac{5}{2}\rangle & |30\tfrac{5}{2}-\tfrac{3}{2}\rangle & |3-1\tfrac{5}{2}-\tfrac{1}{2}\rangle & |3-2\tfrac{5}{2}\tfrac{1}{2} & 3-3\tfrac{5}{2}\tfrac{3}{2}\rangle \\
\left(\begin{array}{ccccc}
-\tfrac{5}{2} & \sqrt{15} & 0 & 0 & 0 \\
\sqrt{15} & 0 & 2\sqrt{6} & 0 & 0 \\
0 & 2\sqrt{6} & \tfrac{1}{2} & \tfrac{3}{2}\sqrt{10} & 0 \\
0 & 0 & \tfrac{3}{2}\sqrt{10} & -1 & 2\sqrt{3} \\
0 & 0 & 0 & 2\sqrt{3} & -\tfrac{9}{2}
\end{array}\right)
\end{array}$$

$$(11\text{-}19\text{h})$$

$$M_F = -\tfrac{5}{2} \qquad
\begin{array}{c}
\\
\langle 30\tfrac{5}{2}-\tfrac{5}{2}| \\
\langle 3-1\tfrac{5}{2}-\tfrac{3}{2}| \\
\langle 3-2\tfrac{5}{2}-\tfrac{1}{2}| \\
\langle 3-3\tfrac{5}{2}\tfrac{1}{2}|
\end{array}
\begin{array}{cccc}
|30\tfrac{5}{2}-\tfrac{5}{2}\rangle & |3-1\tfrac{5}{2}-\tfrac{3}{2}\rangle & |3-2\tfrac{5}{2}-\tfrac{1}{2}\rangle & |3-3\tfrac{5}{2}\tfrac{1}{2}\rangle \\
\left(\begin{array}{cccc}
0 & \sqrt{15} & 0 & 0 \\
\sqrt{15} & \tfrac{3}{2} & 2\sqrt{5} & 0 \\
0 & 2\sqrt{5} & 1 & \tfrac{3}{2}\sqrt{6} \\
0 & 0 & \tfrac{3}{2}\sqrt{6} & -\tfrac{3}{2}
\end{array}\right)
\end{array}$$

$$(11\text{-}19\text{i})$$

$$M_F = -\tfrac{7}{2} \qquad
\begin{array}{c}
\\
\langle 3-1\tfrac{5}{2}-\tfrac{5}{2}| \\
\langle 3-2\tfrac{5}{2}-\tfrac{3}{2}| \\
\langle 3-3\tfrac{5}{2}-\tfrac{1}{2}|
\end{array}
\begin{array}{ccc}
|3-1\tfrac{5}{2}-\tfrac{5}{2}\rangle & |3-2\tfrac{5}{2}-\tfrac{3}{2}\rangle & |3-3\tfrac{5}{2}-\tfrac{1}{2}\rangle \\
\left(\begin{array}{ccc}
\tfrac{5}{2} & \tfrac{5}{2}\sqrt{2} & 0 \\
\tfrac{5}{2}\sqrt{2} & 3 & 2\sqrt{3} \\
0 & 2\sqrt{3} & \tfrac{3}{2}
\end{array}\right)
\end{array} \qquad (11\text{-}19\text{j})$$

$$M_F = -\tfrac{9}{2} \qquad
\begin{array}{c}
\\
\langle 3-2\tfrac{5}{2}-\tfrac{5}{2}| \\
\langle 3-3\tfrac{5}{2}-\tfrac{3}{2}|
\end{array}
\begin{array}{cc}
|3-2\tfrac{5}{2}-\tfrac{5}{2}\rangle & |3-3\tfrac{5}{2}-\tfrac{3}{2}\rangle \\
\left(\begin{array}{cc}
5 & \tfrac{1}{2}\sqrt{30} \\
\tfrac{1}{2}\sqrt{30} & \tfrac{9}{2}
\end{array}\right)
\end{array}$$

$$(11\text{-}19\text{k})$$

$$M_F = -\tfrac{11}{2} \qquad
\begin{array}{c}
\\
\langle 3-3\tfrac{5}{2}-\tfrac{5}{2}|
\end{array}
\begin{array}{c}
|3-3\tfrac{5}{2}-\tfrac{5}{2}\rangle \\
\left(\begin{array}{c}
\tfrac{15}{2}
\end{array}\right)
\end{array} \qquad (11\text{-}19\text{l})$$

The above matrices can be checked by diagonalization, and the resultant eigenvalues should be proportional to the value K defined in (11-9) for all cases, where F has its maximal value of M_F. The eigenvalues for the above matrices are,

$$M_F = \tfrac{11}{2} \quad \tfrac{15}{2} \tag{11-20a}$$
$$M_F = \tfrac{9}{2} \quad \tfrac{15}{2}, \quad 2 \tag{11-20b}$$
$$M_F = \tfrac{7}{2} \quad \tfrac{15}{2}, \quad 2, \quad -\tfrac{5}{2} \tag{11-20c}$$
$$M_F = \tfrac{5}{2} \quad \tfrac{15}{2}, \quad 2, \quad -\tfrac{5}{2}, \quad -6 \tag{11-20d}$$

$$M_F = \tfrac{3}{2} \quad \tfrac{15}{2}, \quad 2, \quad -\tfrac{5}{2}, \quad -6, \quad -\tfrac{17}{2} \qquad (11\text{-}20e)$$

$$M_F = \tfrac{1}{2} \quad \tfrac{15}{2}, \quad 2, \quad -\tfrac{5}{2}, \quad -6, \quad -\tfrac{17}{2}, \quad -10 \qquad (11\text{-}20f)$$

$$M_F = -\tfrac{1}{2} \quad -\tfrac{17}{2}, \quad 2, \quad -\tfrac{5}{2}, \quad -10, \quad -6, \quad \tfrac{15}{2} \qquad (11\text{-}20g)$$

$$M_F = -\tfrac{3}{2} \quad -\tfrac{17}{2}, \quad 2, \quad -\tfrac{5}{2}, \quad -6, \quad \tfrac{15}{2} \qquad (11\text{-}20h)$$

$$M_F = -\tfrac{5}{2} \quad 2, \quad -\tfrac{5}{2}, \quad -6, \quad \tfrac{15}{2} \qquad (11\text{-}20i)$$

$$M_F = -\tfrac{7}{2} \quad 2, \quad -\tfrac{5}{2}, \quad \tfrac{15}{2} \qquad (11\text{-}20j)$$

$$M_F = -\tfrac{9}{2} \quad 2, \quad \tfrac{15}{2} \qquad (11\text{-}20k)$$

$$M_F = -\tfrac{11}{2} \quad \tfrac{15}{2} \qquad (11\text{-}20l)$$

In a crystal, splittings that arise from the internal electric fields are substantially greater than the hyperfine splittings. This means that it is more realistic to calculate the hyperfine matrix elements in the JM_JIM_I scheme where the electronic and nuclear angular momenta are uncoupled. In this case, the diagonal matrix elements are given by (6-4a),

$$\langle \alpha SLJM_JIM_I|\mathbf{H}_{hfs}|\alpha'S'L'JM_JIM_I\rangle = AM_JM_I \qquad (11\text{-}21)$$

and the off-diagonal matrix elements by (6-4b),

$$\langle \alpha SLJM_JII_z|\mathbf{H}_{hfs}|\alpha'S'L'JM_J\pm1IM_I\mp1\rangle$$
$$= \tfrac{1}{2}A\sqrt{(J\mp M_J)(J\pm M_J+1)(I\pm M_I)(I\mp M_I+1)} \qquad (11\text{-}22)$$

In order to evaluate the hyperfine splitting one needs to construct the crystal field states as linear combinations of the $|JM_J\rangle$ components. In the present example that involves construction of the states that are symmetrized with respect to the octahedral point group, as discussed in chapter 9.

11.2 AN OCTAHEDRAL CRYSTAL FIELD

To a first approximation, the Pr^{3+} ion sees a predominantly octahedral (O) crystal field which is represented by the crystal field potential

$$V_{cryst} = A_4\left[C_0^{(4)} - \frac{\sqrt{70}}{7}\left(C_3^{(4)} - C_{-3}^{(4)}\right)\right]$$
$$+ A_6\left[C_0^{(6)} + \frac{\sqrt{210}}{24}\left(C_3^{(6)} - C_{-3}^{(6)}\right) + \frac{\sqrt{231}}{24}\left(C_6^{(6)} + C_{-6}^{(6)}\right)\right] \qquad (11\text{-}23)$$

The first part of the potential associated with the fourth-order invariant B_4 defines an operator (as in (9-16)),

$$X_4 = 594\left[C_0^{(4)} - \frac{\sqrt{70}}{7}\left(C_3^{(4)} - C_{-3}^{(4)}\right)\right] \qquad (11\text{-}24)$$

where the common factor of 594 allows one to obtain simple matrix elements and consequently integer eigenvalues. The matrix elements of crystal field potential presented in (9-17a)-(9-17c) were evaluated in a $|J M_J\rangle$ basis, and diagonalized to find the octa-hedrally symmetrized states (9-19a)- (9-19g). These symmetry adapted functions are identified here by the appropriate irreducible representations of the octahedral group,

$$|(^3\Gamma_2)\rangle = \frac{1}{3}\left[-\sqrt{2}|^3F_{3,3}\rangle - \sqrt{5}|^3F_{3,0}\rangle + \sqrt{2}|^3F_{3,-3}\rangle\right] \tag{11-25a}$$

$$|(^3\Gamma_4)a\rangle = \frac{1}{\sqrt{2}}\left[|^3F_{3,3}\rangle + |^3F_{3,-3}\rangle\right] \tag{11-25b}$$

$$|(^3\Gamma_4)b\rangle = \frac{1}{\sqrt{6}}\left[-\sqrt{5}|^3F_{3,1}\rangle + |^3F_{3,-2}\rangle\right] \tag{11-25c}$$

$$|(^3\Gamma_4)c\rangle = \frac{1}{\sqrt{6}}\left[\sqrt{5}|^3F_{3,-1}\rangle + |^3F_{3,2}\rangle\right] \tag{11-25d}$$

$$|(^3\Gamma_5)a\rangle = \frac{1}{6}\left[-\sqrt{10}|^3F_{3,3}\rangle + 4|^3F_{3,0}\rangle + \sqrt{10}|^3F_{3,-3}\rangle\right] \tag{11-25e}$$

$$|(^3\Gamma_5)b\rangle = \frac{1}{\sqrt{6}}\left[|^3F_{3,1}\rangle + \sqrt{5}|^3F_{3,-2}\rangle\right] \tag{11-25f}$$

$$|(^3\Gamma_5)c\rangle = \frac{1}{\sqrt{6}}\left[|^3F_{3,-1}\rangle - \sqrt{5}|^3F_{3,2}\rangle\right] \tag{11-25g}$$

These functions are used for the evaluation of the matrix elements of the magnetic dipole interactions that modify the energy levels of $4f^2$ configuration in octahedral field.

Introduction of the sixth-order invariant in (9-20)

$$X_6 = 6864\left[C_0^{(6)} + \frac{\sqrt{210}}{24}\left(C_3^{(6)} - C_{-3}^{(6)}\right) + \frac{\sqrt{231}}{24}\left(C_6^{(6)} + C_{-6}^{(6)}\right)\right] \tag{11-26}$$

allows one to write the complete octahedral crystal field as

$$V_{Oh} = X_4 A^4 + X_6 A^6, \tag{11-27}$$

and to obtain the complete crystal field matrices which, in terms of the octahedrally symmetrized states of (11-25a-g), are diagonal, namely

$$
\begin{array}{c}
\quad\quad\quad |(^3\Gamma_4)a\rangle \quad\quad\quad |(^3\Gamma_4)b\rangle \quad\quad\quad |(^3\Gamma_4)c\rangle \\
\begin{array}{c}\langle(^3\Gamma_4)a| \\ \langle(^3\Gamma_4)b| \\ \langle(^3\Gamma_4)c|\end{array}
\begin{pmatrix}
-3A^4 - 405A^6 & 0 & 0 \\
0 & -3A^4 - 405A^6 & 0 \\
0 & 0 & -3A^4 - 405A^6
\end{pmatrix}
\end{array} \tag{11-28a}
$$

$$
\begin{array}{c}
\quad\quad\quad |(^3\Gamma_5)a\rangle \quad\quad\quad |(^3\Gamma_5)b\rangle \quad\quad\quad |(^3\Gamma_5)c\rangle \\
\begin{array}{c}\langle(^3\Gamma_5)a| \\ \langle(^3\Gamma_5)b| \\ \langle(^3\Gamma_5)c|\end{array}
\begin{pmatrix}
9A^4 + 225A^6 & 0 & 0 \\
0 & 9A^4 + 225A^6 & 0 \\
0 & 0 & 9A^4 + 225A^6
\end{pmatrix}
\end{array} \tag{11-28b}
$$

$$
\begin{array}{c}
\quad |(^3\Gamma_1)\rangle \\
\langle(^3\Gamma_1)| \left(-18A^4 + 540A^6\right)
\end{array} \tag{11-28c}
$$

Thus, starting from the initial level of calculations at which the free ionic system functions represented by purely atomic terms are applied, the spin orbit interaction is included to give the functions defined within the improved intermediate coupling scheme, still valid for the spherical symmetry of a free ion. The perturbing influence of the environment of the lanthanide ion distorts the spherical symmetry giving rise to the corrections to the energy that are the eigenvalues of the matrices (11-28a-c). As a consequence, the wave functions, which are the eigenvectors of these matrices and which are presented in (11-25a-g), are linear combinations that are symmetry adapted and possess appropriate transformation properties. In the particular example analyzed here, the seven components of the atomic term $^3F_3$ for $M_J = 0, \pm1, \pm2, \pm3$ are combined together to give the functions that are invariant under the symmetry operations of the octahedral group.

For the simplicity of notation denoting a frequently appearing factor as

$$M_\pm = \sqrt{(\tfrac{5}{2} \pm M_I)(\tfrac{7}{2} \mp M_I)} \tag{11-29}$$

finally, using the general expressions for the matrix elements of H_{hfs} in (11-19) and (11-20), it is rather straightforward to obtain the values of matrix elements that are determined by M_J and M_I,

$$\langle(^3\Gamma_2)IM_I|H_{hfs}|(^3\Gamma_2)IM_I\rangle \qquad = 0 \tag{11-30a}$$

$$\langle(^3\Gamma_2)IM_I|H_{hfs}|(^3\Gamma_5)aIM_I\rangle \qquad = -2M_I$$

$$\langle(^3\Gamma_2)IM_I|H_{hfs}|(^3\Gamma_5)bIM_I - 1\rangle \ = \sqrt{2}M_+$$

$$\langle(^3\Gamma_2)IM_I|H_{hfs}|(^3\Gamma_5)cIM_I + 1\rangle \ = -\sqrt{2}M_- \tag{11-30b}$$

$$\langle(^3\Gamma_2)IM_I|H_{hfs}|(^3\Gamma_4)aIM_I\rangle \qquad = 0$$

$$\langle(^3\Gamma_2)IM_I|H_{hfs}|(^3\Gamma_4)bIM_I\rangle \qquad = 0$$

$$\langle(^3\Gamma_2)IM_I|H_{hfs}|(^3\Gamma_4)cIM_I\rangle \qquad = 0 \tag{11-30c}$$

$$\langle(^3\Gamma_5)aIM_I|H_{hfs}|(^3\Gamma_5)aIM_I\rangle \quad = 0$$

$$\langle(^3\Gamma_5)bIM_I|H_{hfs}|(^3\Gamma_5)bIM_I\rangle \quad = \frac{M_I}{2}$$

$$\langle(^3\Gamma_5)cIM_I|H_{hfs}|(^3\Gamma_5)cIM_I\rangle \quad = -\frac{M_I}{2} \tag{11-30d}$$

$$\langle(^3\Gamma_5)aIM_I|H_{hfs}|(^3\Gamma_5)bIM_I - 1\rangle = \frac{\sqrt{2}}{4}M_+$$

$$\langle(^3\Gamma_5)aIM_I|H_{hfs}|(^3\Gamma_5)cIM_I + 1\rangle = \frac{\sqrt{2}}{4}M_-$$

$$\langle(^3\Gamma_5)bIM_I|H_{hfs}|(^3\Gamma_5)cIM_I\rangle \qquad = 0 \tag{11-30e}$$

$$\langle(^3\Gamma_5)aIM_I|H_{hfs}|(^3\Gamma_4)aIM_I\rangle \qquad = -\sqrt{5}M_I$$

$$\langle(^3\Gamma_5)aIM_I|H_{hfs}|(^3\Gamma_4)bIM_I - 1\rangle = \frac{\sqrt{10}}{4}M_+$$

$$\langle(^3\Gamma_5)aIM_I|H_{hfs}|(^3\Gamma_4)cIM_I + 1\rangle = -\frac{\sqrt{10}}{4}M_-$$

$$\langle(^3\Gamma_5)bIM_I|H_{hfs}|(^3\Gamma_4)aIM_I + 1\rangle = -\frac{\sqrt{10}}{4}M_-$$

$$\langle(^3\Gamma_5)bI\,M_I|H_{hfs}|(^3\Gamma_4)bI\,M_I\rangle = -\frac{\sqrt{5}}{2}M_I$$

$$\langle(^3\Gamma_5)bI\,M_I|H_{hfs}|(^3\Gamma_4)cI\,M_I-1\rangle = \frac{\sqrt{10}}{2}M_+$$

$$\langle(^3\Gamma_5)cI\,M_I|H_{hfs}|(^3\Gamma_4)aI\,M_I-1\rangle = \frac{\sqrt{10}}{4}M_+$$

$$\langle(^3\Gamma_5)cI\,M_I|H_{hfs}|(^3\Gamma_4)bI\,M_I+1\rangle = \frac{\sqrt{10}}{2}M_-$$

$$\langle(^3\Gamma_5)cI\,M_I|H_{hfs}|(^3\Gamma_4)cI\,M_I\rangle = -\frac{\sqrt{5}}{2}M_I \qquad (11\text{-}30\text{f})$$

$$\langle(^3\Gamma_4)aI\,M_I|H_{hfs}|(^3\Gamma_4)aI\,M_I\rangle = 0$$

$$\langle(^3\Gamma_4)bI\,M_I|H_{hfs}|(^3\Gamma_4)bI\,M_I\rangle = -\frac{3M_I}{2}$$

$$\langle(^3\Gamma_4)cI\,M_I|H_{hfs}|(^3\Gamma_4)cI\,M_I\rangle = +\frac{3M_I}{2}$$

$$\langle(^3\Gamma_4)aI\,M_I|H_{hfs}|(^3\Gamma_4)bI\,M_I-1\rangle = \frac{3\sqrt{2}}{4}M_+$$

$$\langle(^3\Gamma_4)aI\,M_I|H_{hfs}|(^3\Gamma_4)cI\,M_I+1\rangle = \frac{3\sqrt{2}}{4}M_-$$

$$\langle(^3\Gamma_4)bI\,M_I|H_{hfs}|(^3\Gamma_4)cI\,M_I\rangle = 0 \qquad (11\text{-}30\text{g})$$

We note that under the octahedral group H_{hfs} transforms as Γ_4. Furthermore, it is seen from (11-30a), (11-30d) and (11-30g) that the Γ_2 state has no first-order magnetic hyperfine structure whereas the splitting for the Γ_4 states is three times larger than for the Γ_5 states.

11.3   OCTAHEDRAL MAGNETIC HYPERFINE MATRIX ELEMENTS

Using (11-30a,d,e,g) we can readily calculate the matrix elements within each octahedral state to give the matrices and their respective eigenvalues, presented in parenthesis at the right hand side of each matrix. The eigenvectors, whose designation $I = \frac{5}{2}$ is suppressed, are displayed below each matrix. Thus, in the particular case of the states $^3\Gamma_5$ and $^3\Gamma_4$ that originate from the atomic term $^3F_3$ of electron configuration $4f^2$ and the nuclear spin $I = \frac{5}{2}$ of Pr^{3+} ion in the octahedral field, we have the following solutions,

for the electronic states $^3\Gamma_5$,

$$
\begin{array}{c c}
 & \begin{array}{cc} |a\frac{5}{2}\frac{5}{2}\rangle & |b\frac{5}{2}\frac{3}{2}\rangle \end{array} \\
(^3\Gamma_5) & \\
\begin{array}{c} \langle a\frac{5}{2}\frac{5}{2}| \\ \langle b\frac{5}{2}\frac{3}{2}| \end{array} & \begin{pmatrix} 0 & \frac{\sqrt{10}}{4} \\ \frac{\sqrt{10}}{4} & \frac{3}{4} \end{pmatrix}
\end{array}
\qquad (-\tfrac{1}{2}, \qquad \tfrac{5}{4})
$$

$$|-\tfrac{1}{2}\rangle_1^{(-3)} = \tfrac{1}{\sqrt{7}}[\sqrt{2}|a\tfrac{5}{2}\rangle - \sqrt{5}|b\tfrac{3}{2}\rangle]$$

$$|\tfrac{5}{4}\rangle_1^{(-3)} = \frac{1}{\sqrt{7}}[\sqrt{5}|a\tfrac{5}{2}\rangle + \sqrt{2}|b\tfrac{3}{2}\rangle] \tag{11-31a}$$

$$
\begin{array}{cccc}
(^3\Gamma_5) & |a\tfrac{5}{2}\tfrac{3}{2}\rangle & |b\tfrac{5}{2}\tfrac{1}{2}\rangle & |c\tfrac{5}{2}\tfrac{5}{2}\rangle \\
\langle a\tfrac{5}{2}\tfrac{3}{2}| & 0 & 1 & \frac{\sqrt{10}}{4} \\
\langle b\tfrac{5}{2}\tfrac{1}{2}| & 1 & \frac{1}{4} & 0 \\
\langle c\tfrac{5}{2}\tfrac{5}{2}| & \frac{\sqrt{10}}{4} & 0 & -\frac{5}{4}
\end{array}
\qquad (-\tfrac{7}{4}, \quad -\tfrac{1}{2}, \quad \tfrac{5}{4})
$$

$$|-\tfrac{1}{2}\rangle_2^{(-3)} = \frac{1}{\sqrt{35}}[3|a\tfrac{3}{2}\rangle - 4|b\tfrac{1}{2}\rangle + \sqrt{10}|c\tfrac{5}{2}\rangle]$$

$$|\tfrac{5}{4}\rangle_2^{(-3)} = \frac{1}{\sqrt{21}}[\sqrt{10}|a\tfrac{3}{2}\rangle + \sqrt{10}|b\tfrac{1}{2}\rangle + |c\tfrac{5}{2}\rangle]$$

$$|-\tfrac{7}{4}\rangle_1^{(-3)} = \frac{1}{\sqrt{15}}[-2|a\tfrac{3}{2}\rangle + |b\tfrac{1}{2}\rangle + \sqrt{10}|c\tfrac{5}{2}\rangle] \tag{11-31b}$$

$$
\begin{array}{cccc}
(^3\Gamma_5) & |a\tfrac{5}{2}\tfrac{1}{2}\rangle & |b\tfrac{5}{2}-\tfrac{1}{2}\rangle & |c\tfrac{5}{2}\tfrac{3}{2}\rangle \\
\langle a\tfrac{5}{2}\tfrac{1}{2}| & 0 & \frac{3\sqrt{2}}{4} & 1 \\
\langle b\tfrac{5}{2}-\tfrac{1}{2}| & \frac{3\sqrt{2}}{4} & -\frac{1}{4} & 0 \\
\langle c\tfrac{5}{2}\tfrac{3}{2}| & 1 & 0 & -\frac{3}{4}
\end{array}
\qquad (-\tfrac{7}{4}, \quad -\tfrac{1}{2}, \quad \tfrac{5}{4})
$$

$$|-\tfrac{1}{2}\rangle_3^{(-3)} = \frac{1}{\sqrt{35}}[|a\tfrac{1}{2}\rangle - 3\sqrt{2}|b-\tfrac{1}{2}\rangle + 4|c\tfrac{3}{2}\rangle]$$

$$|\tfrac{5}{4}\rangle_3^{(-3)} = \frac{1}{\sqrt{7}}[2|a\tfrac{1}{2} > +\sqrt{2}|b-\tfrac{1}{2}\rangle + |c\tfrac{3}{2}\rangle]$$

$$|-\tfrac{7}{4}\rangle_2^{(-3)} = \frac{1}{\sqrt{5}}[-\sqrt{2}|a\tfrac{1}{2}\rangle + |b-\tfrac{1}{2}\rangle + \sqrt{2}|c\tfrac{3}{2}\rangle] \tag{11-31c}$$

$$
\begin{array}{cccc}
(^3\Gamma_5) & |a\tfrac{5}{2}-\tfrac{1}{2}\rangle & |b\tfrac{5}{2}-\tfrac{3}{2}\rangle & |c\tfrac{5}{2}\tfrac{1}{2}\rangle \\
\langle a\tfrac{5}{2}-\tfrac{1}{2}| & 0 & 1 & \frac{3\sqrt{2}}{4} \\
\langle b\tfrac{5}{2}-\tfrac{3}{2}| & 1 & -\frac{3}{4} & 0 \\
\langle c\tfrac{5}{2}\tfrac{1}{2}| & \frac{3\sqrt{2}}{4} & 0 & -\frac{1}{4}
\end{array}
\qquad (-\tfrac{7}{4}, \quad -\tfrac{1}{2}, \quad \tfrac{5}{4})
$$

$$|-\tfrac{1}{2}\rangle_4^{(-3)} = \frac{1}{\sqrt{35}}[|a-\tfrac{1}{2}\rangle + 4|b-\tfrac{3}{2}\rangle - 3\sqrt{2}|c\tfrac{1}{2}\rangle]$$

$$|\tfrac{5}{4}\rangle_4^{(-3)} = \frac{1}{\sqrt{7}}[2|a-\tfrac{1}{2} > +|b-\tfrac{3}{2}\rangle + \sqrt{2}|c\tfrac{1}{2}\rangle]$$

$$|-\tfrac{7}{4}\rangle_3^{(-3)} = \frac{1}{\sqrt{5}}[-\sqrt{2}|a-\tfrac{1}{2}\rangle + \sqrt{2}|b-\tfrac{3}{2}\rangle + |c\tfrac{1}{2}\rangle] \tag{11-31d}$$

$$
\begin{array}{cccc}
(^3\Gamma_5) & |a\tfrac{5}{2}-\tfrac{3}{2}\rangle & |b\tfrac{5}{2}-\tfrac{5}{2}\rangle & |c\tfrac{5}{2}-\tfrac{1}{2}\rangle \\
\langle a\tfrac{5}{2}-\tfrac{3}{2}| & 0 & \frac{\sqrt{10}}{4} & 1 \\
\langle b\tfrac{5}{2}-\tfrac{5}{2}| & \frac{\sqrt{10}}{4} & -\frac{5}{4} & 0 \\
\langle c\tfrac{5}{2}-\tfrac{1}{2}| & 1 & 0 & \frac{1}{4}
\end{array}
\qquad (-\tfrac{7}{4}, \quad -\tfrac{1}{2}, \quad \tfrac{5}{4})
$$

$$|-\tfrac{1}{2}\rangle_5^{(-3)} = \tfrac{1}{\sqrt{35}}[-3|a-\tfrac{3}{2}\rangle - \sqrt{10}|b-\tfrac{5}{2}\rangle + 4|c-\tfrac{1}{2}\rangle]$$

$$|\tfrac{5}{4}\rangle_5^{(-3)} = \tfrac{1}{\sqrt{21}}[\sqrt{10}|a-\tfrac{3}{2}\rangle + |b-\tfrac{5}{2}\rangle + \sqrt{10}|c-\tfrac{1}{2}\rangle]$$

$$|-\tfrac{7}{4}\rangle_4^{(-3)} = \tfrac{1}{\sqrt{15}}[-2|a-\tfrac{3}{2}\rangle + \sqrt{10}|b-\tfrac{5}{2}\rangle + |c-\tfrac{1}{2}\rangle] \qquad (11\text{-}31\text{e})$$

$$(^3\Gamma_5) \qquad |a\tfrac{5}{2}-\tfrac{5}{2}\rangle \quad |c\tfrac{5}{2}-\tfrac{3}{2}\rangle$$

$$\begin{matrix} \langle a\tfrac{5}{2}-\tfrac{5}{2}| \\ \langle c\tfrac{5}{2}-\tfrac{3}{2}| \end{matrix} \begin{pmatrix} 0 & \tfrac{\sqrt{10}}{4} \\ \tfrac{\sqrt{10}}{4} & \tfrac{3}{4} \end{pmatrix} \qquad (-\tfrac{1}{2}, \quad \tfrac{5}{4})$$

$$|-\tfrac{1}{2}\rangle_6^{(-3)} = \tfrac{1}{\sqrt{7}}[-\sqrt{5}|a-\tfrac{5}{2}\rangle + \sqrt{2}|c-\tfrac{3}{2}\rangle]$$

$$|\tfrac{5}{4}\rangle_6^{(-3)} = \tfrac{1}{\sqrt{7}}[\sqrt{2}|a-\tfrac{5}{2}\rangle + \sqrt{5}|c-\tfrac{3}{2}\rangle] \qquad (11\text{-}31\text{f})$$

$$(^3\Gamma_5) \qquad |b\tfrac{5}{2}\tfrac{5}{2}\rangle$$

$$\langle b\tfrac{5}{2}\tfrac{5}{2}| \begin{pmatrix} \tfrac{5}{4} \end{pmatrix} \qquad (\tfrac{5}{4})$$

$$|\tfrac{5}{4}\rangle_7^{(-3)} = |b\tfrac{5}{2}\rangle \qquad (11\text{-}31\text{g})$$

$$(^3\Gamma_5) \qquad |c\tfrac{5}{2}-\tfrac{5}{2}\rangle$$

$$\langle c\tfrac{5}{2}-\tfrac{5}{2}| \begin{pmatrix} \tfrac{5}{4} \end{pmatrix} \qquad (\tfrac{5}{4})$$

$$|\tfrac{5}{4}\rangle_8^{(-3)} = |c-\tfrac{5}{2}\rangle \qquad (11\text{-}31\text{h})$$

and for the electronic states $^3\Gamma_4$

$$(^3\Gamma_4) \qquad |a\tfrac{5}{2}\tfrac{5}{2}\rangle \quad |b\tfrac{5}{2}\tfrac{3}{2}\rangle$$

$$\begin{matrix} \langle a\tfrac{5}{2}\tfrac{5}{2}| \\ \langle b\tfrac{5}{2}\tfrac{3}{2}| \end{matrix} \begin{pmatrix} 0 & \tfrac{3\sqrt{10}}{4} \\ \tfrac{3\sqrt{10}}{4} & -\tfrac{9}{4} \end{pmatrix} \qquad (-\tfrac{15}{4}, \quad \tfrac{3}{2})$$

$$|\tfrac{3}{2}\rangle_1^{(9)} = \tfrac{1}{\sqrt{7}}[\sqrt{5}|a\tfrac{5}{2}\rangle + \sqrt{2}|b\tfrac{3}{2}\rangle]$$

$$|-\tfrac{15}{4}\rangle_1^{(9)} = \tfrac{1}{\sqrt{7}}[\sqrt{2}|a\tfrac{5}{2}\rangle - \sqrt{5}|b\tfrac{3}{2}\rangle] \qquad (11\text{-}32\text{a})$$

$$(^3\Gamma_4) \qquad |a\tfrac{5}{2}\tfrac{3}{2}\rangle \quad |b\tfrac{5}{2}\tfrac{1}{2}\rangle \quad |c\tfrac{5}{2}\tfrac{5}{2}\rangle$$

$$\begin{matrix} \langle a\tfrac{5}{2}\tfrac{3}{2}| \\ \langle b\tfrac{5}{2}\tfrac{1}{2}| \\ \langle c\tfrac{5}{2}\tfrac{5}{2}| \end{matrix} \begin{pmatrix} 0 & 3 & \tfrac{3\sqrt{10}}{4} \\ 3 & -\tfrac{3}{4} & 0 \\ \tfrac{3\sqrt{10}}{4} & 0 & \tfrac{15}{4} \end{pmatrix} \qquad (-\tfrac{15}{4}, \quad \tfrac{3}{2}, \quad \tfrac{21}{4})$$

$$|\tfrac{3}{2}\rangle_2^{(9)} = \tfrac{1}{\sqrt{35}}[3|a\tfrac{3}{2}\rangle + 4|b\tfrac{1}{2}\rangle - \sqrt{10}|c\tfrac{5}{2}\rangle]$$

$$|-\tfrac{15}{4}\rangle_2^{(9)} = \tfrac{1}{\sqrt{21}}[-\sqrt{10}|a\tfrac{3}{2}\rangle + \sqrt{10}|b\tfrac{1}{2}\rangle + |c\tfrac{5}{2}\rangle]$$

$$|\tfrac{21}{4}\rangle_1^{(9)} = \tfrac{1}{\sqrt{15}}[2|a\tfrac{3}{2}\rangle + |b\tfrac{1}{2}\rangle + \sqrt{10}|c\tfrac{5}{2}\rangle] \qquad (11\text{-}32\text{b})$$

$$({}^3\Gamma_4) \qquad |a\tfrac{5}{2}\tfrac{1}{2}\rangle \quad |b\tfrac{5}{2}-\tfrac{1}{2}\rangle \quad |c\tfrac{5}{2}\tfrac{3}{2}\rangle$$

$$\begin{array}{c} \langle a\tfrac{5}{2}\tfrac{1}{2}| \\ \langle b\tfrac{5}{2}-\tfrac{1}{2}| \\ \langle c\tfrac{5}{2}\tfrac{3}{2}| \end{array} \begin{pmatrix} 0 & \frac{9\sqrt{2}}{4} & 3 \\ \frac{9\sqrt{2}}{4} & \frac{3}{4} & 0 \\ 3 & 0 & \frac{9}{4} \end{pmatrix} \qquad (-\tfrac{15}{4},\ \tfrac{3}{2},\ \tfrac{21}{4})$$

$$|\tfrac{3}{2}\rangle_3^{(9)} = \tfrac{1}{\sqrt{35}}[|a\tfrac{1}{2}\rangle + 3\sqrt{2}|b-\tfrac{1}{2}\rangle - 4|c\tfrac{3}{2}\rangle]$$

$$|-\tfrac{15}{4}\rangle_3^{(9)} = \frac{1}{\sqrt{7}}[-2|a\tfrac{1}{2}\rangle + \sqrt{2}|b-\tfrac{1}{2}\rangle + |c\tfrac{3}{2}\rangle]$$

$$|\tfrac{21}{4}\rangle_2^{(9)} = \frac{1}{\sqrt{5}}[\sqrt{2}|a\tfrac{1}{2}\rangle + |b-\tfrac{1}{2}\rangle + \sqrt{2}|c\tfrac{3}{2}\rangle] \qquad (11\text{-}32\mathrm{c})$$

$$({}^3\Gamma_4) \qquad |a\tfrac{5}{2}-\tfrac{1}{2}\rangle \quad |b\tfrac{5}{2}-\tfrac{3}{2}\rangle \quad |c\tfrac{5}{2}\tfrac{1}{2}\rangle$$

$$\begin{array}{c} \langle a\tfrac{5}{2}-\tfrac{1}{2}| \\ \langle b\tfrac{5}{2}-\tfrac{3}{2}| \\ \langle c\tfrac{5}{2}\tfrac{1}{2}| \end{array} \begin{pmatrix} 0 & 3 & \frac{9\sqrt{2}}{4} \\ 3 & \frac{9}{4} & 0 \\ \frac{9\sqrt{2}}{4} & 0 & \frac{3}{4} \end{pmatrix} \qquad (-\tfrac{15}{4},\ \tfrac{3}{2},\ \tfrac{21}{4})$$

$$|\tfrac{3}{2}\rangle_4^{(9)} = \tfrac{1}{\sqrt{35}}[|a-\tfrac{1}{2}\rangle - 4|b-\tfrac{3}{2}\rangle + 3\sqrt{2}|c\tfrac{1}{2}\rangle]$$

$$|-\tfrac{15}{4}\rangle_4^{(9)} = \frac{1}{\sqrt{7}}[-2|a-\tfrac{1}{2}\rangle + |b-\tfrac{3}{2}\rangle + \sqrt{2}|c\tfrac{1}{2}\rangle]$$

$$|\tfrac{21}{4}\rangle_3^{(9)} = \frac{1}{\sqrt{5}}[\sqrt{2}|a-\tfrac{1}{2}\rangle + \sqrt{2}|b-\tfrac{3}{2}\rangle + |c\tfrac{1}{2}\rangle] \qquad (11\text{-}32\mathrm{d})$$

$$({}^3\Gamma_4) \qquad |a\tfrac{5}{2}-\tfrac{3}{2}\rangle \quad |b\tfrac{5}{2}-\tfrac{5}{2}\rangle \quad |c\tfrac{5}{2}-\tfrac{1}{2}\rangle$$

$$\begin{array}{c} \langle a\tfrac{5}{2}-\tfrac{3}{2}| \\ \langle b\tfrac{5}{2}-\tfrac{5}{2}| \\ \langle c\tfrac{5}{2}-\tfrac{1}{2}| \end{array} \begin{pmatrix} 0 & \frac{3\sqrt{10}}{4} & 3 \\ \frac{3\sqrt{10}}{4} & \frac{15}{4} & 0 \\ 3 & 0 & -\frac{3}{4} \end{pmatrix} \qquad (-\tfrac{15}{4},\ \tfrac{3}{2},\ \tfrac{21}{4})$$

$$|\tfrac{3}{2}\rangle_5^{(9)} = \tfrac{1}{\sqrt{35}}[3|a-\tfrac{3}{2}\rangle - \sqrt{10}|b-\tfrac{5}{2}\rangle + 4|c-\tfrac{1}{2}\rangle]$$

$$|-\tfrac{15}{4}\rangle_5^{(9)} = \frac{1}{\sqrt{21}}[-\sqrt{10}|a-\tfrac{3}{2}\rangle + |b-\tfrac{5}{2}\rangle + \sqrt{10}|c-\tfrac{1}{2}\rangle]$$

$$|\tfrac{21}{4}\rangle_4^{(9)} = \frac{1}{\sqrt{15}}[2|a-\tfrac{3}{2}\rangle + \sqrt{10}|b-\tfrac{5}{2}\rangle + |c-\tfrac{1}{2}\rangle] \qquad (11\text{-}32\mathrm{e})$$

$$({}^3\Gamma_4) \qquad |a\tfrac{5}{2}-\tfrac{5}{2}\rangle \quad |c\tfrac{5}{2}-\tfrac{3}{2}\rangle$$

$$\begin{array}{c} \langle a\tfrac{5}{2}-\tfrac{5}{2}| \\ \langle c\tfrac{5}{2}-\tfrac{3}{2}| \end{array} \begin{pmatrix} 0 & \frac{3\sqrt{10}}{4} \\ \frac{3\sqrt{10}}{4} & -\frac{9}{4} \end{pmatrix} \qquad (-\tfrac{15}{4},\ \tfrac{3}{2})$$

$$|\tfrac{3}{2}\rangle_6^{(9)} = \tfrac{1}{\sqrt{7}}[\sqrt{5}|a-\tfrac{5}{2}\rangle + \sqrt{2}|c-\tfrac{3}{2}\rangle]$$

$$|-\tfrac{15}{4}\rangle_6^{(9)} = \frac{1}{\sqrt{7}}[\sqrt{2}|a-\tfrac{5}{2}\rangle - \sqrt{5}|c-\tfrac{3}{2}\rangle] \qquad (11\text{-}32\mathrm{f})$$

$$(^3\Gamma_4) \quad |b\tfrac{5}{2}\tfrac{5}{2}\rangle$$
$$\langle b\tfrac{5}{2}\tfrac{5}{2}| \left(-\tfrac{15}{4} \right) \quad (-\tfrac{15}{4})$$
$$|-\tfrac{15}{4}\rangle_7^{(9)} = |b\tfrac{5}{2}\rangle \tag{11-32g}$$

$$(^3\Gamma_4) \quad |c\tfrac{5}{2} - \tfrac{5}{2}\rangle$$
$$\langle c\tfrac{5}{2} - \tfrac{5}{2}| \left(-\tfrac{15}{4} \right) \quad (-\tfrac{15}{4})$$
$$|-\tfrac{15}{4}\rangle_8^{(9)} = |c - \tfrac{5}{2}\rangle \tag{11-32h}$$

The eigenvectors associated with each matrix form an orthonormal set, but they are determined only to within an overall phase. Hence caution must be exercised in using them to calculate matrix elements that involve eigenvectors associated with different orthonormal sets.

Furthermore, note that the above eigenvalues are associated with considerable degeneracies, more than might at first be expected. In particular,

$$^3\Gamma_2 \qquad\qquad 6(0) \tag{11-33a}$$
$$^3\Gamma_4 \qquad 6(\tfrac{3}{2}) \qquad 8(-\tfrac{15}{4}) \qquad 4(\tfrac{21}{4}) \tag{11-33b}$$
$$^3\Gamma_5 \qquad 6(-\tfrac{1}{2}) \qquad 8(\tfrac{5}{4}) \qquad 4(-\tfrac{7}{4}) \tag{11-33c}$$

where there are just three distinct eigenvalues for the Γ_4 and Γ_5 irreducible representations. Noting that

$$\Gamma_4 \times (\Gamma_7 + \Gamma_8) = \Gamma_6 + 2\Gamma_7 + 3\Gamma_8 \tag{11-34a}$$
$$\Gamma_5 \times (\Gamma_7 + \Gamma_8) = 2\Gamma_6 + \Gamma_7 + 3\Gamma_8 \tag{11-34b}$$

we might have expected two sets of six distinct eigenvalues arising from the Γ_4 and Γ_5 irreps. Nor does the eigenvalue spectrum involve equal spacings. Some of the extra degeneracy would be lifted, if the magnetic hyperfine interaction between the states of the three representations, (Γ_2, Γ_4, Γ_5) were included, as seen by inspection of (11-31b) and (11-31f). However, a more important consideration is the influence of the trigonal C_{3v} crystal field.

At the end of chapter 9 it was concluded that in our example the crystal field is predominantly octahedral with a smaller trigonal component. In accordance with the $O \rightarrow C_{3v}$ branching rules

$$\Gamma_2 \rightarrow \Gamma_2$$
$$\Gamma_4 \rightarrow \Gamma_2 + \Gamma_3$$
$$\Gamma_5 \rightarrow \Gamma_1 + \Gamma_3 \tag{11-35}$$

the octahedral Γ_2 irreducible representation remains non-degenerate, while the Γ_4 and Γ_5 split into a singlet and a doublet. The trigonal splitting octahedral Γ_5 irrep is five times larger than for the Γ_4 irrep. Inspection of the magnetic hyperfine matrices (11-31a-h) and (11-32a-h) shows that for both of the octahedral irreps Γ_4 and Γ_5

there is no coupling between the b and c components, but there is coupling with the a components. If the trigonal crystal field interaction is significantly greater than that of the magnetic hyperfine interaction, then the off-diagonal matrix elements of the magnetic hyperfine interaction may be ignored. In such a situation only the diagonal magnetic hyperfine matrix elements in (11-31a-h) and (11-32a-h) need to be considered. It is readily seen that the diagonal elements, and their degeneracies are

$$^3\Gamma_4(0)6, \ (\tfrac{15}{4})2, \ (\tfrac{9}{4})2, \ (\tfrac{3}{4})2, \ (-\tfrac{3}{4})2, \ (-\tfrac{9}{4})2,$$

$$(-\tfrac{15}{4})2 \tag{11-36a}$$

$$^3\Gamma_5(0)6, \ (\tfrac{5}{4})2, \ (\tfrac{3}{4})2, \ (\tfrac{1}{4})2, \ (-\tfrac{1}{4})2, \ (-\tfrac{3}{4})2,$$

$$(-\tfrac{5}{4})2 \tag{11-36b}$$

$$\Gamma_2(0)6 \tag{11-36c}$$

Taking into account that under $SO(3) \to C_{3v}$,

$$[5/2] \to \Gamma_4 + \Gamma_5 + 2\Gamma_6, \tag{11-37}$$

and that for C_{3v}

$$\Gamma_1 \times 2(\Gamma_4 + \Gamma_5) + \Gamma_6 = 2(\Gamma_4 + \Gamma_5) + \Gamma_6$$
$$\Gamma_2 \times 2(\Gamma_4 + \Gamma_5) + \Gamma_6 = 2(\Gamma_4 + \Gamma_5) + \Gamma_6$$
$$\Gamma_3 \times 2(\Gamma_4 + \Gamma_5) + \Gamma_6 = (\Gamma_4 + \Gamma_5) + 5\Gamma_6, \tag{11-38}$$

it is not surprising that the eigenvalue spectrum involves, to first order, 6-fold eigenvalues that show no hyperfine splitting. In addition, it is also not surprising to obtain two groups of six two-fold degenerate equally spaced eigenvalues, each set being derived from the interaction of the nuclear magnetic moment with a Γ_3 C_{3v} crystal field level. In accord with observation, the total width of the six hyperfine levels derived from the Γ_3 C_{3v} irrep, which originate from the octahedral Γ_4 irrep, is 3/2 times larger than that derived from the octahedral Γ_5 irrep. Departures from equal spacing probably come as consequences of off-diagonal hyperfine matrix elements rather than of more exotic conjectured effects.

REFERENCES

1. Chukalina E P, Popova M N, Antic-Fidancev E and Chaminade J P, (1999) Hyperfine Structure in Optical Spectra of $CsCdBr_3$: Pr^{3+} *Phys. Lett.,* **A 258** 375-8.
2. Popova M N, Chukalina E P, Malkin B Z, Iskhakova A I, Antic-Fidancev E, Porcher P and Chaminade J P, High-resolution Infrared Absorption Spectra, Crystal-field Levels, and Relaxation Processes in $CsCdBr_3$: Pr^{3+} (2001) *Phys. Rev.,* **B 63** 075103-1 -9.
3. Agladze N I and Popova M N, (1985) Hyperfine Structure in Optical Spectra of $LiYF_4$: Ho *Sol. St. Comm.,* **55** 1097.
4. Agladze N I, Vinogradov E A and Popova M N, (1986) Manifestation of Quadrupole Hyperfine Interaction and of Interlevel Interaction in the Optical Spectrum of the $LiYF_4$: Ho Crystal *J. Exp. Theor. Phys.* **64** 716.

5. Agladze N I, Popova M N, Zhizhin G N, Egorov V J and Petrova M A, (1991) Isotope Structure in Optical Spectra of LiYF$_4$: Ho^{3+} *Phys. Rev. Lett.* **66** 477.

6. Agladze N I, Popova M N, Zhizhin G N and Becucci M, (1993) Study of Isotope Composition in Crystals by High Resolution Spectroscopy of Monoisotope Impurity *J. Exper. Theor. Phys.* **76** 1110.

7. Agladze N I, Popova M N, Koreiba M A and Malkin B Z, (1993) Isotope Effects in the Lattice Structure and Vibrational and Optical Spectra of ^{6}Li$_x$^{7}Li$_{1-x}$YF$_4$: Ho Crystals *J. Exper. Theor. Phys.* **77** 1021.

12 Magnetic Interactions in *f*-electron Systems

Real mathematical theorems will require the same stamina whether you measure the effort in months or years. You can forget the idea, if you ever had it, that all you require is a bit of natural genius and that then you can wait for inspiration to strike. There is simply no substitute for hard work and perseverance.

Andrew Wiles, 13 July 2001

In this chapter we start with a brief discussion of units and with some remarks on the structure of f^N electron configurations. The particular emphasis is on the calculation of the low lying states of f^6 configuration as realized in neutral samarium, $(Sm\,I)$, and in triply ionized europium, $(Eu\,IV$ or $Eu^{3+})$. Special attention is paid to the ground multiplet ^7F. In the search for the explanation of its structure, a detailed account and calculation of the Zeeman effect with particular emphasis on the two lowest levels, $^7F_0$ and $^7F_1$, of the f^6 configuration for both weak and strong magnetic fields in the presence of magnetic hyperfine interactions are given.

There is currently considerable interest in the possibility of realizing quantum computers with rare earth ion doped crystals[1,2]. The rare earth ion Eu^{3+} doped in crystals of $Y_2Si\,O_5$ is of particular interest[3]. The precursor to this work goes back to the theory of nuclear magnetic resonance in Eu^{3+} developed in the late 1950's by R J Elliott[4] and studies of the anomalous quadrupole coupling in europium ethylsulphate[5]. The role of hyperfine and magnetic interactions in determining the very small splittings in the ground state, and the exceedingly low probabilities associated with the $^7F_0 \rightarrow$ $^5D_0$ transition play a crucial role in the possibility of creating rare earth ion based quantum computers.

In order to perform some practical calculations that can be compared with experimental measurements we must give some consideration to units. The units appearing in experimental papers usually reflect the type of experimental equipment being used. Thus an optical spectroscopist might report wavelengths in Ångstroms $(1\text{Å} = 10^{-8}cm)$ or in nanometers $(1nm = 10^{-1}\text{Å} = 10^{-9}cm)$, while a radioastronomer might report wavelengths in centimeters or meters. Or again an optical spectroscopist may choose to report "energies" of atomic levels in *wave numbers*, the reciprocal of wavelength using cm^{-1} with the electron spin resonance (ESR) spectroscopist using MHz $(1MHz = 10^6 Hz)$ or possibly GHz $(1GHz = 10^9 Hz)$. We may wish that all used a common system of units, but things are different when one works in the real world rather than Utopia; therefore we must be able to switch between units. In particular note that

$$1eV = 8.0655 \times 10^3 cm^{-1} = 2.418 \times 10^5 MHz = 1.602 \times 10^{-19}J$$
$$1cm^{-1} = 1.2398 \times 10^{-4}eV = 2.998 \times 10^4 MHz$$
$$1MHz = 4.136 \times 10^{-6}eV = 3.336 \times 10^{-5}cm^{-1}$$

In much of our discussion of magnetic field effects we give magnetic fields in the unit Tesla (T) and make use of the Bohr magneton $\mu_B = \frac{e\hbar}{2m_e}$. In that case we frequently put

$$\mu_B = 0.46686cm^{-1}T^{-1} \tag{12-1}$$

12.1   THE f^N ELECTRON CONFIGURATIONS

The lanthanides and actinides[6] are characterized by the systematic filling of the $4f^N$ and $5f^N$ shells respectively where $N = 0, 1, \ldots, 14$. The enumeration of the states of the f^N configurations has been outlined by Judd[7] using the group schemes introduced by Racah[8]. There the states are labeled by the irreducible representations of the group chain

$$U_{14} \supset SU_2 \times SU_7 \supset SO_7 \supset G_2 \supset SO_3 \tag{12-2}$$

In SmI the lowest electron configuration is $4f^66s^2$, whereas in $EuIV$ it is just $4f^6$. Since the $6s^2$ shell is closed the SL terms of the two configurations are identical. A complete listing of the group labeled terms is given in Table 12-1. It follows from Hund's rules that the ground term is ^7F, and the ground state is $^7F_0$.

TABLE 12-1
Group Classification of the States of the f^6
Electron Configuration

SU_7	SO_7	G_2	^{2S+1}L
$\{1^6\}$	[100]	(10)	^7F
$\{21^4\}$	[210]	(21)	^5DFGHKL
		(20)	^5DGI
		(11)	^5PH
	[111]	(20)	^5DGI
		(10)	^5F
		(00)	^5S
$\{2^21^2\}$	[221]	(31)	$^3PDF_2GH_2I_2K_2LMNO$
		(30)	$^3PFGHIKM$
		(21)	^3DFGHKL
		(20)	^3DGI
		(11)	^3PH
		(10)	^3F
	[211]	(30)	$^3PFGHIKM$
		(21)	^3DFGHKL
		(20)	^3DGI
		(11)	^3PH
		(10)	^3F
	[110]	(11)	^3PH
		(10)	^3F

TABLE 12-1 *Continued*
Group Classification of the States of the f^6
Electron Configuration

SU_7	SO_7	G_2	^{2S+1}L
$\{2^3\}$	[222]	(40)	$^1SDFG_2HI_2KL_2MNQ$
		(30)	$^1PFGHIKM$
		(20)	^1DGI
		(10)	^1F
		(00)	^1S
	[220]	(22)	$^1SDGHILN$
		(21)	^1DFGHKL
		(20)	^1DGI
	[200]	(20)	^1DGI
	[000]	(00)	^1S

The experimentally determined low lying energy levels of $Sm\ I$ and $Eu\ IV$ may be extracted from the NIST database[9]. In Table 12-2 the levels of the ^7F multiplet and the lowest ^5D multiplet for the $4f^66s^2$ configuration in $Sm\ I$ are presented (note that the $4f^65d6s$ levels that commence at $\sim 10800cm^{-1}$ are omitted). The Lande

TABLE 12-2
The Energy Levels of the ^7F Multiplet
and the Lowest ^5D Multiplet of Sm I

Configuration	Term J	Level	Lande g
$4f^66s^2$	$^7F_0$	0.00	
	$^7F_1$	292.58	1.49839
	$^7F_2$	811.92	1.49779
	$^7F_3$	1489.55	1.49707
	$^7F_4$	2273.09	1.49625
	$^7F_5$	3125.46	1.49532
	$^7F_6$	4020.66	1.49417
	$^5D_0$	...	...
	$^5D_1$	15914.55	...
	$^5D_2$	17864.29	...
	$^5D_3$	20195.76	...
	$^5D_4$	...	...

TABLE 12-3
Low-Lying Terms of Eu IV

Configuration	Term J	Level
$4f^6$	$^7F_0$	0
	$^7F_1$	[370]
	$^7F_2$	[1040]
	$^7F_3$	[1890]
	$^7F_4$	[2860]
	$^7F_5$	[3910]
	$^7F_6$	[4940]
	$^5D_0$	[17270]
	$^5D_1$	[19030]
	$^5D_2$	[21510]
	$^5D_3$	[24390]
	$^5D_4$	[27640]

$g-$factors displayed there were determined by atomic beam measurements[10], and the positions of the $^5D_0$ and $^5D_4$ are, as yet, undetermined.

The energy levels of the ^7F multiplet and the lowest ^5D multiplet of the $4f^6$ configuration of $Eu\ IV$ are listed in Table 12-3. However, these data are *not* the energies of ionized Eu^{3+}, but are the levels deduced from experimental studies of trivalent europium in crystals; and hence must be viewed as approximate. This is indicated by enclosing the wavenumbers in square brackets.

12.2   CALCULATION OF THE FREE ION ENERGY LEVELS OF *SM I*

Conway and Wybourne[11] used the complete Coulomb and spin-orbit interaction matrices for the f^6 configuration to calculate the free ion energy levels of the ^7F multiplet for $Sm\ I$. In some senses their calculation could be thought as rather crude because they chose to fix the ratios of the Slater integrals F_4/F_2 and F_6/F_2 to those of a $4f$ hydrogenic eigenfunction,

$$F_4/F_2 = 0.13805 \qquad F_6/F_2 = 0.015108 \qquad (12\text{-}3)$$

As a consequence the elements of the energy matrices are expressed in terms of the two integrals F_2 and ζ_{4f}. These two integrals were then treated as freely variable parameters to be determined from the experimental data. In spite of the apparent crudity of the calculation, they found the mean error between the experimental and calculated energy levels was $< 0.2cm^{-1}$. Using the resultant eigenvectors, they were able to calculate intermediate coupling corrected Lande $g-$values that, with the appropriate relativistic corrections[12], agreed with the experimental values to within the fifth decimal place. A reanalysis of the experimental data using atomic beam measurements[10] produced an even smaller discrepancy between the calculated and experimental values.

The atomic beam measurements were made on the two stable isotopes of samarium, ^{147}Sm and ^{149}Sm. Both isotopes have $I = \frac{7}{2}$. Before discussing the hyperfine structure in detail let us consider the Zeeman effect for the levels of the ^7F multiplet in the two coupling schemes $J I F M_F$ and $J M_J I M_I M_F$ in sufficient detail to include interactions that do not preserve J or F as good quantum numbers.

12.3   THE ZEEMAN EFFECT IN *Sm I* (WITHOUT NUCLEAR SPIN EFFECTS)

It is seen from Table 12-2 that the ground state of *Sm I* is $^7F_0$, and the $^7F_1$ level occurs at $292.58cm^{-1}$. It is probable then that in magnetic field there is a coupling of these two levels. The Zeeman part of the Hamiltonian has a form of (6-5)

$$H_{mag} = \mu_B B_z \left(L_0^{(1)} + g_s S_0^{(1)} + g_I' I_0^{(1)} \right) \tag{12-4}$$

For the completeness of the analysis, the nuclear Zeeman term is retained here, though it is usually about three orders of magnitude smaller than the electronic term. As a first exercise the hyperfine interaction is ignored and the behavior of the two lowest ^7F terms in a magnetic field B_z is considered in a $|JM\rangle$ basis. Thus there are four basis states originating from $^7F_0$,

$$|^7F00\rangle, \quad |^7F10\rangle, \quad |^7F11\rangle, \quad |^7F1-1\rangle$$

It is straightforward to evaluate the diagonal and off-diagonal Zeeman matrix elements. Thus, from (2-53), the diagonal element is determined by

$$\langle \alpha SLJM|H_{mag}|\alpha SLJM\rangle = B_z \mu_B M g(SLJ)$$

with the Lande $g-$ factor evaluated from (2-54)

$$g(^7F_1) = \frac{1+g_s}{2} \tag{12-5}$$

For the required off-diagonal matrix element, in accordance with (2-58), we have

$$\langle ^7F00|H_{mag}|^7F10\rangle = B_z \mu_B 2(g_s - 1) \tag{12-6}$$

As seen in Table 12-2 the $^7F_1$ and $^7F_0$ levels in $Sm I$ are separated by $202.58cm^{-1}$. This splitting comes almost entirely from the spin-orbit interaction. Within the ^7F multiplet the spin-orbit interaction matrix elements are given by[7]

$$\langle f^6\,^7F_J|H_{so}|f^6\,^7F_J\rangle = \frac{\zeta_f}{12}[J(J+1) - L(L+1) - S(S+1)] \tag{12-7}$$

and since $S = L = 3$, we have

$$\Delta E = E(^7F_1) - E(^7F_0) = \frac{\zeta_f}{6} = 292.58cm^{-1} \tag{12-8}$$

and hence $\zeta_f \sim 1755.5cm^{-1}$. In relation to the ground state $^7F_0$, the energy of the $|^7F1 \pm 1\rangle$ states is given by

$$E(^7F1 \pm 1) = \frac{\zeta_f}{6} \pm \frac{1+g_s}{2}\mu_B B_z \tag{12-9}$$

This means that the Zeeman splitting in this case is $\sim \pm 0.7cm^{-1}$ with the magnetic field in Tesla.

The $|^7F00\rangle$ and $|^7F10\rangle$ states are coupled by the magnetic field, and hence a complete treatment requires diagonalization of the matrix

$$\begin{array}{cc} & \begin{array}{cc} |^7F00\rangle & |^7F10\rangle \end{array} \\ \begin{array}{c} \langle^7F00| \\ \langle^7F10| \end{array} & \begin{pmatrix} 0 & 2(g_s - 1)\mu_B B_z \\ 2(g_s - 1)\mu_B B_z & \frac{\zeta_f}{6} \end{pmatrix} \end{array} \tag{12-10}$$

The eigenvalues are evaluated from the secular equation

$$\lambda^2 - \frac{\zeta_f}{6}\lambda - 4(g_s - 1)^2 \mu_B^2 B_z^2 = 0 \tag{12-11}$$

and they are

$$\lambda_\pm = \frac{\zeta_f}{12}\left[1 \pm \sqrt{1 + [24(g_s - 1)\frac{\mu_B B_z}{\zeta_f}]^2}\right] \tag{12-12}$$

The net effect is for the ground state to be decreased in energy, and the other level with $M_J = 0$ to be increased in energy as the two levels mutually repel. For the Bohr magneton from (12-1), $\mu_B \sim 0.47cm^{-1}T^{-1}$, with $\zeta_{4f} \sim 1756cm^{-1}$ for Sm I, and with $(g_s - 1) \sim 1$ we have from (12-12)

$$\lambda_\pm \sim 146.3\left[1 \pm \sqrt{1 + 4.1 \times 10^{-5}B_z^2}\right]$$

which gives two solutions

$$\lambda_+ \sim 292.6 + 2.05 \times 10^{-5}B_z^2 \ cm^{-1}$$
$$\lambda_- \sim -2.05 \times 10^{-5}B_z^2 \ cm^{-1}$$

In a typical Zeeman experiment the magnetic field $B_z \sim 1T$, and thus the Zeeman shift for the two $M_J = 0$ states is very small. The eigenvectors associated with the two levels have the form

$$|0\rangle = a|^7F00\rangle + b|^7F10\rangle \tag{12-13a}$$
$$|1\rangle = -b|^7F00\rangle + a|^7F10\rangle \tag{12-13b}$$

As a consequence the ground state has a very small $J = 1$ character, and it remains non-degenerate, as expected.

12.4   THE ZEEMAN EFFECT IN *Sm I*, INCLUDING NUCLEAR SPIN

If the nuclear spin is included, the calculations are performed in either a $|JIFM_F\rangle$ basis, with the electronic and nuclear momenta coupled, or a $|JM_JIM_IM_F\rangle$ basis, with uncoupled momenta. Which basis to choose depends on the relative strengths of the magnetic and hyperfine interactions. As discussed in chapter 6, for a weak magnetic field the $|JIFM_F\rangle$ basis is usually the simplest, whereas for a strong magnetic field the calculations within the $|JM_JIM_IM_F\rangle$ basis are often simpler.

As found earlier in (6-8), the diagonal matrix element is expressed as

$$\langle \alpha JIFM_F|H_{mag}|\alpha'JIFM_F\rangle = \mu_B B_z M_F g_F \qquad (12\text{-}14)$$

with

$$g_F = g_J \frac{[F(F+1) - I(I+1) + J(J+1)]}{2F(F+1)}$$

$$+ g_I \frac{[F(F+1) + I(I+1) - J(J+1)]}{2F(F+1)} \qquad (12\text{-}15)$$

If the spin-orbit interaction is taken into account, and as a result the electronic part of the functions are expressed in the intermediate coupling scheme, g_J in (12-15) has to be replaced by its intermediate coupling values. As a crucial part of present calculations it is recommended to exercise the derivations of various matrix elements at the end of chapter 6.

12.5   SOME MAPLE ZEEMAN EFFECT PROGRAMMES

The results of the above exercises can easily be written as simple MAPLE programmes for evaluation of the matrix elements for H_{mag} in the $|JIFM_F\rangle$ scheme.

```
#Diagonal Zeeman Effect (6-8) and (12-14)#
gf:=proc(J,i,F,M)
local result;
result:=simplify(M*gj*((F*(F+1)-i*(i+1)
+J*(J+1))/(2*F*(F+1))) + M*gi*((F*(F+1)
+i*(i+1)-J*(J+1))/(2*F*(F+1))));
end;

#Off-diagonal in F (6-50)#
gof:=proc(J,i,F,M)
local result;
result:=simplify((gj-gi)*sqrt((F+1)^2 - M^2)*
sqrt((((F+J+i+2)*(F+J-i+1)*(F-J+i+1)*(i+J-F))/(4*(F+1)^2
*(2*F+1)*(2*F+3)))));
end;

#Off-diagonal in J (6-51)#
goj:=proc(J,i,F,M,S,L)
```

```
local result;
result:=simplify((((M*(1-gs))/(4*F*(F+1)*(J+1)))
*sqrt((F+i+J+2)*(F+J-i+1)*(J+i-F+1)*(i+F-J)*(J+L+S+2)
*(J+L-S+1) *(J-L+S+1)*(S+L-J))/sqrt((2*J+1)*(2*J+3)));
end;
```

```
#Off-diagonal in J and F (6-52)#
gojf:=proc(J,i,F,M,S,L)
local result;
result:=simplify((((gs-1)/(4*(F+1)*(J+1)))
*sqrt((F+M+1)*(F-M+1))
*sqrt((F+J+i+2)*(F+J+i+3)*(F+J-i+1)*(F+J-i+2))
*sqrt((J+L+S+2)*(J+L-S+1)*(J-L+S+1)*(S+L-J))/sqrt((2*F+1)
*(2*F+3)*(2*J+1)*(2*J+3)));
end;
```

```
#Off-diagonal in J and F → F-1 (6-54)#
gojfd:=proc(J,i,F,M,S,L)
local result;
result:=simplify((((((gs-1)*sqrt(F^2-M^2))/(4*F*(J+1)))*
sqrt((i+F-J-1)*(i+F-J)*(i+J-F+1)
*(i+J-F+2)*(J+L+S+2)*(J+L-S+1)*(J-L+S+1)*(S+L-J)))
/(sqrt((2*J+1)*(2*J+3)*(2*F-1)
*(2*F+1))));
end;
```

The above MAPLE programmes make the calculation of the Zeeman matrices a trivial process. In the case of a nuclear spin of $I = \frac{7}{2}$ we obtain the typical magnetic interaction matrices expressed in terms of the magnetic field B_z. Here the matrices for the positive values of M_F are listed, leaving it as an exercise to find what changes must be made in order to obtain the matrices for the negative values of M_F.

Magnetic interaction matrices are as follows,

$$
\begin{array}{cc}
M_F = \frac{9}{2} & |(1, \frac{7}{2})\frac{9}{2}, \frac{9}{2}\rangle \\
\langle(1, \frac{7}{2})\frac{9}{2}, \frac{9}{2}| & \left(g_J + \frac{7}{2}g_I' \right)
\end{array}
\tag{12-16a}
$$

$$
\begin{array}{cccc}
M_F = \frac{7}{2} & |(0, \frac{7}{2})\frac{7}{2}, \frac{7}{2}\rangle & |(1, \frac{7}{2})\frac{7}{2}, \frac{7}{2}\rangle & |(1, \frac{7}{2})\frac{9}{2}, \frac{7}{2}\rangle \\
\langle(0, \frac{7}{2})\frac{7}{2}, \frac{7}{2}| \Bigg(& \frac{7}{2}g_I' & 2(1-g_s)\frac{\sqrt{7}}{3} & 2(g_s-1)\frac{\sqrt{2}}{3} \\
\langle(1, \frac{7}{2})\frac{7}{2}, \frac{7}{2}| & 2(1-g_s)\frac{\sqrt{7}}{3} & \frac{2}{9}g_J + \frac{59}{18}g_I' & \frac{\sqrt{14}}{9}(g_J - g_I') \\
\langle(1, \frac{7}{2})\frac{9}{2}, \frac{7}{2}| & 2(g_s-1)\frac{\sqrt{2}}{3} & \frac{\sqrt{14}}{9}(g_J - g_I') & \frac{7}{9}g_J + \frac{49}{18}g_I' \Bigg)
\end{array}
\tag{12-16b}
$$

$$M_F = \tfrac{5}{2} \quad \begin{pmatrix} & |0,\tfrac{7}{2})\tfrac{7}{2},\tfrac{5}{2}\rangle & |1,\tfrac{7}{2})\tfrac{9}{2},\tfrac{5}{2}\rangle & |1,\tfrac{7}{2})\tfrac{7}{2},\tfrac{5}{2}\rangle & |1,\tfrac{7}{2})\tfrac{5}{2},\tfrac{5}{2}\rangle \\ \langle 0,\tfrac{7}{2})\tfrac{7}{2},\tfrac{5}{2}| & \tfrac{5}{2}g_I & \tfrac{\sqrt{14}}{3}(g_s-1) & -\tfrac{10\sqrt{7}}{21}(g_s-1) & \tfrac{\sqrt{42}}{7}(g_s-1) \\ \langle 1,\tfrac{7}{2})\tfrac{9}{2},\tfrac{5}{2}| & \tfrac{\sqrt{14}}{3}(g_s-1) & \tfrac{5}{9}g_J+\tfrac{35}{18}g_I' & \tfrac{7\sqrt{2}}{18}(g_J-g_I') & 0 \\ \langle 1,\tfrac{7}{2})\tfrac{7}{2},\tfrac{5}{2}| & -\tfrac{10\sqrt{7}}{21}(g_s-1) & \tfrac{7\sqrt{2}}{18}(g_J-g_I') & \tfrac{10}{63}g_J+\tfrac{295}{126}g_I' & \tfrac{3\sqrt{6}}{14}(g_J-g_I') \\ \langle 1,\tfrac{7}{2})\tfrac{5}{2},\tfrac{5}{2}| & \tfrac{\sqrt{42}}{7}(g_s-1) & 0 & \tfrac{3\sqrt{6}}{14}(g_J-g_I') & -\tfrac{5}{7}g_J+\tfrac{45}{14}g_I' \end{pmatrix}$$ (12-16c)

$$M_F = \tfrac{3}{2} \quad \begin{pmatrix} & |0,\tfrac{7}{2})\tfrac{7}{2},\tfrac{3}{2}\rangle & |1,\tfrac{7}{2})\tfrac{9}{2},\tfrac{3}{2}\rangle & |1,\tfrac{7}{2})\tfrac{7}{2},\tfrac{3}{2}\rangle & |1,\tfrac{7}{2})\tfrac{5}{2},\tfrac{3}{2}\rangle \\ \langle 0,\tfrac{7}{2})\tfrac{7}{2},\tfrac{3}{2}| & \tfrac{3}{2}g_I' & \sqrt{2}(g_s-1) & -\tfrac{2\sqrt{7}}{7}(g_s-1) & \tfrac{\sqrt{70}}{7}(g_s-1) \\ \langle 1,\tfrac{7}{2})\tfrac{9}{2},\tfrac{3}{2}| & \sqrt{2}(g_s-1) & \tfrac{1}{6}(2g_J+7g_I') & \tfrac{\sqrt{14}}{6}(g_J-g_I') & 0 \\ \langle 1,\tfrac{7}{2})\tfrac{7}{2},\tfrac{3}{2}| & -\tfrac{2\sqrt{7}}{7}(g_s-1) & \tfrac{\sqrt{14}}{6}(g_J-g_I') & \tfrac{1}{42}(4g_J+59g_I') & \tfrac{3\sqrt{10}}{14}(g_J-g_I') \\ \langle 1,\tfrac{7}{2})\tfrac{5}{2},\tfrac{3}{2}| & \tfrac{\sqrt{70}}{7}(g_s-1) & 0 & \tfrac{3\sqrt{10}}{14}(g_J-g_I') & \tfrac{3}{14}(-2g_J+9g_I') \end{pmatrix}$$ (12-16d)

$$M_F = \tfrac{1}{2} \quad \begin{pmatrix} & |0,\tfrac{7}{2})\tfrac{7}{2},\tfrac{1}{2}\rangle & |1,\tfrac{7}{2})\tfrac{9}{2},\tfrac{1}{2}\rangle & |1,\tfrac{7}{2})\tfrac{7}{2},\tfrac{1}{2}\rangle & |1,\tfrac{7}{2})\tfrac{5}{2},\tfrac{1}{2}\rangle \\ \langle 0,\tfrac{7}{2})\tfrac{7}{2},\tfrac{1}{2}| & \tfrac{1}{2}g_I' & \tfrac{2\sqrt{5}}{3}(g_s-1) & -\tfrac{2\sqrt{7}}{21}(g_s-1) & \tfrac{2\sqrt{21}}{7}(g_s-1) \\ \langle 1,\tfrac{7}{2})\tfrac{9}{2},\tfrac{1}{2}| & \tfrac{2\sqrt{5}}{3}(g_s-1) & g_J+\tfrac{7}{18}g_I' & \tfrac{\sqrt{35}}{9}(g_J-g_I') & 0 \\ \langle 1,\tfrac{7}{2})\tfrac{7}{2},\tfrac{1}{2}| & -\tfrac{2\sqrt{7}}{21}(g_s-1) & \tfrac{\sqrt{35}}{9}(g_J-g_I') & \tfrac{1}{126}(4g_J+59g_I') & \tfrac{3\sqrt{3}}{7}(g_J-g_I') \\ \langle 1,\tfrac{7}{2})\tfrac{5}{2},\tfrac{1}{2}| & \tfrac{2\sqrt{21}}{7}(g_s-1) & 0 & \tfrac{3\sqrt{3}}{7}(g_J-g_I') & \tfrac{1}{14}(-4g_J+9g_I') \end{pmatrix}$$ (12-16e)

In the absence of nuclear spin the ground state ($^7F_0$) is non-degenerate but with the occurrence of a nuclear spin I the ground state has a degeneracy of $2I + 1$; this degeneracy can be lifted by an external magnetic field. Can the hyperfine interaction, at higher than first-order, lift the degeneracy or could an external electric field, perhaps in a crystal, lift the degeneracy, at least partially? What about relativistic effects? The objective of the further discussion is to supply quantitative answers to these questions.

12.6 ZEEMAN MATRICES IN A $|J\,M_J\,I\,M_I\,M_F\rangle$ BASIS

It is possible also to calculate the Zeeman matrix elements in a $|J M_J I M_I M_F\rangle$ basis. With the aid of (6-12), as an exercise, derive the following results for a particular example discussed here

$$M_F = \tfrac{9}{2} \quad |11,\tfrac{7}{2}\tfrac{7}{2};\tfrac{9}{2}\rangle$$
$$\langle 11,\tfrac{7}{2}\tfrac{7}{2};\tfrac{9}{2}| \left(g_J + \tfrac{7}{2}g_I' \right)$$ (12-17a)

$$M_F = \tfrac{7}{2} \qquad |00, \tfrac{7}{2}\tfrac{7}{2}; \tfrac{7}{2}\rangle \qquad |10, \tfrac{7}{2}\tfrac{7}{2}; \tfrac{7}{2}\rangle \qquad |11, \tfrac{7}{2}\tfrac{5}{2}; \tfrac{7}{2}\rangle$$

$$\begin{array}{l}\langle 00, \tfrac{7}{2}\tfrac{7}{2}; \tfrac{7}{2}| \\ \langle 10, \tfrac{7}{2}\tfrac{7}{2}; \tfrac{7}{2}| \\ \langle 11, \tfrac{7}{2}\tfrac{5}{2}; \tfrac{7}{2}|\end{array}\begin{pmatrix} \tfrac{7}{2}g'_I & 2(g_s - 1) & 0 \\ 2(g_s - 1) & \tfrac{7}{2}g'_I & 0 \\ 0 & 0 & g_J + \tfrac{5}{2}g'_I \end{pmatrix} \qquad \text{(12-17b)}$$

$$M_F = \tfrac{5}{2} \qquad |00, \tfrac{7}{2}\tfrac{5}{2}; \tfrac{5}{2}\rangle \quad |10, \tfrac{7}{2}\tfrac{5}{2}; \tfrac{5}{2}\rangle \quad |11, \tfrac{7}{2}\tfrac{3}{2}; \tfrac{5}{2}\rangle \quad |1-1, \tfrac{7}{2}\tfrac{7}{2}; \tfrac{5}{2}\rangle$$

$$\begin{array}{l}\langle 00, \tfrac{7}{2}\tfrac{5}{2}; \tfrac{5}{2}| \\ \langle 10, \tfrac{7}{2}\tfrac{5}{2}; \tfrac{5}{2}| \\ \langle 11, \tfrac{7}{2}\tfrac{3}{2}; \tfrac{5}{2}| \\ \langle 1-1, \tfrac{7}{2}\tfrac{7}{2}; \tfrac{5}{2}|\end{array}\begin{pmatrix} \tfrac{5}{2}g'_I & 2(g_s - 1) & 0 & 0 \\ 2(g_s - 1) & \tfrac{5}{2}g'_I & 0 & 0 \\ 0 & 0 & g_J + \tfrac{3}{2}g'_I & 0 \\ 0 & 0 & 0 & -g_J + \tfrac{7}{2}g'_I \end{pmatrix}$$

$$\text{(12-17c)}$$

$$M_F = \tfrac{3}{2} \qquad |00, \tfrac{7}{2}\tfrac{3}{2}; \tfrac{3}{2}\rangle \quad |10, \tfrac{7}{2}\tfrac{3}{2}; \tfrac{3}{2}\rangle \quad |11, \tfrac{7}{2}\tfrac{1}{2}; \tfrac{3}{2}\rangle \quad |1-1, \tfrac{7}{2}\tfrac{5}{2}; \tfrac{3}{2}\rangle$$

$$\begin{array}{l}\langle 00, \tfrac{7}{2}\tfrac{3}{2}; \tfrac{3}{2}| \\ \langle 10, \tfrac{7}{2}\tfrac{3}{2}; \tfrac{3}{2}| \\ \langle 11, \tfrac{7}{2}\tfrac{1}{2}; \tfrac{3}{2}| \\ \langle 1-1, \tfrac{7}{2}\tfrac{5}{2}; \tfrac{3}{2}|\end{array}\begin{pmatrix} \tfrac{3}{2}g'_I & 2(g_s - 1) & 0 & 0 \\ 2(g_s - 1) & \tfrac{3}{2}g'_I & 0 & 0 \\ 0 & 0 & g_J - \tfrac{1}{2}g'_I & 0 \\ 0 & 0 & 0 & -g_J + \tfrac{3}{2}g'_I \end{pmatrix}$$

$$\text{(12-17d)}$$

$$M_F = \tfrac{1}{2} \qquad |00, \tfrac{7}{2}\tfrac{1}{2}; \tfrac{1}{2}\rangle \quad |10, \tfrac{7}{2}\tfrac{1}{2}; \tfrac{1}{2}\rangle \quad |11\tfrac{7}{2} - \tfrac{1}{2}; \tfrac{1}{2}\rangle \quad |1-1, \tfrac{7}{2}\tfrac{3}{2}; \tfrac{1}{2}\rangle$$

$$\begin{array}{l}\langle 00, \tfrac{7}{2}\tfrac{1}{2}; \tfrac{1}{2}| \\ \langle 10, \tfrac{7}{2}\tfrac{1}{2}; \tfrac{1}{2}| \\ \langle 11\tfrac{7}{2} - \tfrac{1}{2}; \tfrac{1}{2}| \\ \langle 1-1, \tfrac{7}{2}\tfrac{3}{2}; \tfrac{1}{2}|\end{array}\begin{pmatrix} \tfrac{1}{2}g'_I & 2(g_s - 1) & 0 & 0 \\ 2(g_s - 1) & \tfrac{1}{2}g'_I & 0 & 0 \\ 0 & 0 & g_J - \tfrac{1}{2}g'_I & 0 \\ 0 & 0 & 0 & -g_J + \tfrac{3}{2}g'_I \end{pmatrix}$$

$$\text{(12-17e)}$$

Check that the traces of the matrices of the two sets (12-16) and (12-17) are in one-to-one correspondence. Are the respective sets of eigenvalues the same in spite of the different bases?

REFERENCES

1. Nilsson M, Rippe L, Ohlsson N, Christiansson T and Kroll S, (2002) Initial Experiments Concerning Quantum Information Processing in Rare-earth-ion Doped Crystals *Physica Scripta* **T102** 178.

2. Wesenberg J and Molmer K, (2003) *Robust Quantum Gates and a Bus Architecture for Quantum Computing with Rare-earth-ion Doped Crystals* arXiv:quant-ph/0301036 v 1 9 Jan 2003.
3. Equall R W, Sun Y, Cone R L and Macfarlane R M, (1994) Ultraslow Optical Dephasing in Eu^{3+} : Y_2SiO_5 *Phys. Rev. Lett.,* **72** 2179.
4. Elliott R J, (1957) Theory of Nuclear Magnetic Resonance in Eu^{3+} *Proc. Phys. Soc. (London)* **B 70** 119.
5. Judd B R, Lovejoy C A and Shirley D A, (1962) Anomalous Quadrupole Coupling in Europium Ethylsulfate *Phys. Rev.,* **128** 1733.
6. Wybourne B G, (1965) *Spectroscopic Properties of Rare Earths* Wiley-Interscience Publication, John Wiley and Sons Inc., New York.
7. Judd B R (1998) *Operator Techniques in Atomic Spectroscopy* Princeton: Princeton University Press.
8. Racah G (1949) Theory of Complex Spectra IV *Phys. Rev.,* **76** 1352.
9. http://physics.nist.gov/cgi-bin/AtData/main_asd.
10. Childs W J and Goodman L S, (1972) Reanalysis of the Hyperfine Structure of the $4f^66s^2$ ^7F Multiplet in $^{147,149}Sm$, Including Measurements for the $^7F_6$ State *Phys. Rev.,* **A 6** 2011.
11. Conway J G and Wybourne B G, (1963) Low-lying Energy Levels of Lanthanide Atoms and Intermediate Coupling *Phys. Rev.,* **130** 2325.
12. Judd B R and Lindgren I, (1961) Theory of the Zeeman Effect in the Ground Multiplets of Rare-earth Atoms *Phys. Rev.,* **122** 1802.

13 Magnetic Hyperfine Interactions in Lanthanides

> *The present generation has no right to complain of the great discoveries already made, as if they left no room for further enterprise. They have only given Science a wider boundary, and we have not only to reduce to order the regions that have been conquered, but to keep up constant operations on the frontier, on a continually increasing scale.*

J C Maxwell, Inaugural Lecture

The conclusions on the effect of an external magnetic field on the $^7F_0$ and $^7F_1$ levels of the $4f^6$ configuration are extended here by the role of magnetic hyperfine interaction on these two levels of two particular stable isotopes, ^{147}Sm, ^{149}Sm that possess the nuclear spin $I = \frac{7}{2}$.

13.1 MAGNETIC HYPERFINE MATRIX ELEMENTS IN $J M_J I M_I$ COUPLING

In (5-31) we have for the magnetic hyperfine interaction, H_{hfs}, for an electron configuration ℓ^N

$$H_{hfs} = a_\ell \sum_{i=1}^{N} \left[\mathbf{l}_i - \sqrt{10}(\mathbf{s}\mathbf{C}^{(2)})_i^{(1)} \right] \cdot \mathbf{I} \tag{13-1}$$

with

$$a_\ell = 2\mu_B \mu_n g_I < r^{-3} > = \frac{2\mu_B \mu_n \mu_I < r^{-3} >}{I} \tag{13-2}$$

where $< r^{-3} >$ is the expectation value of the inverse-cube radius of the electron orbital. Noting (13-1), we define again,

$$\mathbf{N}^{(1)} = \sum_{i=1}^{N} \mathbf{N}_i^{(1)} = \sum_{i=1}^{N} \left(\mathbf{l}_i - \sqrt{10}(\mathbf{s}\mathbf{C}^{(2)})_i^{(1)} \right) \tag{13-3}$$

Let us consider the scalar tensor product

$$\mathbf{N}^{(1)} \cdot \mathbf{I}^{(1)} = N_0^{(1)} I_0^{(1)} - \left(N_1^{(1)} I_{-1}^{(1)} + N_{-1}^{(1)} I_1^{(1)} \right) \tag{13-4}$$

and the evaluation of the matrix elements in $J M_J I M_I$ coupling. Since the tensors are of rank 1, the matrix elements vanish unless $J' = J$ or $J' = J \pm 1$. Applying the

Wigner-Eckart theorem we may evaluate (13-4) as

$$\langle \alpha J M_J I M_I M_F | \mathbf{N}^{(1)} \cdot \mathbf{I}^{(1)} | \alpha' J' M_J' I M_I' M_F' \rangle$$

$$= \delta_{M_F, M_F'} \sum_{q=-1}^{1} (-1)^q \langle \alpha J M_J I M_I M_F | N_q^{(1)} I_{-q}^{(1)} | \alpha' J' M_J' I M_I' M_F' \rangle$$

$$= \delta_{M_F, M_F'} \sum_{q=-1}^{1} (-1)^q (-1)^{J-M_J} \begin{pmatrix} J & 1 & J' \\ -M_J & q & M_J - q \end{pmatrix} \langle \alpha J \| N^{(1)} \| \alpha' J' \rangle$$

$$\times (-1)^{I-M_I} \begin{pmatrix} I & 1 & I \\ -M_I & -q & M_I + q \end{pmatrix} \langle I \| I^{(1)} \| I \rangle \qquad (13\text{-}5)$$

where only the matrix elements diagonal in the nuclear spin I are considered. The last reduced matrix element in (13-5) is simply

$$\langle I \| I^{(1)} \| I \rangle = \sqrt{I(I+1)(2I+1)} \qquad (13\text{-}6)$$

We now use (13-5) to obtain expressions for the various particular matrix elements.

The matrix elements that are diagonal in all the quantum numbers other than, possibly, α, α', using (2-48c) and expanding the two $3 - j$ symbols, are expressed in the following way

$$\langle \alpha J M_J I M_I M_F | \mathbf{N}^{(1)} \cdot \mathbf{I}^{(1)} | \alpha' J M_J I M_I M_F \rangle = M_I M_J \frac{\langle \alpha J \| N^{(1)} \| \alpha' J \rangle}{\sqrt{J(J+1)(2J+1)}} \qquad (13\text{-}7a)$$

which is the same result as in (6-4a), where the element is defined in the terms of the magnetic hyperfine structure constant from (5-35d).

The list of possible matrix elements for particular values of angular momenta is extended here by the expressions that are diagonal in the M quantum numbers, but with $J' = J + 1$ which, after taking into account (2-56), have the form

$$\langle \alpha J M_J I M_I M_F | \mathbf{N}^{(1)} \cdot \mathbf{I}^{(1)} | \alpha' J + 1 M_J I M_I M_F \rangle$$

$$= -M_I \sqrt{\frac{(J + M_J + 1)(J - M_J + 1)}{(J + 1)(2J + 1)(2J + 3)}} \langle \alpha J \| N^{(1)} \| \alpha' J + 1 \rangle \qquad (13\text{-}7b)$$

The matrix elements while diagonal in J and M_F and off-diagonal in the quantum numbers, M_J, M_I have $q = \pm 1$ (whereas the previous all involved $q = 0$). Evaluating the $3j$−symbols leads to the expressions

$$\langle \alpha J M_J I M_I M_F | \mathbf{N}^{(1)} \cdot \mathbf{I}^{(1)} | \alpha' J M_J \pm 1 I M_I \mp 1 M_F \rangle =$$

$$\mp \frac{1}{2} \sqrt{\frac{(J \mp M_J)(J \pm M_J + 1)}{J(J+1)(2J+1)}} \sqrt{(I \pm M_I)(I \mp M_I + 1)} \langle \alpha J \| N^{(1)} \| \alpha' J \rangle \qquad (13\text{-}7c)$$

and the matrix elements with $J' = J + 1$ become

$$\langle \alpha J M_J I M_I M_F | \mathbf{N}^{(1)} \cdot \mathbf{I}^{(1)} | \alpha' J + 1 M_J \pm 1 I M_I \mp 1 M_F \rangle$$

$$= \pm \frac{1}{2} \sqrt{\frac{(J \pm M_J + 1)(J \pm M_J + 2)(I \mp M_I + 1)(I \pm M_I)}{(J + 1)(2J + 1)(2j + 3)}} \langle \alpha J \| N^{(1)} \| \alpha' J + 1 \rangle$$

$$(13\text{-}7d)$$

In the above the special $3j$—symbols are used

$$\begin{pmatrix} J & J & 1 \\ M & -M - 1 & 1 \end{pmatrix} = (-1)^{J-M} \sqrt{\frac{(J - M)(J - M + 1)}{2J(J + 1)(2J + 1)}} \qquad (13\text{-}8a)$$

and

$$\begin{pmatrix} J + 1 & J & 1 \\ M & -M - 1 & 1 \end{pmatrix} = (-1)^{J-M-1} \sqrt{\frac{(J - M)(J - M + 1)}{(2J + 1)(J + 1)(2J + 3)}} \qquad (13\text{-}8b)$$

To complete the calculation we need to evaluate the reduced matrix elements $\langle \alpha J \| N^{(1)} \| \alpha' J' \rangle$. This means that, in accordance with (13-1), two reduced matrix elements have to be evaluated, and after making the description of each electronic state complete, they are

$$\langle \alpha S L J \| L^{(1)} \| \alpha' S' L' J' \rangle \qquad (13\text{-}9a)$$

and

$$-\sqrt{10} \langle \alpha S L J \| \sum_i (s^{(1)} C^{(2)})_i^{(1)} \| \alpha' S' L' J' \rangle \qquad (13\text{-}9b)$$

The first matrix element (13-9a) may be evaluated using (2-46), and with the explicit value of the $6j$—symbol, there are two cases

$$\langle \alpha S L J \| L^{(1)} \| \alpha' S' L' J \rangle$$

$$= \delta_{\alpha,\alpha'} \delta_{S,S'} \delta_{L,L'} \frac{1}{2} [J(J + 1) + L(L + 1) - S(S + 1)] \sqrt{\frac{2J + 1}{J(J + 1)}} \qquad (13\text{-}10a)$$

and

$$\langle \alpha S L J \| L^{(1)} \| \alpha' S' L' J + 1 \rangle =$$

$$\delta_{\alpha,\alpha'} \delta_{S,S'} \delta_{L,L'} \frac{1}{2} \sqrt{\frac{(S + L + J + 2)(J + L - S + 1)(J - L + S + 1)(S + L - J)}{(J + 1)}} \qquad (13\text{-}10b)$$

The second matrix element (13-9b) can be evaluated using (2-43) to give

$$\langle \alpha S L J \| - \sqrt{10} \sum_i (s^{(1)} C^{(2)})_i^{(1)} \| \alpha' S' L' J \rangle = -\sqrt{2(2J + 1)(2J' + 1)}$$

$$\times \begin{Bmatrix} S & S' & 1 \\ L & L' & 2 \\ J & J' & 1 \end{Bmatrix} \langle s \| s^{(1)} \| s \rangle \langle \ell \| C^{(2)} \| \ell \rangle \langle \alpha S L \| V^{(12)} \| \alpha' S' L' \rangle$$

$$(13\text{-}11)$$

where the double tensor operators $\mathbf{V}^{(\kappa,k)}$ are defined by the one-electron reduced matrix elements

$$\langle s\ell \| v^{(\kappa,k)} \| s'\ell'\rangle = \delta_{s,s'}\delta_{\ell,\ell'}\sqrt{(2\kappa+1)(2k+1)} \qquad (13\text{-}12)$$

The properties of double tensor operators are discussed in detail in Chapter 6 of Judd's book[1]. In particular, for the $f^6\,{}^7F$ multiplet the reduced many-electron matrix element have the value

$$\langle {}^7F\|V^{(12)}\|{}^7F\rangle = -\sqrt{\frac{70}{3}} \qquad (13\text{-}13)$$

With the above results it is not difficult to deduce that

$$\langle f^6\,{}^7F_1\|N^{(1)}\|f^6\,{}^7F_1\rangle = \sqrt{\frac{2}{3}} \qquad (13\text{-}14\text{a})$$

$$\langle f^6\,{}^7F_0\|N^{(1)}\|f^6\,{}^7F_0\rangle = 0 \qquad (13\text{-}14\text{b})$$

$$\langle f^6\,{}^7F_0\|N^{(1)}\|f^6\,{}^7F_1\rangle = \frac{5\sqrt{3}}{3} \qquad (13\text{-}14\text{c})$$

13.2   MAGNETIC HYPERFINE MATRIX ELEMENTS FOR THE ^7F $J = 0, 1$ LEVELS

We are now in a position to be able to calculate the magnetic hyperfine matrix elements for the ^7F $J = 0, 1$ levels in the $|JM_J IM_I M_F\rangle$ basis. In a similar manner as was done for the Zeeman effect in chapter 12 we obtain the matrices as ($M_F = M_J + M_I$),

$$
\begin{array}{cc}
M_F = \frac{9}{2} & |11,\tfrac{7}{2}\tfrac{7}{2};\tfrac{9}{2}\rangle \\
\langle 11,\tfrac{7}{2}\tfrac{7}{2};\tfrac{9}{2}| & \left(\dfrac{7}{6} \right)
\end{array}
\qquad (13\text{-}15\text{a})
$$

$$
\begin{array}{c}
M_F = \frac{7}{2} \\[4pt]
\langle 00,\tfrac{7}{2}\tfrac{7}{2};\tfrac{7}{2}| \\
\langle 10,\tfrac{7}{2}\tfrac{7}{2};\tfrac{7}{2}| \\
\langle 11,\tfrac{7}{2}\tfrac{5}{2};\tfrac{7}{2}|
\end{array}
\begin{array}{ccc}
|00,\tfrac{7}{2}\tfrac{7}{2};\tfrac{7}{2}\rangle & |10,\tfrac{7}{2}\tfrac{7}{2};\tfrac{7}{2}\rangle & |11,\tfrac{7}{2}\tfrac{5}{2};\tfrac{7}{2}\rangle \\
\left(\begin{array}{ccc}
0 & -\dfrac{35}{6} & \dfrac{5\sqrt{14}}{6} \\
-\dfrac{35}{6} & 0 & -\dfrac{\sqrt{14}}{6} \\
\dfrac{5\sqrt{14}}{6} & -\dfrac{\sqrt{14}}{6} & \dfrac{5}{6}
\end{array} \right)
\end{array}
\qquad (13\text{-}15\text{b})
$$

$$
\begin{array}{c}
M_F = \frac{5}{2} \\[4pt]
\langle 00,\tfrac{7}{2}\tfrac{5}{2};\tfrac{5}{2}| \\
\langle 10,\tfrac{7}{2}\tfrac{5}{2};\tfrac{5}{2}| \\
\langle 11,\tfrac{7}{2}\tfrac{3}{2};\tfrac{5}{2}| \\
\langle 1-1,\tfrac{7}{2}\tfrac{7}{2};\tfrac{5}{2}|
\end{array}
\begin{array}{cccc}
|00,\tfrac{7}{2}\tfrac{5}{2};\tfrac{5}{2}\rangle & |10,\tfrac{7}{2}\tfrac{5}{2};\tfrac{5}{2}\rangle & |11,\tfrac{7}{2}\tfrac{3}{2};\tfrac{5}{2}\rangle & |1-1,\tfrac{7}{2}\tfrac{7}{2};\tfrac{5}{2}\rangle \\
\left(\begin{array}{cccc}
0 & -\dfrac{25}{6} & \dfrac{5\sqrt{6}}{3} & -\dfrac{5\sqrt{14}}{6} \\
-\dfrac{25}{6} & 0 & \dfrac{\sqrt{6}}{3} & \dfrac{\sqrt{14}}{6} \\
\dfrac{5\sqrt{6}}{3} & \dfrac{\sqrt{6}}{3} & \dfrac{1}{2} & 0 \\
-\dfrac{5\sqrt{14}}{6} & \dfrac{\sqrt{14}}{6} & 0 & -\dfrac{7}{6}
\end{array} \right)
\end{array}
\qquad (13\text{-}15\text{c})
$$

$$M_F = \tfrac{3}{2} \qquad |00, \tfrac{7}{2}\tfrac{3}{2}; \tfrac{3}{2}\rangle \quad |10, \tfrac{7}{2}\tfrac{3}{2}; \tfrac{3}{2}\rangle \quad |11, \tfrac{7}{2}\tfrac{1}{2}; \tfrac{3}{2}\rangle \quad |1-1, \tfrac{7}{2}\tfrac{5}{2}; \tfrac{3}{2}\rangle$$

$$
\begin{array}{c}
\langle 00, \tfrac{7}{2}\tfrac{3}{2}; \tfrac{3}{2}| \\[4pt]
\langle 10, \tfrac{7}{2}\tfrac{3}{2}; \tfrac{3}{2}| \\[4pt]
\langle 11, \tfrac{7}{2}\tfrac{1}{2}; \tfrac{3}{2}| \\[4pt]
\langle 1-1, \tfrac{7}{2}\tfrac{5}{2}; \tfrac{3}{2}|
\end{array}
\left(
\begin{array}{cccc}
0 & -\tfrac{5}{2} & \tfrac{5\sqrt{30}}{6} & -\tfrac{5\sqrt{6}}{3} \\[6pt]
-\tfrac{5}{2} & 0 & -\tfrac{\sqrt{30}}{6} & \tfrac{\sqrt{6}}{3} \\[6pt]
\tfrac{5\sqrt{30}}{6} & -\tfrac{\sqrt{30}}{6} & \tfrac{1}{6} & 0 \\[6pt]
-\tfrac{5\sqrt{6}}{3} & \tfrac{\sqrt{6}}{3} & 0 & -\tfrac{5}{6}
\end{array}
\right)
$$

$$(13\text{-}15\mathrm{d})$$

$$M_F = \tfrac{1}{2} \qquad |00, \tfrac{7}{2}\tfrac{1}{2}; \tfrac{1}{2}\rangle \quad |10, \tfrac{7}{2}\tfrac{1}{2}; \tfrac{1}{2}\rangle \quad |11\tfrac{7}{2} - \tfrac{1}{2}; \tfrac{1}{2}\rangle \quad |1-1, \tfrac{7}{2}\tfrac{3}{2}; \tfrac{1}{2}\rangle$$

$$
\begin{array}{c}
\langle 00, \tfrac{7}{2}\tfrac{1}{2}; \tfrac{1}{2}| \\[4pt]
\langle 10, \tfrac{7}{2}\tfrac{1}{2}; \tfrac{1}{2}| \\[4pt]
\langle 11\tfrac{7}{2} - \tfrac{1}{2}; \tfrac{1}{2}| \\[4pt]
\langle 1-1, \tfrac{7}{2}\tfrac{3}{2}; \tfrac{1}{2}|
\end{array}
\left(
\begin{array}{cccc}
0 & -\tfrac{5}{6} & \tfrac{10\sqrt{2}}{3} & -\tfrac{5\sqrt{30}}{6} \\[6pt]
-\tfrac{5}{6} & 0 & -\tfrac{2\sqrt{2}}{3} & \tfrac{\sqrt{30}}{6} \\[6pt]
\tfrac{10\sqrt{2}}{3} & -\tfrac{2\sqrt{2}}{3} & -\tfrac{1}{6} & 0 \\[6pt]
-\tfrac{5\sqrt{30}}{6} & \tfrac{\sqrt{30}}{6} & 0 & -\tfrac{1}{2}
\end{array}
\right)
$$

$$(13\text{-}15\mathrm{e})$$

To obtain correct numerical results the matrix elements must be multiplied by a_ℓ as defined in (13-2). Note that there is no first-order hyperfine splitting for the $^7 F_0$ ground state. This is not surprising because the matrix elements of any non-scalar interaction must vanish between states with $J = 0$. However, there is a question whether it is possible to obtain the splitting by including the second-order interactions via the $^7 F_1$ level. Would such splitting be measurable?

The second-order perturbation on the $^7 F_0$ level by magnetic hyperfine interaction with the $^7 F_1$ level is approximately determined by a product of two matrix elements

$$E(^7 F00\tfrac{7}{2}M_I; M_F) = -\frac{1}{\Delta} \sum_{M_J, M_I'} \left(a_\ell \langle ^7 F00\tfrac{7}{2}M_I; M_F | H_{hfs} |^7 F1 M_J \tfrac{7}{2}M_I'; M_F \rangle \right)^2$$

$$(13\text{-}16)$$

where $\Delta = E(^7 F_1) - E(^7 F_0)$. The summation in (13-16) proceeds by summing the squares of the matrix elements for the first row of each of the matrices for each value of M_F. However, each summation results in the value

$$a_\ell^2 \frac{175}{4} \qquad (13\text{-}17)$$

and hence the shift is the *same* for all the values of M_F. This means that this mechanism cannot lift the ground state degeneracy!

Experimentally[2], it is found for $^{147} Sm\ I$ that $a_\ell \sim -140 MHz$, which allows one to estimate the size of the shift of the ground state of this atom. Using (13-17), and $\Delta \sim 292 cm^{-1}$ in (13-16), the size of this shift is as follows

$$E(^7 F00\tfrac{7}{2}M_I; M_F) \sim -\left(140 \times \frac{175}{4}\right)^2 \times (292 \times 2.998 \times 10^4)^{-1}$$

$$= -4.3 MHz = -1.4 \times 10^{-4} cm^{-1} \qquad (13\text{-}18)$$

13.3   COMBINED MAGNETIC AND HYPERFINE FIELDS IN $Sm\ I$

Let us now consider the combined action of an external magnetic field B_z and the magnetic hyperfine interaction on the $^7F_0$ ground state of $Sm\ I$. We have seen that neither, by itself, can produce a splitting at second-order. In second-order we need to evaluate the cross terms

$$E(^7F00\tfrac{7}{2}M_I; M_F) = -\delta_{M_I, M_F}\frac{2a_\ell\mu_B B_z}{\Delta}$$
$$\times\langle^7F00\tfrac{7}{2}M_I; M_F|\mathbf{N}^{(1)}\cdot\mathbf{I}^{(1)}|^7F10M_IM_F\rangle$$
$$\times\langle^7F10M_IM_F|L + g_sS|^7F00\tfrac{7}{2}M_I; M_F\rangle \quad (13\text{-}19)$$

The first matrix element follows from specialization of (13-7b) to give

$$\langle^7F00\tfrac{7}{2}M_I; M_F|\mathbf{N}^{(1)}\cdot\mathbf{I}^{(1)}|^7F10M_IM_F\rangle = -\frac{5}{3}M_F \quad (13\text{-}20)$$

and the second from (13-6)

$$\langle^7F10M_IM_F|L + g_sS|^7F00\tfrac{7}{2}M_I; M_F\rangle = 2(g_s - 1)\mu_B B_z \quad (13\text{-}21)$$

and hence (13-19) evaluates as

$$E(^7F00\tfrac{7}{2}M_I; M_F) = \frac{\delta_{M_I, M_F}}{\Delta}M_F 2a_\ell\mu_B B_z\frac{5}{3}2(g_s - 1) \quad (13\text{-}22)$$

Here we notice that the perturbation is directly proportional to the M_F quantum number and to the applied magnetic field. Thus the combined effect is to produce a small hyperfine splitting of the ground state - how small? Assuming an external field of $1T$ we find from (13-22) that

$$E(^7F00\tfrac{7}{2}M_I; M_F) \sim 1.5M_F MHz \sim 5\times 10^{-5}M_F cm^{-1} \quad (13\text{-}23)$$

13.4   COMBINED MAGNETIC HYPERFINE AND CRYSTAL FIELDS

In the quest for the explanation of the hyperfine structure of the ground state of $4f^6$ configuration of Sm^{2+} and Eu^{3+}, the approach is extended by the case of an ion inserted in a crystal field. Thus the combined actions of the magnetic hyperfine interactions and the crystal field are taken into account at the second order.

    In order to perform this investigation in a systematic way it is worthwhile to recapitulate the physical reality of the lanthanide ion in a crystal analyzing the total Hamiltonian which represents the major interactions,

$$H = H_0 + H_{corr} + H_{so} + H_{hfs} + H_{EM} + H_{mag} + H_{cryst} \quad (13\text{-}24)$$

where H_0 is the central field approximation energy operator, H_{corr} represents the electron correlation effects, H_{so} is the spin-orbit interaction, H_{hfs} and H_{EM} represent the magnetic and electric multipole hyperfine interactions, respectively. The external fields are represented by the interactions with the magnetic field H_{mag} and crystal field potential H_{cryst}.

The first three operators in (13-24) included together as H_{free} describe the electronic structure of a free ion. The remaining operators represent all important physical mechanisms that have to be included in a reliable description of a many electron system. However, based on the physical reality of each system it is possible to arrange these terms in a series of decreasing importance, which in fact is already applied in (13-24).

To avoid the task of evaluation of the radial integrals that determine the matrix elements of all energy operators, it is possible to apply the fitting procedure that reproduces the measured energies that are, from a theoretical point of view, defined in terms of various parameters (radial integrals). As a result of such calculations performed with H_{free}, the energy levels are described by the functions in the intermediate coupling scheme. In this procedure the Slater radial integrals (or equivalently the Racah radial parameters) and the spin orbit coupling constant associated with the integral of r^{-3} are the adjusted parameters. The Hamiltonian H_{free} describes any many electron system within the single configuration approximation. This limitation is broken by H_{corr}, or simply by introducing additional parameters that originate from configuration interaction, including Judd parameters that are due to the three-particle interactions. It is also possible to perform the fitting procedure for the free system hamiltonian extended by the crystal field potential, if the ion is doped into a crystal. This scheme introduces additional parameters, the crystal field parameters, and it leads to the energy levels described by the irreducible representation of a certain point group. Obviously, the results of such calculations contain the impact of various interactions within the electronic structure of an ion, while they do not include the properties of the nucleus and its interactions. In particular, the magnetic and electric multipole hyperfine interactions may be treated as perturbations that create additional corrections to the electronic energy evaluated within the described procedure. The illustration of this kind of investigation is presented in a recent paper [3], where the magnetic interactions combined with the external magnetic field are discussed in the case of Tm^{3+} ion in $Y_3Al_5O_{12}$ host.

Instead of evaluating the whole matrices with the elements of various interactions, we apply here the perturbation approach in order to monitor certain physical terms and their contributions to the energy. The perturbation expansion is performed for the Hamiltonian

$$H = H_{free} + \lambda H_{cryst} \qquad (13\text{-}25)$$

which includes the fact that the ion is perturbed by the surrounding environment. Consequently, limiting the analysis to the corrections to the energy of lowest order, the energy levels are now described by the function improved by the first order correction

$$\Phi_i = \Psi_i^0 + \lambda \Psi_i^1 \qquad (13\text{-}26)$$

and

$$\Psi^1 = \sum_{\chi^0 \neq \Psi^0} {}' \frac{\langle \chi^0 \mid V_{cryst} \mid \Psi_i^0 \rangle}{E_\Psi^0 - E_\chi^0} \chi^0 \qquad (13\text{-}27)$$

where $\Psi^0 \equiv \Psi$ and $\chi^0 \equiv \chi$ are the solutions of the zero order eigenvalue problem, and the sum is over all the other states of $4f^N$ configuration for which the energy denominator is not vanishing.

The energy of magnetic hyperfine interactions is now determined by the matrix elements that are of the second order,

$$\Gamma_{hfs}^2 = \lambda[\langle\Psi|H_{hfs}|\Psi^1\rangle + \langle\Psi^1|H_{hfs}|\Psi\rangle] \tag{13-28}$$

where the first order correction to the wave function is defined in (13-27), and the magnetic hyperfine interaction is defined in (13-1).

The specification of the wave functions in the matrix elements of (13-28) should be extended by appropriate quantum numbers of the nuclear part of the system. However, since only the electronic part is modified by the crystal field potential, and the total functions are expressed in the uncoupled scheme of electronic and nuclear momenta, further analysis concerns only the electronic part.

The analog of (13-19) that represents the combined interactions of the magnetic hyperfine origin and those caused by the crystal field potential, in the case of the first component of (13-28), has the form

$$\gamma_{hfs}^\Psi(1) = \sum_{\chi \neq \Psi}^{all} \sum_{kq} B_q^k \langle 4f|r^k|4f\rangle \langle\Psi|H_{hfs}|\chi\rangle\langle\chi|C_q^{(k)}|\Psi\rangle/(E_\Psi - E_\chi) \tag{13-29}$$

In practical numerical calculations

$$A_q^k = B_q^k \langle 4f|r^k|4f\rangle \tag{13-30}$$

and the values of the crystal field parameters are obtained from the fitting procedure. The energy correction in (13-29) represents the particular case of interactions between various energy states, Ψ and χ, of the ground configuration of the lanthanide ion, $4f^N$, taken into account via the crystal field potential. The selection rules for the non-vanishing matrix elements of the spherical tensor in the second matrix element limit the expansion of the crystal field potential to the even rank k components. These are the same terms of V_{cryst} that contribute to the energy at the first order.

In the particular example of the term $^7F_2$ of $4f^6$ of an ion in a crystal field, the second-order energy correction caused by H_{hfs} is a sum of two contributions arising from two possible intermediate states χ, namely

$$\gamma_{hfs}^{^7F_2}(1) = \sum_{kq}^{all} B_q^k \langle 4f|r^k|4f\rangle[\langle^7F_2|H_{hfs}|^7F_1\rangle\langle^7F_1|C_q^{(k)}|^7F_2\rangle/(E_{^7F_2} - E_{^7F_1})$$

$$+ \langle^7F_2|H_{hfs}|^7F_3\rangle\langle^7F_3|C_q^{(k)}|^7F_2\rangle/(E_{^7F_2} - E_{^7F_3})] \tag{13-31}$$

The intermediate states $^7F_1$ and $^7F_3$ are the terms of the $4f^6$ configuration with the values of J allowed by the selection rules for the non-vanishing matrix element of H_{hfs}. Taking into account the possible ranks of the crystal field potential k, finally the correction has the form of three additive terms

$$
\begin{aligned}
\gamma_{hfs}^{^7F_2}(1) = \sum_q \Big[& A_q^2 \langle ^7F_2|H_{hfs}|^7F_1\rangle\langle ^7F_1|C_q^{(2)}|^7F_2\rangle/(E_{^7F_2} - E_{^7F_1}) \\
+ & A_q^2 \langle ^7F_2|H_{hfs}|^7F_3\rangle\langle ^7F_3|C_q^{(2)}|^7F_2\rangle/(E_{^7F_2} - E_{^7F_3}) \\
+ & A_q^4 \langle ^7F_2|H_{hfs}|^7F_3\rangle\langle ^7F_3|C_q^{(4)}|^7F_2\rangle/(E_{^7F_2} - E_{^7F_3})\Big]
\end{aligned}
\tag{13-32}
$$

where the product of structural parameter and the radial integral is replaced by the crystal field parameters defined in (13-30). The evaluation of appropriate matrix elements in (13-31) is trivial since those with H_{hfs} are determined by the expression (6-3). At the same time, the matrix elements of the spherical tensors are defined by the unit tensor operators,

$$
\langle n\ell^N \Psi|C_q^{(k)}|n\ell^N\chi\rangle = \langle\ell||C^{(k)}||\ell\rangle\langle n\ell^N\Psi|U_q^{(k)}|n\ell^N\chi\rangle
\tag{13-33}
$$

The matrix element of $U_q^{(k)}$ represents the genealogy of the states involved, and their values are readily available in the Tables of Nielson and Koster[4]. Obviously if the H_{so} is included in H_{free}, the wave functions of the $4f^N$ configuration are expressed in the intermediate coupling scheme. Thus, the matrix elements have to be evaluated with appropriate linear combinations of purely $S - L$-states.

　　　The second-order crystal field contributions to the magnetic hyperfine interaction energy vanish for the particular case of $^7F_0$ state of the $4f^6$ configuration. In this particular case, since H_{hfs} is essentially a tensor operator with rank 1, it requires the intermediate state χ in (13-29) assigned to $J = 1$, which in turn determines the rank of crystal field potential $k = 1$. This condition is in contradiction with the even parity ranks of tensor operators of crystal field potential that act within the $4f^N$ configuration. This means that in the particular case of the ground state $^7F_0$ of $4f^6$ configuration there is no additional information gained from the analysis of the crystal field and magnetic hyperfine interactions combined at the second order.

　　　Inspection of the tensorial structure of the magnetic hyperfine interaction operators provides a possibility for finding new second-order corrections to the energy. These are the terms which originate from the interactions between various configurations. These new terms break down the limitations of the single configuration approximation of the original approach discussed above.

　　　For the purpose of further analysis the definition of H_{hfs} from (13-1) is rewritten here in the generalized tensorial form

$$
V_{hfs} = \mathcal{A}r^{-3}(L^{(1)} - \sqrt{15}\langle\ell||\mathbf{C}^{(2)}||\ell'\rangle W^{(12)1}(sn\ell, sn'\ell')) \cdot \mathbf{I}^{(1)}
\tag{13-34}
$$

where the numerical factor contains $\mathcal{A} = 2\mu_B\mu_n g_I$, and the radial part is treated separately. $W^{(12)1}$ is the unit double tensor operator with an even rank in the orbital part, which represents the inter-shell interactions, and is defined by the reduced matrix element

$$
\langle s\ell''||w^{(\kappa_1 k_1)}(s\ell, s\ell')||s\ell'''\rangle = \delta(\ell'', \ell)\,\delta(\ell', \ell''')
\tag{13-35}
$$

The presence of the radial part in the original operators makes it possible to evaluate the matrix elements between the states of two different electron configurations. In particular, the non-vanishing interactions via both components of V_{hfs} in (13-34) are obtained when the matrix elements are evaluated between the states of $4f^N$ and the states of singly excited configurations in which one electron is promoted from the $4f$ shell to the excited states $n'f$, for all $n' > 4$. This fact is denoted in (13-34) by an additional specification of the double tensor operator.

As seen from the reduced matrix element of spherical tensor in (13-34) from all excited one-electron states only those described by the functions of $\ell' = odd$ are allowed. This means that it is possible to evaluate the perturbing influence of the excited configurations $4f^{N-1}n'f$ taken into account at least via the second part of V_{hfs} defined in (13-34).

In a similar way, the inter-shell unit tensor operator acting only within the orbital space is defined, and as a consequence the first part of the V_{hfs} has a general effective form

$$r^{-3}L^{(1)} \Rightarrow \langle n\ell|r^{-3}|n'\ell\rangle\langle\ell||\ell^{(1)}||\ell'\rangle u_q^{(1)}(n\ell, n'\ell') \tag{13-36}$$

where again $\ell' \equiv \ell \equiv f$; thus, V_{hfs} is the inter-shell analog of H_{hfs}.

The general expression for the second-order corrections to the energy caused by magnetic hyperfine interactions with the simultaneous inclusion of the perturbing influence of crystal field potential is a sum of two terms that differ by the order of the perturbing operators in the products of the matrix elements. For the excited configurations that satisfy the parity selection rules we have,

$$\Gamma_{hfs}^{\Psi}(inter)$$

$$= \sum_{n',\chi}^{even}\sum_{kq} \frac{B_q^k}{(E_\Psi - E_\chi)} \left[\langle 4f^N \Psi|V_{hfs}|4f^{N-1}n'f\chi\rangle\langle 4f^{N-1}n'f\chi|r^k C_q^{(k)}|4f^N \Psi\rangle \right.$$

$$\left. + \langle 4f^N \Psi|r^k C_q^{(k)}|4f^{N-1}n'f\chi\rangle\langle 4f^{N-1}n'f\chi|V_{hfs}|4f^N \Psi\rangle \right] \tag{13-37}$$

where the expansion of the crystal field potential is limited to the even values of k. Furthermore, since χ is the energy state of excited configurations, the energy denominators are always non-vanishing, and therefore there is no limitation on the summation in (13-37).

Note that here there are the structural parameters B_q^k and the radial parts of the operators are treated separately, since they are involved in the off diagonal radial integrals. Collecting r^{-3} from V_{hfs} and r^k from the crystal field potential, the radial term of correction $\Gamma_{hfs}^{\Psi}(inter)$ has the form

$$R_{hfs}^k(n'; \Psi\chi) = \frac{\langle 4f|r^{-3}|n'f\rangle\langle n'f|r^k|4f\rangle}{(E_\Psi - E_\chi)} \tag{13-38}$$

It should be remembered that the functions Ψ and χ describe the energy states of two different configurations, and therefore the energy denominator never vanishes.

It is not easy to evaluate the energy correction presented in (13-37). In order to see the source of the difficulty in performing numerical calculations the general expression is applied for the example of $^7F_2$ state of $4f^6$ configuration. In this particular case we have for the first sequence of operators in (13-37),

$$\gamma_{hfs}^{^7F_2}(inter) = \sum_{n'}\sum_{kq}^{even} B_q^k$$
$$\times\left[R_{hfs}^k(n';{}^7F_2\,{}^7F_1)\langle 4f^6\,{}^7F_2|H_{hfs}|4f^5n'f\,{}^7F_1\rangle\langle 4f^5n'f\,{}^7F_1|C_q^{(k)}|4f^6\,{}^7F_2\rangle\right.$$
$$\left.+R_{hfs}^k(n';{}^7F_2\,{}^7F_3)\langle 4f^6\,{}^7F_2|H_{hfs}|4f^5n'f\,{}^7F_3\rangle\langle 4f^5n'f\,{}^7F_3|C_q^{(k)}|4f^6\,{}^7F_2\rangle\right]$$

$$(13\text{-}39)$$

The most troublesome point of practical evaluation of this energy correction is the summation over n' which means to take into account all possible single excitations $4f^{N-1}n'f$, for discrete one-electron states and including the continuum.

The states of the intermediate configuration $4f^5n'f$ are assigned to $J' = J \pm 1$, and therefore there are two distinct terms contributing to the correction. In the case of the first term, $J = 1$, the rank of the crystal field potential has just one value $k = 2$, and therefore this term is multiplied by the structural parameter B_q^2. For the matrix elements that involve $^7F_3$ of $4f^5n'f$, there are two allowed values of $k = 2, 4$; this is the contribution from the part of the crystal field potential associated with B_q^2 and B_q^4. For each term a different radial integral is required. In fact the latter differ from each other just by the energy denominators. It is instructive to analyze these energy denominators in detail to derive some conclusions that provide simplification of the expressions, and make the practical calculations not only possible but also simpler.

As a result of this analysis we see that in (13-39) there are three kinds of radial terms,

$$R_{hfs}^2(n';{}^7F_2,{}^7F_1) \Rightarrow R_{hfs}^2(4f^6\,{}^7F_2 - 4f^5n'f\,{}^7F_1)$$
$$R_{hfs}^2(4f^6\,{}^7F_2 - 4f^5n'f\,{}^7F_3)$$
$$R_{hfs}^4(4f^6\,{}^7F_2 - 4f^5n'f\,{}^7F_3) \qquad (13\text{-}40)$$

The first two terms contain the same radial integrals but various energy denominators, which are determined by the difference between the energy of the states of ground and excited configurations. Realizing that the excited configurations of the lanthanide ions are energetically distant from the ground configuration, which is especially true in the case of $4f^{N-1}n'f$ (which in the energy scale is not the first excited configuration), it is possible to assume that all these differences are approximately the same. This assumption is widely used in the theoretical description of the spectroscopic properties of lanthanides, and it defines the basic approximation of the Judd-Ofelt theory of electric dipole $f \longleftrightarrow f$ transitions discussed in chapter 17. This approximation, although criticized in the literature as a drastic assumption which is not always valid, makes practical calculations possible due to the rather extensive simplification of various expressions.

If in addition the zero order problem of H_{free} is solved for the average energy of configuration and the excited one-electron states are generated with the frozen core, then the radial terms defined in (13-38) are simplified to the following term

$$R^k_{hfs}(n'; \Psi \chi) \equiv R^k_{hfs}(4f^N \Psi - 4f^{N-1}n'f\chi)$$

$$\simeq R^k_{hfs}(n'f) = \frac{\langle 4f | r^{-3} | n'f \rangle \langle n'f | r^k | 4f \rangle}{(\epsilon_{4f} - \epsilon_{n'f})} \tag{13-41}$$

where ϵ_{4f} and $\epsilon_{n'f}$ are the orbital energies of one-electron states of given symmetry. As a consequence, since the radial terms (mainly their energy denominators) are independent of all the other quantum numbers but n', it is possible to perform the closure over the intermediate states χ in the products of matrix elements in (13-37).

Note that in fact the frozen core realization of zero order problem of H_{free} is implicitly assumed in (13-38), since there are no overlap integrals between the one-electron states that are occupied in both configurations. They are assumed to be the same for ground and excited configurations, and only the excited one-electron state $n'f$ is adjusted to the frozen core orbitals to satisfy all symmetry and orthogonality requirements.

The detailed procedure of deriving the effective operators that act within the ground configuration of the lanthanides is based on the contraction of creation and annihilation operators, which in pairs represent each tensor operator; this is described systematically in chapter 17, which is devoted to the Judd-Ofelt theory of $f \longleftrightarrow f$ transitions. For the sake of clarity of the present discussion only the outline is presented here. In the case of the contraction of two single particle inter-shell tensor operators that act within the orbital space we use the following rule of coupling

$$[u^{(1)}_i(4f, n'\ell') \times u^{(t)}_i(n'\ell', 4f)]^{(k)}_q = (-1)^k [k]^{1/2} \begin{Bmatrix} t & k & 1 \\ f & \ell' & f \end{Bmatrix} u^{(k)}_{i,q}(4f, 4f) \tag{13-42}$$

This is the case of the matrix elements of (13-37) with the first part of V_{hfs} taken into account. For the inter-shell double tensor operators, as those that define the second part of V_{hfs}, the rule of performing closure is more complex, since these are the objects acting within the spin-orbital space (for details see chapter 18), namely

$$w^{(\kappa_1 k_1)x}(s\ell, s\ell') \, w^{(\kappa_2 k_2)y}(s\ell', s\ell)$$

$$= \sum_{z\zeta} \sum_{\kappa_3 k_3} (-1)^{x-y-\zeta+\kappa_3+k_3+2s} \, [z, x, y]^{1/2} \, [\kappa_3, k_3] \begin{pmatrix} x & y & z \\ \varrho & \eta & -\zeta \end{pmatrix}$$

$$\begin{Bmatrix} \kappa_2 & \kappa_3 & \kappa_1 \\ \frac{1}{2} & \frac{1}{2} & \frac{1}{2} \end{Bmatrix} \begin{Bmatrix} k_2 & k_3 & k_1 \\ \ell & \ell' & \ell \end{Bmatrix} \begin{Bmatrix} \kappa_1 & k_1 & x \\ \kappa_2 & k_2 & y \\ \kappa_3 & k_3 & z \end{Bmatrix} w^{(\kappa_3 k_3)z}_\zeta(\ell\ell) \tag{13-43}$$

In terms of the effective operators the second-order contributions to the energy of magnetic hyperfine interactions taken into account in the description of the lanthanide ion in crystal have the form

$$
\Gamma^2_{hfs}(inter) = -4\sqrt{21} \sum_{kq}^{even} B^k_q R^k_{hfs} \langle \ell || \mathbf{C}^{(k)} || \ell \rangle \sum_{\lambda\mu}^{odd} (-1)^{\lambda-\mu} \begin{pmatrix} 1 & k & \lambda \\ \zeta & q & -\mu \end{pmatrix}
$$
$$
\times \left[(-1)^q [\lambda] \begin{Bmatrix} k & \lambda & 1 \\ \ell & \ell & \ell \end{Bmatrix} \langle 4f^N \Psi | U^{(\lambda)}_\mu | 4f^N \Psi \rangle \right.
$$
$$
\left. - \sum_x^{even} [\lambda]^{\frac{1}{2}} [x] \begin{Bmatrix} k & x & 2 \\ \ell & \ell & \ell \end{Bmatrix} \begin{Bmatrix} x & \lambda & 1 \\ 1 & 2 & k \end{Bmatrix} \langle 4f^N \Psi | W^{(1x)\lambda}_\mu | 4f^N \Psi \rangle \right]
$$

$$(13\text{-}44)$$

where the radial integrals, as the only dependent terms on the principal quantum number of excited one-electron states, are summed over n',

$$
R^k_{hfs} = \sum_{n'}^{exc} R^k_{hfs}(n'f) = \frac{\langle 4f | r^{-3} | n'f \rangle \langle n'f | r^k | 4f \rangle}{(\epsilon_{4f} - \epsilon_{n'f})}
\qquad (13\text{-}45)
$$

Instead of performing the summation over the complete radial basis sets of one-electron excited states as in (13-45), it is possible to introduce new functions, the so-called perturbed functions that are the solutions of a differential equation. As a result, without special effort it is possible to evaluate the values of appropriate radial integrals (the details of this procedure are presented in chapter 17).

The limitations on certain indices in (13-44) come from different origins. The rank of the tensor operators of the crystal field potential has even values, since this is the requirement for the non-vanishing matrix element of spherical tensor in (13-33). When both sequences of the operators $V_{hfs}V_{cryst}$ and $V_{cryst}V_{hfs}$ are included (as in (13-38)), the final expressions for the effective operator with the first part of V_{hfs} differ from each other only by a phase factor, and only if $\lambda = odd$ the contributions do not vanish. In the case of the second part of V_{hfs}, when both terms arising from two sequences of operators are compared, it is concluded that $x = even$. Consequently, since the effective operators analyzed here determine the energy, they have to be hermitian; as a consequence also in this case λ is odd ($x + 1 + \lambda$ must be even).

It is interesting to note that these second-order contributions are new, and they are not included even in the fitting procedure based on the standard single configuration approximation. As a consequence there is no danger in including the same effect twice when evaluating these corrections to the energy. When the energy of the whole Hamiltonian, extended by the crystal field potential, is evaluated in a semi-empirical way, together with the free ionic system parameters, the crystal field parameters with even ranks are determined via the reproduction of the measured energies. In such a procedure these even rank parameters, although the same as in the corrections of (13-44), are associated with the unit tensor operators U with even ranks. Here, when the hyperfine interactions are combined with the crystal field perturbing influence, the crystal field parameters (in fact the structural parameters since the radial integrals are treated separately) are associated with the unit tensor operators of odd ranks. Similarly, when the spin orbit interaction is a part of the total Hamiltonian and the spin orbit coupling parameter is determined via the fitting procedure, it is associated with the matrix elements of the double tensor operator $W^{(11)0}$, while here we have terms with $W^{(1,even)odd}$.

The tensorial structure of the effective operators in (13-44) shows that there are no corrections to the energy of the $^7F_0$ state of the $4f^6$ configuration. Indeed, the first part of the contribution vanishes because the lowest rank of unit tensor operator U is 1, while the triangular condition for the matrix element with the states of $J = 0$ requires the rank 0; the second part also vanishes for of the same reason. There is hope that possibly the electric multipole hyperfine interactions discussed in chapter 16 provide new corrections to the energy of $^7F_0$ of $4f^6$ configuration. However, before extending our discussion by new effects, at first the higher order contributions to the magnetic hyperfine interactions should at least be checked.

13.5   OTHER PHYSICAL MECHANISMS AND HIGHER-ORDER CORRECTIONS

When the energy of many electron system is determined through the fitting procedure, the resulting parameters that represent the interactions between various configurations via the electron correlation operator contain the impact due to the two- and three-particle operators acting within the orbital space. In this sense the reproduction of the measured energies is based on the standard approach in which in part electron correlation effects are included in spite of the fact that the zero order problem is defined within the single configuration approximation.

Inclusion of the spin-orbit interaction provides the fitted value of the spin-orbit coupling parameter associated with the double tensor operator $w^{(11)0}$, that is a scalar product of two vectors in both, orbital and spin sub-spaces.

If direct calculations are to be performed, there is a problematic situation, which is mainly caused by the size of the system, especially when the electron correlation effects have to be included to meet a certain accuracy of the results. There is a powerful tool developed for the description of atomic/ionic systems that is based on Multiconfiguration Hartree Fock approach, which in its best numerical realization was constructed by Charlotte Froese Fischer [5]. Although some calculations have already been performed for the energy of free lanthanides ions [6], to describe their properties using the MCHF functions is almost unrealistic. This is evident when analyzing the explosion of various terms contributing to the magnetic hyperfine interaction energy defined only for a single ground configuration, for example. Choosing a simpler tool for inclusion of electron correlation effects, which is based on the Configuration Interaction approach, simplifies the task but not far enough to make the calculations practical.

In order to see the concept and its complexity, we use the perturbation approach to verify which contributions may be expected when the physical mechanisms that are the most important for the description of the lanthanide ion in crystal are taken into account.

In addition to the combined magnetic hyperfine and crystal field interactions it is instructive to regard the second-order corrections that originate from electron correlation effects. In particular the Coulomb interaction operator V_c, that is two particle in origin, introduces completely different effective operators.

At the second order there is for example a correction, which is determined by the following product of matrix elements,

$$\Gamma_{hfs}^{\Psi}(corr) = \langle 4f^N \Psi | V_{hfs} | 4f^{N-1} n' f \chi \rangle \langle 4f^{N-1} n' f \chi | V_c | 4f^N \Psi \rangle \qquad (13\text{-}46)$$

where V_c is defined in (2-29), and its angular part is expressed by the scalar product of spherical tensors. In the particular case of the matrix element in (13-46) the spherical tensors are replaced by the inter shell unit tensor operators

$$V_c \sim \sum_s \sum_{i<j} \left(u_i^{(s)}(4f, 4f) \cdot u_j^{(s)}(n'f, 4f) \right) \qquad (13\text{-}47)$$

using the definition

$$C^{(s)}(n'\ell', n\ell) = \langle \ell' || C^{(s)} || \ell \rangle u^{(s)}(n'\ell', n\ell) \qquad (13\text{-}48)$$

In the case of the first part of V_{hfs} from (13-36), the coupling of the inter shell unit tenors operators and their contraction in accordance with (13-42) result in the second-order correction to the energy, which is determined by two-particle effective operator with the angular part proportional to (the $3 - j$ coefficients and numerical factors are omitted for simplicity),

$$-[x]\sqrt{1/3} \left\{ \begin{matrix} s & x & 1 \\ f & f & f \end{matrix} \right\} \langle 4f^N \Psi | [u_i^{(x)}(4f, 4f) \times u_j^{(s)}(4f, 4f)]_q^{(1)} | 4f^N \Psi \rangle \quad (13\text{-}49)$$

For the second part of V_{hfs} the coupling and closure procedure are more complex, because the inter-shell operators are double tensor operators, and therefore the commutator in (13-43) has to be used. Consequently, the second-order correction is again determined by a new object, which is associated with the tensorial product of double tensor operators

$$\sim \langle 4f^N \Psi | [w_i^{(1k)x}(4f, 4f) \times w_j^{(0s)s}(4f, 4f)]_q^{(1)} | 4f^N \Psi \rangle \qquad (13\text{-}50)$$

Although these terms complete the list of second-order corrections to the energy caused by V_{hfs}, again they do not contribute to the energy of $^7F_0$ of $4f^6$ configuration.

To conclude the second-order analysis, the answer to the question whether these contributions are indeed new, or are they already included within the adjusted parameters obtained from a fitting procedure should be addressed. As mentioned at the beginning of this discussion, the radial integrals of the two body interactions within the orbital space are among those from the list of standard energy parameters. Therefore it is expected that the contributions determined by (13-49) are taken into account automatically via the fitting procedure. This conclusion obviously indicates that the standard parametrization of the energy structure of many electron system is more general than its basic assumptions suggest. The situation is different in the case of contributions determined by the operators in (13-50), since they are the two particle objects acting within the spin-orbital space, while only one particle double operators $w^{(11)0}$ are included if the spin-orbit coupling parameter is evaluated.

The second-order corrections to the energy contain the impact of the configuration interaction between the ground configuration and singly excited ones taken into account via various physical mechanisms. At the third order of perturbation expansion, when the Coulomb interaction operator is taken as a perturbation (in fact together with the potential of central field approximation to meet the criterion of smallness of the perturbation) it is possible to include the perturbing influence of doubly excited configurations, which play a dominant role in the direct evaluation of the energy of many electron systems based on *MCHF* or *CI*. However, if the partitioning of the hamiltonian that describes the system is properly performed, and the basic requirements for the applicability of the perturbation approach are satisfied, it is expected that the series of energy corrections is convergent. Thus the third-order corrections should be, *a priori*, smaller than those of the second order discussed above, but it does not imply that they are small enough to be neglected.

Because of new selection rules introduced to the theoretical analysis, it is interesting to inspect the tensorial structure of the following third-order term that represents the inter-shell interactions via V_{hfs}, V_{cryst} and V_c,

$$\Lambda_{hfs}^{\Psi} = \langle 4f^N \Psi | V_{hfs} | 4f^{N-1} n' f \chi \rangle \langle 4f^{N-1} n' f \chi | V_{cryst} | Bb \rangle \langle Bb | V_c | 4f^N \Psi \rangle \quad (13\text{-}51)$$

Bb denotes the states b of excited configurations B, which in the particular case here, due to the two-particle nature of Coulomb interaction, describe the promotion of two electrons from the $4f$ shell,

$$4f^{N-2} n' f n'' f$$
$$4f^{N-2} n' f n'' p$$
$$4f^{N-2} n' f^2 \quad (13\text{-}52)$$

When the first doubly excited configuration is taken into account in the sequence of unit tensor operators that contains the first part of V_{hfs}, there are two pairs of unit tensor operators that have to be contracted. In the following expression the first with the third, and then the second with the fourth operators have to be contracted,

$$\langle 4f^N \Psi | u_i^{(1)}(4f, n'f) u_j^{(k)}(4f, n''\ell'') \left(u_i^{(s)}(n'f, 4f) \cdot u_j^{(k)}(n''\ell'', 4f) \right) | 4f^N \Psi \rangle \quad (13\text{-}53)$$

since, for example, the third operator annihilates the state $4f$ and creates $n'f$, which has to be destructed by the first operator, which indeed annihilates $n'f$ and creates in turn $4f$. In summary, these two pairs of creation and annihilation operators act like the effective operator $u^{(x)}(4f4f)$, which has the unit reduced matrix element between functions of $4f^N$ configuration. The situation is the same in the case of the second pair of tensor operators. Indeed, the second and fourth operators in (13-53) have to act on the same coordinate to be contracted.

Using again the same rules as before, it becomes straightforward at this point to predict that the third-order effective operator is associated with the two-particle operator

$$\langle 4f^N \Psi | [u_i^{(x)}(4f, 4f) \times u_j^{(y)}(4f, 4f)]_q^{(\lambda)} | 4f^N \Psi \rangle \quad (13\text{-}54)$$

which for $\lambda = 0$ requires only the equality $x = y$. Finally, this term defines the third-order correction to the energy of $^7F_0$ of $4f^6$, which is due to magnetic hyperfine interactions. The third-order effective operator originating from the second part of V_{hfs} also contributes to this energy. Similarly as at the second-order of analysis, now the closure procedure is applied for the double tensor operators, and as a result tensorial product of two double unit tensor operators is obtained. It is interesting to note that the third-order contributions associated with (13-54), and with its analog for the spin-orbital space, are the first among all the corrections to the magnetic hyperfine energy analyzed here that represent the perturbing influence of doubly excited configurations.

In order to complete the review of the physical nature of third-order energy corrections, the following possibilities have to be also included,

$$\Lambda_{hfs}^{\Psi} = \langle 4f^N \Psi | V_{hfs} | 4f^{N-1} n' f \chi \rangle \langle 4f^{N-1} n' f \chi | V_{so} | Bb \rangle \langle Bb | V_{cryst} | 4f^N \Psi \rangle$$
$$(13\text{-}55)$$

and

$$\Lambda_{hfs}^{\Psi} = \langle 4f^N \Psi | V_{hfs} | 4f^{N-1} n' f \chi \rangle \langle 4f^{N-1} n' f \chi | V_{so} | Bb \rangle \langle Bb | V_c | 4f^N \Psi \rangle \quad (13\text{-}56)$$

from the structure of which, without even performing a deep analysis, at least by analogy with the previous cases, it is clearly seen that:

- in the first sequence, (13-55) it is possible to include only single excitations (since all operators are single particle), while in the second, (13-56), the configurations B, as in the case of (13-52), may describe double excitations with certain parity requirements satisfied;
- since in both expressions the spin-orbit interaction operator is present, all the operators, including those which are effective, are double tensor operators (the closure defined by (13-43) should be applied);
- the effective operator of (13-55) is a one particle object, while in the case of (13-56) the two particle operator (tensorial product of two operators) contributes to the energy correction;
- there are three possible ranks of the effective double tensor operators in their spin part, namely we have $w^{(\kappa k)\lambda}$ for $\kappa = 0, 1, 2$, where
  for $\kappa = 0$ the double tensor operator is reduced to the orbital object $u^{(k)}$ since $k = \lambda$, and this is the contribution that originates from the first part of V_{hfs};
  for $\kappa = 1$ the double tensor operator is the same as the spin-orbit interaction operator, which is included as a perturbation in the original definition, and in general these terms, together with the former ones, are possibly taken into account if the energy is determined through the semi-empirical procedure;
  for $\kappa = 2$ the effective operator results from the contraction of the double tensor operator of the second part of V_{hfs} and the spin orbit operator; these objects are new, and they are beyond the standard parametrization scheme.

The interplay of all these important physical mechanisms appears at the fourth order, as for example in the following general term,

$$\Theta^{\Psi}_{hfs} = \langle 4f^N \Psi | V_{hfs} | Aa \rangle \langle Aa | V_{cryst} | Bb \rangle \langle Bb | V_{so} | Xx \rangle \langle Xx | V_c | 4f^N \Psi \rangle \qquad (13\text{-}57)$$

where $A \equiv 4f^{N-1}n'f$, and B and X are the excited configurations of appropriate parity.

Further analysis of various contributions to the magnetic hyperfine interactions energy is far beyond the limits of the present discussion. It is worthwhile to summarize at this point that in all new contributions presented here, the expansion of the crystal field potential is limited by the parity requirements to its even part. This means that even extending the analysis to the fourth order, the energy, even when improved by the corrections originating from the subtle magnetic hyperfine interactions, is expressed by even rank crystal field parameters. This conclusion leaves the problem of the theoretical description of spectroscopic properties of lanthanides in crystals still present, since the results of this discussion demonstrate that it is impossible to determine the odd rank crystal field parameters via the fitting procedure applied for the energy. At the same time, the knowledge of these parameters is crucial for the direct evaluation of the amplitude of $f \longleftrightarrow f$ transitions, as discussed in detail in chapter 17.

13.6   EXERCISES

13-1. Give a physical interpretation of the result found in (13-41).

13-2. Explain why if you diagonalize the matrices (13-39a-e), with the elements off-diagonal in J put to zero you get just 4 distinct eigenvalues $(0, -\frac{1}{3}, -\frac{3}{2}, \frac{7}{6})$.

REFERENCES

1. Judd B R, (1998) *Operator Techniques in Atomic Spectroscopy* Princeton: Princeton University Press.
2. Childs W J and Goodman L S, (1972) Reanalysis of the Hyperfine Structure of the $4f^6 6s^2$ ^7F Multiplet in ^{147,149}Sm, Including Measurements for the $^7F_6$ State *Phys. Rev.,* **A 6** 2011.
3. Guillot-Noël O, Goldner Ph, and Antic-Fidancev E, (2005) Analysis of Magnetic Interactions in Rare-earth-doped Crystals for Quantum Manipulation *Phys. Rev.,* **B 71**, 174409
4. Nielson, C W and Koster, G F, (1963) *Spectroscopic Coefficients for the p^n, d^n, and f^n Configurations*, Cambridge: MIT Press.
5. Froese Fischer C, (1977) *The Hartree-Fock Methods for Atoms* Wiley-Interscience Publication, John Wiley and Sons Inc., New York.
6. Cai Z, Umar V M and Froese Fischer C, (1992) Large-scale Relativistic Correlation Calculations: Levels of P^{3+} *Phys. Rev. Lett.,* **68**, 297

14 Electric Quadrupole Hyperfine Interactions

Never having experienced the classical education as fragmentarily delivered by the English public and grammar schools, nor a university grounding in Newtonian science, Faraday had no preconceptions, and was thus uniquely receptive when he first encountered science in London.

James Hamilton, *Faraday, the life* (London: HarperCollins 2002)

Continuing the search discussed in chapter 13, we explore the role of electric quadrupole hyperfine interactions in the $4f^6\ ^7F$ multiplet of the trivalent europium ion. Europium has two stable isotopes ^{151}Eu and ^{153}Eu, each having a nuclear spin of $I = \frac{5}{2}$, and hence each possesses both a nuclear magnetic moment and an electric quadrupole moment.

Using the definition of the electric quadrupole hyperfine interaction operator defined in (6-34), the matrix element evaluated in the coupled momenta scheme is separated, in accordance with the rule of (2-44), into the product of two distinct electronic and nuclear parts,

$$\langle \alpha JIFM_F | H_{EQ} | \alpha' J'IFM_F \rangle$$
$$= -e^2(-1)^{J'+I+F} \begin{Bmatrix} J' & I & F \\ I & J & 2 \end{Bmatrix} \langle \alpha J \| r_e^{-3} C_e^{(2)} \| \alpha' J' \rangle \langle I \| r_n^2 C_n^{(2)} \| I \rangle \quad (14\text{-}1)$$

In the terms of the nuclear quadrupole moment introduced in (6-37),

$$Q = \sqrt{\frac{4I(2I-1)}{(I+1)(2I+1)(2I+3)}} \langle I \| r_n^2 C_n^{(2)} \| I \rangle \quad (14\text{-}2)$$

the matrix element in (14-1) is rewritten as

$$\langle \alpha JIFM_F | H_{EQ} | \alpha' J'IFM_F \rangle$$
$$= -b_\ell(-1)^{J'+I+F} \begin{Bmatrix} J' & I & F \\ I & J & 2 \end{Bmatrix} \sqrt{\frac{(I+1)(2I+1)(2I+3)}{4I(2I-1)}} \langle \ell \| C^{(2)} \| \ell \rangle$$
$$\times \langle \alpha J \| U^{(2)} \| \alpha' J' \rangle \quad (14\text{-}3)$$

where, as defined in (6-39), $b_\ell = e^2 Q \langle r^{-3} \rangle$.

The total angular momentum J in the matrix element of unit tensor operator is a result of coupling of S and L, therefore expanding the identification of the electronic states, we have

$$\langle \alpha S L J \| U^{(2)} \| \alpha' S' L' J' \rangle$$

$$= \delta_{S,S'} (-1)^{S+L'+J} \sqrt{(2J+1)(2J'+1)} \begin{Bmatrix} J & 2 & J' \\ L' & S & L \end{Bmatrix}$$

$$\times \langle \alpha S L \| U^{(2)} \| \alpha' S' L' \rangle \qquad (14\text{-}4)$$

Now the matrix element of unit tensor operator is evaluated in a pure $L - S$ coupling, and for a single electron we define the unit tensor operator $\mathbf{u}^{(k)}$ as previously,

$$\langle \ell \| u^{(k)} \| \ell' \rangle = \delta_{\ell,\ell'} \qquad (14\text{-}5)$$

The reduced matrix elements of $U^{(2)}$ are tabulated by Nielson and Koster[1], and in particular

$$\langle f^6 \, {}^7 F \| U^{(2)} \| f^6 \, {}^7 F \rangle = -1 \qquad (14\text{-}6)$$

As a consequence, for various components of ${}^7 F$ we have

$$
\begin{array}{c c c c}
U^{(2)} & |{}^7 F_0\rangle & |{}^7 F_1\rangle & |{}^7 F_2\rangle \\
\langle {}^7 F_0| & 0 & 0 & -\frac{\sqrt{7}}{7} \\
\langle {}^7 F_1| & 0 & \frac{3\sqrt{14}}{28} & -\frac{\sqrt{42}}{28} \\
\langle {}^7 F_2| & -\frac{\sqrt{7}}{7} & \frac{\sqrt{42}}{28} & \frac{11\sqrt{6}}{84}
\end{array}
\qquad (14\text{-}7)
$$

With these results it is possible to estimate the extent to which the ground state of ${}^{151}Eu^{3+}$ is perturbed by electric quadrupole interaction with the ${}^7 F_2$ state at $\sim 1040 cm^{-1}$ above the ground state. Taking into account that the reduced matrix element of spherical tensor has the value

$$\langle 3 \| C^{(2)} \| 3 \rangle = -\frac{2\sqrt{105}}{15} \qquad (14\text{-}8)$$

the matrix element is as follows

$$\langle (0\tfrac{5}{2})\tfrac{5}{2} | H_{EQ} | (2\tfrac{5}{2})\tfrac{5}{2} \rangle$$

$$= -b_{4f}(-1)^{2+\frac{5}{2}+\frac{5}{2}} \begin{Bmatrix} 2 & \frac{5}{2} & \frac{5}{2} \\ \frac{5}{2} & 0 & 2 \end{Bmatrix} \sqrt{\frac{\frac{7}{2} \cdot 6 \cdot 8}{4 \cdot \frac{5}{2} \cdot 4}} \cdot -\frac{2\sqrt{105}}{15} \cdot -\frac{\sqrt{7}}{7}$$

$$= -\frac{\sqrt{210}}{75} b_{4f} \qquad (14\text{-}9)$$

However, for the crystal field calculations the $|J M_J I M_I\rangle$ basis with uncoupled electron and nuclear momenta is most appropriate. In order to perform such calculations, the results of chapter 6 have to be extended by the matrix elements off-diagonal in J,

$$\langle \alpha J M_J I M_I M_F | H_{EQ} | \alpha' J' M_J \pm q\, I M_I \mp q M_F' \rangle$$

$$= -e^2 \delta_{M_F, M_F'} (-1)^{J - M_J} \begin{pmatrix} J & 2 & J' \\ -M_J & \mp q & M_J \pm q \end{pmatrix} (-1)^{I - M_I}$$

$$\times \begin{pmatrix} I & 2 & I \\ -M_I & \pm q & M_I \mp q \end{pmatrix} \langle \alpha J \| r_e^{-3} C_e^2 \| \alpha' J' \rangle \langle I \| r_n^2 C_n^{(2)} \| I \rangle \quad (14\text{-}10)$$

where the $3j-$ symbols result from the Wigner Eckart theorem applied twice for the electron and nuclear parts of the element separately. The elements obtained in (14-10) are evaluated in the same manner as in the case of the previous coupling scheme, since they are the same. Finally we have

$$\langle \alpha S L J M_J I M_I M_F | H_{EQ} | \alpha' S L' J' M_J \pm q\, I M_I \mp q M_F' \rangle$$

$$= \delta_{M_F, M_F'} (-1)^{J - M_J} \begin{pmatrix} J & 2 & J' \\ -M_J & \mp q & M_J \pm q \end{pmatrix} (-1)^{I - M_I} \begin{pmatrix} I & 2 & I \\ -M_I & \pm q & M_I \mp q \end{pmatrix}$$

$$\times -e^2 \langle r_e^{-3} \rangle Q \sqrt{\frac{(I+1)(2I+1)(2I+3)}{4I(2I-1)}} \langle \ell \| C^{(2)} \| \ell \rangle$$

$$\times (-1)^{S + L' + J} \sqrt{(2J+1)(2J'+1)} \begin{Bmatrix} J & 2 & J' \\ L' & S & L \end{Bmatrix} \langle \alpha S L \| U^{(2)} \| \alpha' S' L' \rangle \quad (14\text{-}11)$$

In the particular case of the example discussed here the interaction between the ground state $^7F_0$ and $^7F_2$ via the electric quadrupole hyperfine interaction operator is determined by the off-diagonal matrix element

$$E(M_I) = \langle 4f^6\, ^7F_0 M_J = 0 \tfrac{5}{2} M_I | H_{EQ} | 4f^6\, ^7F_2 M_J = 0 \tfrac{5}{2} M_I \rangle \quad (14\text{-}12)$$

which, taking into account the following angular momenta coupling coefficients

$$\begin{pmatrix} 2 & 2 & 0 \\ 0 & 0 & 0 \end{pmatrix} = \frac{1}{\sqrt{5}} \quad (14\text{-}13)$$

and

$$(-1)^{\tfrac{5}{2} - M_I} \begin{pmatrix} \tfrac{5}{2} & 2 & \tfrac{5}{2} \\ -M_I & 0 & M_I \end{pmatrix} = \frac{12 M_I^2 - 35}{16\sqrt{105}} \quad (14\text{-}14)$$

has the value

$$E(M_I) = -e^2 Q \langle r_e^{-3} \rangle \frac{\sqrt{3}}{600} [12 M_I^2 - 35] \quad (14\text{-}15)$$

It must be realized that this particular matrix element does not determine by itself the energy correction; but it contributes to such, if the second-order terms are analyzed. This is the case when the second-order perturbation is responsible for the ground state hyperfine splitting that is assisted by the mixing via the crystal field potential, for example. The result obtained in (14-15) demonstrates that independently of the choice of the second physical mechanism that contributes at the second order, the electric quadrupole hyperfine splittings are proportional to M_I^2, and as a consequence, the states with $\pm M_I$ are two-fold degenerate.

We have from (14-15) the ratio

$$\Delta_{calc} = \frac{E(\pm\frac{5}{2}) - E(\pm\frac{3}{2})}{E(\pm\frac{3}{2}) - E(\pm\frac{1}{2})} = 2 \tag{14-16}$$

which may be compared with the experimental ratios in $Y\,AlO_3 : Eu^{3+}$ of [2]

$$\Delta_{expt}(^{151}Eu) = \frac{45.99}{23.03} = 1.997 \tag{14-17a}$$

and

$$\Delta_{expt}(^{153}Eu) = \frac{119.20}{59.65} = 1.998 \tag{14-17b}$$

14.1   DERIVATION OF A TENSORIAL FORM OF H_{EQ}

The above results are encouraging, but we still have to consider the *sign* of the splitting so as to determine the *ordering* of the three hyperfine levels, and finally the *magnitude* of the splittings. We also want to make sure that our results are correct and consistent. Therefore we attempt an alternative derivation of the tensorial form based upon an expression for the electric quadrupole hyperfine interaction due to Abragam and Pryce[3] who give the interaction as

$$H_{EQ} = \frac{e^2 Q}{2I(2I-1)} \sum_i \left\{ \frac{I(I+1)}{r_i^3} - \frac{3(\mathbf{r}\cdot\mathbf{I})^2}{r_i^5} \right\} \tag{14-18}$$

where the summation is over all electrons outside of closed shells. Our task is to first express the operator enclosed in curly brackets in terms of tensor operators. We start with the second term.

$$(\mathbf{r}\cdot\mathbf{I}) = r(\mathbf{C}^{(1)}\cdot\mathbf{I}^{(1)}) = -\sqrt{3}r(\mathbf{C}^{(1)}\mathbf{I}^{(1)})_0^{(0)} \tag{14-19}$$

Next

$$\begin{aligned}
(\mathbf{r}\cdot\mathbf{I})^2 &= 3r^2(\mathbf{C}^{(1)}\mathbf{I}^{(1)})_0^{(0)}(\mathbf{C}^{(1)}\mathbf{I}^{(1)})_0^{(0)} \\
&= 3r^2\left((\mathbf{C}^{(1)}\mathbf{I}^{(1)})^{(0)}(\mathbf{C}^{(1)}\mathbf{I}^{(1)})^{(0)}\right)_0^{(0)} \\
&= 3r^2\sum_k \langle(11)0(11)0; 0|(11)k(11)k; 0\rangle\left((\mathbf{C}^{(1)}\mathbf{C}^{(1)})^{(k)}(\mathbf{I}^{(1)}\mathbf{I}^{(1)})^{(k)}\right)_0^{(0)} \\
&= 3r^2\sum_k (2k+1)\left\{\begin{matrix} 1 & 1 & 0 \\ 1 & 1 & 0 \\ k & k & 0 \end{matrix}\right\}\left((\mathbf{C}^{(1)}\mathbf{C}^{(1)})^{(k)}(\mathbf{I}^{(1)}\mathbf{I}^{(1)})^{(k)}\right)_0^{(0)} \tag{14-20}
\end{aligned}$$

The $9j$–symbol vanishes unless $k = 0, 1, 2$. Furthermore

$$(\mathbf{C}^{(1)}\mathbf{C}^{(1)})^{(0)} = -\frac{1}{\sqrt{3}} \tag{14-21a}$$

$$(\mathbf{C}^{(1)}\mathbf{C}^{(1)})^{(1)} = 0 \tag{14-21b}$$

$$(\mathbf{C}^{(1)}\mathbf{C}^{(1)})^{(2)} = \sqrt{\frac{2}{3}}\mathbf{C}^{(2)} \tag{14-21c}$$

Consider the $k = 0$ term in (14-20). Evaluating the $9j$–symbol and using (14-21a) leads to

$$3r^2 \cdot \frac{1}{3} \cdot -\frac{1}{\sqrt{3}}(\mathbf{I}^{(1)}\mathbf{I}^{(1)})^{(0)} = -\frac{r^2}{\sqrt{3}}(\mathbf{I}^{(1)}\mathbf{I}^{(1)})^{(0)} \tag{14-22}$$

Then

$$\langle I M_I | (\mathbf{I}^{(1)}\mathbf{I}^{(1)})_0^{(0)} | I M_I \rangle = (-1)^{I-M_I}\begin{pmatrix} I & 0 & I \\ -M_I & 0 & M_I \end{pmatrix} \langle I \| (\mathbf{I}^{(1)}\mathbf{I}^{(1)})^{(0)} \| I \rangle$$

$$= \frac{(-1)^{2I}}{\sqrt{2I+1}}\begin{Bmatrix} 1 & 0 & 1 \\ I & I & I \end{Bmatrix} \langle I \| I^{(1)} \| I \rangle^2$$

$$= -\frac{1}{\sqrt{3}(2I+1)}I(I+1)(2I+1)$$

$$= -\frac{I(I+1)}{\sqrt{3}} \tag{14-23}$$

Noting (14-22) we see that the $k = 0$ term exactly cancels the first term in (14-18), thus only the $k = 2$ term need to be considered.

The reduced matrix element of the tensorial product for the rank 2 is determined by a simple expression

$$\langle I \| (I^{(1)}I^{(1)})^{(2)} \| I \rangle = \sqrt{5}(-1)^{2I}\begin{Bmatrix} I & 1 & I \\ 2 & I & 1 \end{Bmatrix} \langle I \| I^{(1)} \| I \rangle^2$$

$$= \sqrt{5}(-1)^{2I}\begin{Bmatrix} I & I & 2 \\ 1 & 1 & I \end{Bmatrix} I(I+1)(2I+1)$$

$$= \sqrt{\frac{I(I+1)(2I+1)(2I+3)(2I-1)}{6}} \tag{14-24}$$

and from the Wigner-Eckart theorem applied to the matrix element that depends on the projections of the nuclear quantum numbers

$$\langle I M_I | (I^{(1)}I^{(1)})_0^{(2)} | I M_I \rangle$$

$$= (-1)^{I-M_I}\begin{pmatrix} I & 2 & I \\ -M_I & 0 & M_I \end{pmatrix} \langle I \| (I^{(1)}I^{(1)})^{(2)} \| I \rangle \tag{14-25}$$

Evaluating the $3j$−symbol and using (14-24) we obtain

$$\langle I M_I | (I^{(1)} I^{(1)})_0^{(2)} | I M_I \rangle = \frac{3 M_I^2 - I(I+1)}{\sqrt{6}} \tag{14-26}$$

Finally, collecting all these partial results, it is possible to evaluate that the $k = 2$ portion of (14-20) becomes

$$r^2 \frac{\sqrt{6}}{3} \sum_q (-1)^q C_q^{(2)} (I^{(1)} I^{(1)})_{-q}^{(2)} \tag{14-27}$$

This means that

$$\left\{ \frac{I(I+1)}{r_i^3} - \frac{3(\mathbf{r} \cdot \mathbf{I})^2}{r_i^5} \right\} = -\langle r_e^{-3} \rangle_i \sqrt{6} \sum_q (-1)^q C_q^{(2)} (I^{(1)} I^{(1)})_{-q}^{(2)} \tag{14-28}$$

Returning this result to (14-18), it is straightforward to evaluate the matrix element of the electric quadrupole hyperfine interactions between the states $^7 F_{00}$ and $^7 F_{20}$,

$$\langle ^7 F 00 I M_I | H_{EQ} |^7 F 20 I M_I \rangle$$

$$= -\frac{e^2 Q \langle r_e^{-3} \rangle}{2I(2I-1)} \sqrt{6} \langle ^7 F 00 | C_0^{(2)} |^7 F 20 \rangle \langle I M_I | (I^{(1)} I^{(1)})_0^{(2)} | I M_I \rangle \tag{14-29}$$

Taking into account the values of the contributing matrix elements evaluated previously, as electron matrix element for the particular states

$$\langle ^7 F 00 | C_0^{(2)} |^7 F 20 \rangle = \frac{2\sqrt{3}}{15} \tag{14-30}$$

and the matrix element in (14-26), with $I = \frac{5}{2}$, we have

$$\langle ^7 F 00 \tfrac{5}{2} M_I | H_{EQ} |^7 F 20 \tfrac{5}{2} M_I \rangle = -e^2 Q \langle r_e^{-3} \rangle \frac{\sqrt{3}}{600} \left[12 M_I^2 - 35 \right] \tag{14-31}$$

This result is in agreement with (14-15). This derivation provides not only an additional exercise in using the tools of the tensor operator algebra, but it also demonstrates how important and also how easy it is to verify the correctness of calculations.

Ultimately our concern is with Eu^{3+} ions in a crystal field defined by the potential V_{cryst} in its usual form

$$V_{cryst} = \sum_{t,p} A_p^t C_p^{(t)} \equiv \sum_{t,p} r^t B_p^t C_p^{(t)} \tag{14-32}$$

where the summation is limited to the particular terms by the symmetry of the environment of the lanthanide ion. It is also limited in an effective way by the selection rules for the non-vanishing matrix elements. Having in mind the second-order corrections

to the energy that are determined by the combined effects of electric multipole interactions and the crystal field, it is useful to evaluate the off-diagonal matrix element
of the spherical tensor

$$\langle \alpha SLJM | C_p^{(t)} | \alpha' SL'J'M' \rangle = (-1)^{J-M} \begin{pmatrix} J & t & J' \\ -M & q & M' \end{pmatrix}$$

$$\times (-1)^{S+L'+J+t} \sqrt{[J,J']} \begin{Bmatrix} J & t & J' \\ L' & S & L \end{Bmatrix}$$

$$\times \langle \alpha SL \| C^{(t)} \| \alpha' SL' \rangle \qquad (14\text{-}33)$$

Inspection of the above leads to the selection rules that must be satisfied if the matrix
element is not to vanish

$$\Delta S = 0, \quad \Delta L <= k, \quad \Delta J <= k, \quad M' = M - q \qquad (14\text{-}34)$$

In (14-29) we saw that the electric quadrupole hyperfine interaction can couple the
$M_J = 0$ states of the $^7F_0$ level to that of the $^7F_2$ level. Clearly any crystal field having
an axial quadrupole term such as $A_0^2 C_0^{(2)}$ can likewise mix those two levels. Thus we
can anticipate a second-order splitting mechanism of the form

$$-2 \frac{\langle ^7F00IM_I | A_0^2 C_0^{(2)} | ^7F20IM_I \rangle \langle ^7F20IM_I | H_{EQ} | ^7F00IM_I \rangle}{E(^7F_2) - E(^7F_0)} \qquad (14\text{-}35)$$

This means that it is expected that the interactions between the excited states of the
$4f^N$ configuration, which are taken into account via the crystal field potential and the
electric quadrupole hyperfine interactions, are responsible for the observed splittings
of the ground state.

REFERENCES

1. Nielson, C W and Koster, G F, (1963) *Spectroscopic Coefficients for the p^n, d^n, and f^n Configurations*, Cambridge: MIT Press.
2. Shelby R M and Macfarlane R M, (1981) Measurement of the Anomalous Nuclear Magnetic Moment of Trivalent Europium *Phys. Rev. Lett.*, **47** 1172.
3. Abragam A and Pryce M H L, (1951) Theory of the Nuclear Hyperfine Structure of Paramagnetic Resonance Spectra in Crystals *Proc. Phys. Soc. (London)* **A** 205 135.

15 Electric Quadrupole Hyperfine Structure in Crystals

Symmetry, as wide or as narrow as you may define its meaning, is one idea by which man through the ages has tried to comprehend and create order, beauty, and perfection.

Hermann Weyl, *Symmetry* (Princeton: Princeton University Press 1952)

The investigations presented now are focused on the evaluation of the *sign* and *magnitude* of the ground state hyperfine splitting of Eu^{3+} in a crystalline environment. The first order analysis, in which just the contributions caused by the electric quadrupole hyperfine interactions are evaluated is extended by the second-order terms that originate from the coupling of two distinct physical mechanisms. Namely, as mentioned at the end of the previous chapter, it is expected that the interactions via the crystal field potential are of the greatest importance.

In order to demonstrate the simplest example, it is assumed that the ion of Eu^{3+} is in a purely axial crystal field. Since an interaction between the $^7F_0$ and $^7F_2$ levels is of the greatest interest, the crystal field is limited to a single term, which traditionally[1] is defined as

$$V_{cryst} = B_0^2 \sum_i \left(3z_i^2 - r_i^2\right) \tag{15-1}$$

where the summation is over all electrons in an open shell, and B_2^0 is a parameter that reflects the structure of the crystal. For the purpose of the evaluation of various matrix elements, this potential is converted into the tensorial form, which enables one to evaluate various matrix elements in a straightforward manner,

$$V_{cryst} = 2B_0^2 \langle r^2 \rangle \sum_i C_{i0}^{(2)} \tag{15-2}$$

Note that in this definition the radial dependence is explicitly presented in spite of the fact that in practical numerical calculations of energy corrections the crystal field parameters A_p^t (compare (13-30)) determined from the fitting procedure that reproduces the measured energies are used. In fact the radial integral in (15-2), which is the expectation value of the radial part of the crystal field potential between the one-electron states of the $4f$ symmetry is correct as long as the energy is evaluated within the single configuration approximation. For the purpose of the analysis presented in the following chapter where the inter-shell interactions via various physical mechanisms are included, to make the definition more general, the radial integral in (15-2) should be explicitly defined as $\langle n\ell | r^t | n'\ell' \rangle$. This definition shows that the matrix elements

177

of the crystal field potential for the states of various configurations do not vanish and also that, in the application here, the tensor operators are inter-shell objects.

As mentioned at the end of the previous chapter, the second-order corrections to the energy, which might possibly explain the observed splittings of the ground state of the $4f^6$ configuration, are in general determined by the product of two matrix elements,

$$\gamma_{EQ}^{\Psi} = \sum_{\chi \neq \Psi} \sum_{tp}^{all} B_p^t \langle 4f|r^t|4f\rangle \langle \Psi|C_p^{(t)}|\chi\rangle \langle \chi|H_{EQ}|\Psi\rangle / (E_\Psi - E_\chi)$$

where the summation is over all excited states of the ground configuration. In particular, for $\Psi \equiv {}^7F_0$, following (14-35), we have,

$$E(I, M_I) = -4B_0^2\langle r^2\rangle \frac{\langle {}^7F00IM_I| \sum_i C_{i0}^{(2)}|{}^7F20IM_I\rangle \langle {}^7F20IM_I|H_{EQ}|{}^7F00IM_I\rangle}{E({}^7F_2) - E({}^7F_0)}$$

(15-3)

The operator in the first matrix element acts only in the space of the electrons and is thus independent of, and diagonal in, the IM_I nuclear quantum numbers; this matrix element was evaluated in (14-30). The second matrix element was evaluated in (14-29) as

$$\langle {}^7F20IM_I|H_{EQ}|{}^7F00IM_I\rangle = -e^2 Q\langle r_e^{-3}\rangle \frac{\sqrt{3}}{15} \frac{[3M_I^2 - I(I+1)]}{I(2I-1)}$$

(15-4)

Thus, the second-order energy correction is determined by

$$E(I, M_I) = \frac{8}{75} \frac{e^2 Q B_0^2 \langle r^2\rangle \langle r_e^{-3}\rangle}{\Delta_2} \frac{[3M_I^2 - I(I+1)]}{I(2I-1)}$$

(15-5)

which is identical with (11) of Elliott[1], where the energy difference is denoted by the symbol

$$\Delta_2 = E\left({}^7F_2\right) - E\left({}^7F_0\right)$$

(15-6)

For the stable europium isotopes $I = \frac{5}{2}$ and the energy correction (15-5) becomes

$$E\left(\tfrac{5}{2}, M_I\right) = \frac{4}{375} \frac{e^2 Q B_0^2 \langle r^2\rangle \langle r_e^{-3}\rangle}{\Delta_2} \left[3M_I^2 - \frac{35}{4}\right]$$

(15-7)

which again agrees with Elliott's result[1] *but* has the opposite sign to the observed splittings. This means that the mechanisms regarded here at the second order cannot, by themselves, explain the ground state hyperfine structure of Eu^{3+}. There is a similar problem in explaining the crystal field splitting[2] of the $|4f^7\,{}^8S_{7/2}\rangle$ ground state of Gd^{3+}.

15.1   EXPLICIT CALCULATION OF THE ELLIOTT'S TERM

Elliot[1] has expressed the energy splitting from (15-5) in the form

$$E(I, M_I) = P\left[M_I^2 - \frac{1}{3}I(I+1)\right] \tag{15-8}$$

with

$$P = \frac{8}{25}e^2 Q B_0^2 \langle r^2 \rangle \frac{\langle r_e^{-3} \rangle}{I(2I-1)\Delta_2} \tag{15-9}$$

To obtain a numerical value for P we need the values of various quantities appearing in (15-9); in particular the radial integral has to be evaluated (or at least estimated). It was done by Elliott for Eu^{3+} in europium ethylsulphate crystals. A value for the structural parameter B_0^2 is obtained, to a reasonable approximation, by noting that experimentally the two $^7F_1$ sublevels, for $M_J = \pm 1$ and $M_J = 0$, are separated by $42cm^{-1}$.

This splitting can be easily calculated, when neglecting possible J−mixing, by using the general expression for the matrix elements of a spherical tensor, and applying the numerical results obtained for the reduced matrix elements of unit tensor operator $U^{(2)}$ that are collected in the matrix (14-7). Obviously the energy separation of these two sublevels is determined directly by the crystal field parameter B_0^2, the value of which is crucial for evaluation of the hyperfine splitting of the ground state in accordance with (15-9). Thus, the energy of both sublevels are determined as follows

$$\langle^7F_1 1|C_0^{(2)}|^7F_1 1\rangle$$

$$= (-1)^{1-1} \begin{pmatrix} 1 & 2 & 1 \\ -1 & 0 & 1 \end{pmatrix} \langle 3\|C^{(2)}\|3\rangle \langle^7F_1\|U^{(2)}\|^7F_1\rangle$$

$$= 1 \times \frac{\sqrt{30}}{30} \times -\frac{2\sqrt{105}}{15} \times \frac{3\sqrt{14}}{28}$$

$$= -\frac{1}{10} \tag{15-10a}$$

$$\langle^7F_1 0|C_0^{(2)}|^7F_1 0\rangle$$

$$= (-1)^{1-0} \begin{pmatrix} 1 & 2 & 1 \\ 0 & 0 & 0 \end{pmatrix} \langle 3\|C^{(2)}\|3\rangle \langle^7F_1\|U^{(2)}\|^7F_1\rangle$$

$$= -1 \times \frac{\sqrt{30}}{15} \times -\frac{2\sqrt{105}}{15} \times \frac{3\sqrt{14}}{28}$$

$$= +\frac{1}{5} \tag{15-10b}$$

The energy separation between them, measured to be $42cm^{-1}$, is theoretically determined by

$$\langle^7F_1 0|V_{cryst}|^7F_1 0\rangle - \langle^7F_1 \pm 1|V_{cryst}|^7F_1 \pm 1\rangle = \frac{3}{5}B_0^2\langle r^2 \rangle \tag{15-11}$$

from which it is deduced that the value of the crystal field parameter, including the appropriate radial integral, is

$$A_0^2 \equiv B_0^2 \langle r^2 \rangle = 70 cm^{-1} \qquad (15\text{-}12)$$

The energy denominator of the second-order correction, which is included within the parameter P in (15-9), is determined experimentally as the energy separation of the $^7F_0$ and $^7F_2$ levels,

$$\Delta_2 = 1015 cm^{-1} \qquad (15\text{-}13)$$

This result made it possible to evaluate indirectly by Bleaney[3] a value of the radial integral

$$\langle r_e^{-3} \rangle = 57 \overset{\circ}{A}^{-3} = 57 \times 10^{-24} cm^{-3} \qquad (15\text{-}14)$$

Furthermore,

$$e^2 = R \, a_0 \qquad (15\text{-}15)$$

where R is the Rydberg constant and a_0 is the Bohr radius. Following tradition the electric quadrupole moment Q is expressed in *barns* with

$$1 barn = 10^{-24} cm^2 \qquad (15\text{-}16)$$

To obtain P in cm^{-1} we take

$$R = 109736 cm^{-1} \quad \text{and} \quad a_0 = 0.5292 \times 10^{-8} cm \qquad (15\text{-}17)$$

Taking the above values together with $I = \frac{5}{2}$ and Q in barns, we obtain from (15-9) Elliott's estimate of

$$\begin{aligned}
P &= +Q \times \frac{4}{125} \times \frac{e^2 B_0^2 \langle r^2 \rangle}{\Delta_2} \times \langle r_e^{-3} \rangle \\
&= +Q \times \frac{4 \times 2 \times 109736 \times 0.5292 \times 10^{-8} \times 70 \times 57}{125 \times 1015} \\
&= +1.46 Q \times 10^{-4} cm^{-1} \\
&= +4.38 Q \, MHz \qquad (15\text{-}18)
\end{aligned}$$

Elliott's estimate for P could be brought up-to-date using modern values, but the change is relatively small. For example, when the radial integral is evaluated directly with the Hartree-Fock one-electron functions generated for the average energy of electron configuration of the free ion, and instead of value determined in (15-14), the value of $55 \times 10^{-24} cm^{-3}$ is used, the final result is reduced to $1.41 Q \times 10^{-4} cm^{-1}$. This means that either some other physical mechanisms have to be taken into account to reproduce the observed splitting, or the investigations should be extended by higher order corrections.

Before exploring the subject further, the impact of spin-orbit interaction of the $^5D_0$ state with the ground state $^7F_0$ is considered. This is exactly the mechanism which is able to mix the states of various multiplicities. In order to analyze this possibility, at first the Coulomb interaction among the three ^5D states of $4f^6$ has to be taken into account, in order to determine the linear combination of the three ^5D states that corresponds to the lowest ^5D term.

The detailed calculation can be found in 8-7 of Judd[4]. The three ^5D states are labeled as

$$|(210)(20)^5 D\rangle, \; |(210)(21)^5 D\rangle, \; |(111)(20)^5 D\rangle \qquad (15\text{-}19)$$

The matrix of elements of Coulomb interaction between these states is as follows

$$\begin{pmatrix} 15E^0 + 6E^1 + \frac{858}{7}E^2 + 11E^3 & \frac{468\sqrt{33}}{7}E^2 & \frac{22\sqrt{14}}{7}E^3 \\ \frac{468\sqrt{33}}{7}E^2 & 15E^0 + 6E^1 - \frac{1131}{7}E^2 + 18E^3 & \frac{12\sqrt{462}}{7}E^3 \\ \frac{22\sqrt{14}}{7}E^3 & \frac{12\sqrt{462}}{7}E^3 & 15E^0 + 9E^1 - 11E^3 \end{pmatrix}$$

$$(15\text{-}20)$$

In order to diagonalize this matrix to determine the coefficients of the linear combination of three components, in principle the Slater integrals have to be evaluated with a chosen radial basis set of one-electron functions. In practice, Judd used $4f-$hydrogenic eigenfunction ratios for the Slater integrals. As a result of such calculations, the eigenfunction for the lowest ^5D term has the form

$$|^5D\rangle = -0.196|(210)(20)^5 D\rangle + 0.770|(210)(21)^5 D\rangle - 0.607|(111)(20)^5 D\rangle \quad (15\text{-}21)$$

This function is subsequently applied to evaluate the impact due to the spin orbit interaction, which couples the excited state ^5D to the ground state $^7F_0$.

15.2   SPIN-ORBIT INTERACTION BETWEEN $^7F_0$ AND THE LOWEST $^5D_0$

The spin-orbit interaction is expressed by a double tensor operator, which is a vector operator in the spin and a vector operator in the orbital spaces; these two objects are coupled together via the scalar product to give the final rank of zero. As a consequence, the matrix elements of the spin orbit interaction operator are diagonal in J and M,

$$\langle \ell^N \alpha S L J M | \sum_{i=1}^{N} (\mathbf{s_i} \cdot \boldsymbol{\ell_i}) | \ell^N \alpha' S' L' J' M'\rangle$$

$$= \delta_{J,J'} \delta_{M,M'} (-1)^{S'+L+J} \frac{\sqrt{\ell(\ell+1)(2\ell+1)}}{\sqrt{6}} \begin{Bmatrix} S & S' & 1 \\ L' & L & J \end{Bmatrix}$$

$$\times \langle \ell^N \alpha S L \| W^{(11)} \| \ell^N \alpha' S' L'\rangle \qquad (15\text{-}22)$$

where $W^{(11)}$ is the double unit tensor operator. In fact this is only the angular part of the spin-orbit interaction, and the total energy correction is determined if (15-22) is multiplied by the appropriate radial integral, or by the spin-orbit coupling factor ζ_{4f}.

For the purpose of present numerical calculations it is recommended that the values of some of the relevant spin-orbit matrix elements collected by Judd [4] (p. 203) be used. In particular one may find there the $\langle {}^7F_1 | H_{so} | {}^5D_1 \rangle$ matrix elements, which indirectly are needed here. In order to evaluate the perturbing influence of the lowest excited state ${}^5D$ upon the energy structure of the ground ${}^7F_0$, the corresponding matrix elements for $J = 0$ are required. However, it is possible to obtain the latter by noting from (15-22) that

$$\frac{\langle {}^7F_0 | H_{so} | {}^5D_0 \rangle}{\langle {}^7F_1 | H_{so} | {}^5D_1 \rangle} = \frac{3\sqrt{2}}{4} \qquad (15\text{-}23)$$

Therefore, we have

$$\langle {}^7F_0 | \left(\begin{array}{ccc} |(210)(20)^5D\rangle & |(210)(21)^5D\rangle & |(111)(20)^5D\rangle \\ -\frac{\sqrt{42}}{7}\zeta_{4f} & -\frac{\sqrt{154}}{7}\zeta_{4f} & 2\sqrt{3}\,\zeta_{4f} \end{array} \right) \qquad (15\text{-}24)$$

As a result, for the lowest ${}^5D_0$ level the off-diagonal matrix element, which contributes to the second-order energy correction, has the value expressed by the spin-orbit coupling factor,

$$\langle {}^7F_0 | H_{so} | {}^5D_0 \rangle$$
$$= \left(-0.196 \times -\frac{\sqrt{42}}{7} + 0.770 \times -\frac{\sqrt{154}}{7} - 0.607 \times 2\sqrt{3} \right) \zeta_{4f}$$
$$\cong -3.3\,\zeta_{4f} \qquad (15\text{-}25)$$

Suppose that the ${}^5D_0$ is above the ${}^7F_0$ level by an amount E and try to estimate the amount of $|{}^5D_0\rangle$ character that gets mixed into the $|{}^7F_0\rangle$ ground state by spin-orbit interaction. Using the previous results the following energy matrix has to be considered

$$\begin{array}{cc} & \begin{array}{cc} |{}^7F_0\rangle & |{}^5D_0\rangle \end{array} \\ \begin{array}{c} \langle {}^7F_0| \\ \langle {}^5D_0| \end{array} & \left(\begin{array}{cc} 0 & -3.3\,\zeta_{4f} \\ -3.3\,\zeta_{4f} & E \end{array} \right) \end{array} \qquad (15\text{-}26)$$

In general, in the case of a rank two matrix of the form

$$\begin{array}{cc} & \begin{array}{cc} |\alpha\rangle & |\beta\rangle \end{array} \\ \begin{array}{c} \langle\alpha| \\ \langle\beta| \end{array} & \left(\begin{array}{cc} 0 & a \\ a & E \end{array} \right) \end{array} \qquad (15\text{-}27)$$

two eigenvalues $\lambda_\pm$ may be found from the requirement that the determinant of the set of linear equations vanishes,

$$\begin{vmatrix} -\lambda & a \\ a & E - \lambda \end{vmatrix} = 0 \qquad (15\text{-}28)$$

This condition leads to the secular equation

$$\lambda^2 - E\lambda - a^2 = 0 \tag{15-29}$$

which possesses two solutions,

$$\lambda_\pm = \frac{1}{2}\left[E \pm \sqrt{E^2 + 4a^2}\right]$$
$$= \frac{1}{2}E\left[1 \pm \sqrt{1 + \frac{4a^2}{E^2}}\right] \tag{15-30}$$

If $E \gg a$ it is possible to approximate the square root to get the two eigenvalues as

$$\lambda_+ = E + \frac{a^2}{E} \quad \text{and} \quad \lambda_- = -\frac{a^2}{E} \tag{15-31}$$

The corresponding eigenvectors are

$$|\lambda_+\rangle = x|\alpha\rangle + y|\beta\rangle \quad \text{and} \quad |\lambda_-\rangle = y|\alpha\rangle - x|\beta\rangle \tag{15-32}$$

with

$$xx^* + yy^* = 1 \tag{15-33}$$

where it may be chosen that the coefficients x, y are real. The components (x, y) are determined from the requirement that

$$\begin{pmatrix} 0 & a \\ a & E \end{pmatrix}\begin{pmatrix} x \\ y \end{pmatrix} = \lambda_\pm \begin{pmatrix} x \\ y \end{pmatrix} \tag{15-34}$$

which gives

$$y = \lambda_\pm \frac{x}{a} \tag{15-35}$$

Because of the condition in (15-33),

$$x = e^{i\theta}\frac{a}{\sqrt{a^2 + \lambda_\pm^2}} \tag{15-36}$$

The fixed phase angle θ to the value for which $e^{i\theta} = +1$ leads to the two eigenvectors

$$|\lambda_-\rangle = \frac{1}{\sqrt{E^2 + a^2}}(E|\alpha\rangle - a|\beta\rangle) \tag{15-37a}$$

$$|\lambda_+\rangle = \frac{1}{\sqrt{E^2 + a^2}}(a|\alpha\rangle + E|\beta\rangle) \tag{15-37b}$$

Using these simple manipulations of the 2×2 problem, it is straightforward to return to the main stream of the discussion that is devoted to the strength of the admixing the excited energy states to the ground level via the spin orbit interaction. Namely, choosing $E = 18000cm^{-1}$ and $\zeta_{4f} = 1015cm^{-1}$ we obtain from (15-32) the eigenvalues

$$E(^7F_0) = -603cm^{-1} \quad \text{and} \quad E(^5D_0) = +18603cm^{-1} \tag{15-38}$$

and from (15-37a,b) the eigenvectors assigned to these energy levels,

$$|E(^7F_0)\rangle = 0.984|^7F_0\rangle + 0.177|^5D_0\rangle \tag{15-39a}$$

$$|E(^5D_0)\rangle = -0.177|^7F_0\rangle + 0.984|^5D_0\rangle \tag{15-39b}$$

These calculations should be regarded as an illustrative example and certainly not as an optimized numerical procedure. The main conclusion is that the mixing of the $^7F_0$ and $^5D_0$ states via the spin-orbit interaction is relatively small and is unable to explain the observations.

REFERENCES

1. Elliott R J, (1957) Anomalous Quadrupole Coupling in Europium Ethylsulfate *Proc. Phys. Soc. (London),* **B 70** 119.
2. Wybourne B G, (1966) Theory of Nuclear Magnetic Resonance in Eu^{3+} *Phys. Rev.,* **148** 317.
3. Bleaney B, (1955) Nuclear Moments of the Lanthanons from Paramagnetic Resonance *Proc. Phys. Soc. (London),* **A68** 937.
4. Judd B R, (1998) *Operator Techniques in Atomic Spectroscopy* Princeton: Princeton University Press.

16 The Electric Multipole Coupling Mechanism in Crystals

Physics is becoming so unbelievably complex that it is taking longer and longer to train a physicist. It is taking so long, in fact, to train a physicist to the place where he understands the nature of physical problems that he is already too old to solve them.

Eugene Wigner

In this chapter the early attempt of Judd, Lovejoy and Shirley[1] to reconcile the discrepancy between Elliott's mechanism[2] for the ground state of europium ethylsulphate and the experimental observation that the splitting had about twice the predicted magnitude and the opposite sign is presented. In the search for new corrections to the energy that originate from the electric multipole interactions, the approach is extended by the inter-shell interactions.

Judd *et al* fitted their experimental data to a predominantly quadrupolar Hamiltonian

$$\mathcal{H} = P\left[M_I^2 - \frac{1}{3}I(I+1)\right] \tag{16-1}$$

and treated P as a parameter obtained[1] for the two isotopes Eu^{152} and Eu^{154} the following values

$$P_{152} = -(6.7 \pm 0.5) \times 10^{-4} cm^{-1} \tag{16-2a}$$

$$P_{154} = -(8.3 \pm 0.7) \times 10^{-4} cm^{-1} \tag{16-2b}$$

These results are clearly of the opposite sign to that predicted by Elliott. The breakdown of the Russell-Saunders coupling, considered in the previous chapter (where the matrix elements were evaluated with the functions in the intermediate coupling scheme and even with the inclusion of the perturbing impact of crystal field potential), is at least two orders of magnitude too small to account for the negative sign of P. Therefore a theoretical description has to be enriched by some new physical mechanisms, which potentially may provide new energy corrections. The first candidate is the Coulomb interaction between various electron configurations, including the excited ones. This step is a natural extension, which breaks down the limitations of single configuration approximation of the standard approach.

16.1   CONFIGURATION INTERACTION MECHANISMS

Given the failure to explain the observed quadrupole splitting in terms of interactions confined to the states of the $4f^6$ configuration, it is natural to investigate the possibility of explaining the discrepancy by including the interactions via various operators with other electron configurations. This concept gives a chance to include the inter-shell interactions similarly as discussed in the case of the magnetic hyperfine interactions in the previous chapter.

Among the possible physical mechanisms that may admix to the wave function of the $4f^N$ configurations new components are the Coulomb interaction, spin-orbit interaction and crystal field potential, to mention the most important ones among those included in the Hamiltonian in (13-24).

The Coulomb inter electronic interaction is defined in the standard way

$$V_c = \sum_{i>j} \frac{e^2}{r_{ij}} \tag{16-3}$$

and in our investigations its tensorial form with the angular part from (13-47) is used together with an appropriate radial integral.

Judd *et al* note that since the crystal field parameters are usually deduced from experiment rather than evaluated by explicit calculation, they already accommodate the contributions from mechanisms such as

$$\langle 4f^6 5s^2 \, ^7F_0|V_c|4f^6 5s 5d \, ^7F_0\rangle \langle 4f^6 5s 5d \, ^7F_0|V_{cryst}|4f^6 5s^2 \, ^7F_2\rangle$$
$$\times \langle 4f^6 5s^2 \, ^7F_2|V_{EQ}|4f^6 5s^2 \, ^7F_0\rangle \, [\Delta_1 \Delta_2]^{-1} \tag{16-4}$$

where

$$\Delta_1 = E(4f^6 5s 5d \, ^7F_0) - E(4f^6 5s^2 \, ^7F_0) \tag{16-5a}$$
$$\Delta_2 = E(4f^6 5s^2 \, ^7F_2) - E(4f^6 5s^2 \, ^7F_0) \tag{16-5b}$$

In the nomenclature of the perturbation theory, the contributions in (16-4) are determined by the third-order terms (two energy denominators). In order to complete the list of the impact due to these three physical mechanisms, one has to take into account the remaining sequences of operators in the triple products of the matrix elements. In the case presented in (16-4) there are inter-shell interactions via the Coulomb interaction and V_{cryst} included, while V_{EQ} acts within the ground configuration $4f^N$. The intermediate states between the first two operators describe the energy levels of the excited configurations in which an electron from the closed shell of $5s$ symmetry is promoted to the $5d$ excited one-electron state. Since the parity of excited configuration is the same as the parity of the ground configuration, the expansion of the Coulomb interaction and also the expansion of the crystal field potential are limited to the terms of even ranks; similarly as in the case of the expansion of V_{EM} defined in general as

$$V_{EM} = \sum_k \frac{r_n^k}{r_e^{k+1}} \left(\mathbf{C}_e^{(k)} \cdot \mathbf{C}_n^{(k)} \right) \tag{16-6}$$

As mentioned, there are other third-order terms contributing to the energy, and among them there is an expression in which the electric quadrupole hyperfine interaction operator acts within the excited configurations, as below

$$\langle 4f^6 5s^2\ {}^7F_0|V_c|4f^6 5s5d\ {}^7F_0\rangle\langle 4f^6 5s5d\ {}^7F_0|V_{EQ}|4f^6 5s^2\ {}^7F_2\rangle$$
$$\times\langle 4f^6 5s^2\ {}^7F_2|V_{cryst}|4f^6 5s^2\ {}^7F_0\rangle\,[\Delta_1\Delta_2]^{-1} \tag{16-7}$$

Continuing the analysis of (16-4), it is seen that the first pair of matrix elements along with the energy denominator Δ_1 amounts to a shielding correction to the matrix element

$$\langle 4f^6\ {}^7F_0|V_{EQ}|4f^6\ {}^7F_2\rangle.$$

Judd *et al* note that such a mechanism cannot account for the required change of sign. Instead they suggest that the most likely configuration interaction mechanism must be of the form

$$-2\frac{\langle 4f^6\ {}^7F_0|V_{cryst}|A\ {}^7L_2\rangle\langle A\ {}^7L_2|V_{EQ}|4f^6\ {}^7F_0\rangle}{E(A)} \tag{16-8}$$

where A denotes an excited electron configuration of *even* parity at an energy $E(A)$ above the ground state, and the hyperfine interactions are limited to the quadrupole term H_{EQ} ($k = 2$ in (16-6)). Of course, several configurations A of even parity in comparison to the parity of $4f^6$ may contribute. In our discussion attention is focussed on the ratio

$$R = \frac{\langle 4f^6\ {}^7F_0|V_{cryst}|A\ {}^7L_2\rangle\langle A\ {}^7L_2|V_{EQ}|4f^6\ {}^7F_0\rangle \times \Delta_2}{\langle 4f^6\ {}^7F_0|V_{cryst}|4f^6\ {}^7F_2\rangle\langle 4f^6\ {}^7F_2|V_{EQ}|4f^6\ {}^7F_0\rangle \times E(A)} \tag{16-9}$$

which describes the relative importance of two mechanisms, interactions between the excited states of the ground configuration (in the denominator), and perturbing influence of singly excited configurations (in the numerator). In both cases the interactions are taken into account via V_{cryst} and V_{EQ}.

The operators, V_{cryst} and V_{EQ}, are single particle objects, and hence at second-order we can restrict the excited configurations A to those involving the excitation of a single electron from an occupied shell, $n\ell$, to a one-particle state $n'\ell'$. Canceling out the common angular dependencies of the matrix elements in R gives the ratio as

$$R = \mathcal{A}\frac{\langle n\ell|r^2|n'\ell'\rangle\langle n'\ell'|r^{-3}|n\ell\rangle \times \Delta_2}{\langle 4f|r^2|4f\rangle\langle 4f|r^{-3}|4f\rangle \times E(A)} \tag{16-10}$$

where $\mathcal{A}$ contains the remaining angular factors that do not cancel, and it vanishes if the triad $(\ell, \ell', 2)$ does not satisfy the usual triangular condition.

It is important to note that the nuclear factors appearing in R as defined in (16-9) necessarily cancel. The same happens with the crystal field parameters that are common for both matrix elements. Hence for an axial field the factor contains the following matrix elements,

$$A = \frac{\langle 4f^6 \; ^7F_0 \| C^{(2)} \| A \; ^7L_2 \rangle \langle A \; ^7L_2 \| C_e^{(2)} \| 4f^6 \; ^7F_0 \rangle}{\langle 4f^6 \; ^7F_0 \| C^{(2)} \| 4f^6 \; ^7F_2 \rangle \langle 4f^6 \; ^7F_2 \| C_e^{(2)} \| 4f^6 \; ^7F_0 \rangle} \qquad (16\text{-}11)$$

where indeed only the electronic part of the matrix elements is included, and therefore in the analysis that follows the notation $C_e^{(2)} \equiv C^{(2)}$ needs to be kept.

The spherical tensor operators $C^{(k)}$ are orbital operators, and their matrix elements are diagonal in all spin quantum numbers. Following the presentation of Judd et al,[1] a particular case of an excitation from a closed shell is now analyzed. This means that electron configuration A describes the excitation of a single electron from a closed shell, $n\ell^{4\ell+2}$, to give excited configuration $n\ell^{4\ell+1}n'\ell'$. This possibility of excitation requires a full description of energy states in (16-11). For the case of the energy levels of the ground configuration we have

$$|4f^6 \; ^7F(n\ell^{4\ell+2})^1 S; \; ^7F_0 0\rangle \qquad (16\text{-}12a)$$

and a typical excited state is the form

$$|4f^6 \; ^7F(n\ell^{4\ell+1}n'\ell')^1 D; \; ^7L_2 0\rangle \qquad (16\text{-}12b)$$

with L restricted to

$$F \times D = P + D + F + G + H, \quad \text{i.e. } L = 1, \ldots, 5 \qquad (16\text{-}13)$$

The dependence of the matrix elements in (16-11) upon the total angular momentum quantum numbers $J M_J$ is the same in the numerator and denominator, therefore we have

$$A = \left| \frac{\langle 4f^6 \; F(\ldots)S; F \| C^{(2)} \| 4f^6 \; F(\ldots)^* D; L \rangle}{\langle 4f^6 \; F(\ldots)S; F \| C^{(2)} \| 4f^6 \; F(\ldots)S; F \rangle} \right|^2 \qquad (16\text{-}14)$$

where for brevity the closed shell is indicated as $(\ldots)$ and single particle excitation as $(\ldots)^*$.

To verify that (16-14) follows from (16-11), an Exercise 16-1 is assigned in this chapter. With the experience gained through the course of the presentation it is easy to derive that (Exercise 16-2) the denominator of (16-14) has the value,

$$\left| \langle 4f^6 \; F(\ldots)S; F \| C^{(2)} \| 4f^6 \; F(\ldots)S; F \rangle \right|^2$$
$$= \left| \langle 3 \| C^{(2)} \| 3 \rangle \langle 4f^6 \; ^7F \| U^{(2)} \| 4f^6 \; ^7F \rangle \right|^2$$
$$= \langle 3 \| C^{(2)} \| 3 \rangle^2$$
$$= \frac{28}{15} \qquad (16\text{-}15)$$

The numerator of (16-14) is determined by the matrix element of $C^{(2)}$ between functions that belong to different configurations,

$$\langle F(\ldots)S; F\|C^{(2)}\|F(\ldots)^*D; L\rangle$$

$$= (-1)^{3+2+3+2}\sqrt{7(2L+1)}\begin{Bmatrix} 3 & 2 & L \\ 2 & 3 & 0 \end{Bmatrix}\langle(n\ell^{4\ell+2})S\|C^{(2)}\|(n\ell^{4\ell+1}n'\ell')D\rangle$$

$$= -(-1)^L\sqrt{\frac{2L+1}{5}}\langle\ell\|C^{(2)}\|\ell'\rangle\langle(n\ell^{4\ell+2})S\|U^{(2)}(\ell\ell')\|(n\ell^{4\ell+1}n'\ell')D\rangle \quad (16\text{-}16)$$

where the unit tensor operator is defined as before,

$$\langle n\ell\|u^{(2)}(\ell\ell')\|n'\ell'\rangle = 1 \quad (16\text{-}17)$$

This gives

$$\langle(n\ell^{4\ell+2})S\|U^{(2)}(\ell\ell')\|(n\ell^{4\ell+1}n'\ell')D\rangle = \sqrt{2} \quad (16\text{-}18)$$

and the matrix element in (16-16) becomes

$$= -(-1)^L\sqrt{\frac{2(2L+1)}{5}}\langle\ell\|C^{(2)}\|\ell'\rangle \quad (16\text{-}19)$$

In the particular case of excitation $5p \to 6p$, from the closed shell $5p^6$ to the first excited one-electron state of p symmetry, the very excitation regarded by Judd et al.,[1] with $\ell = 1$ in (16-19), and including (16-15) and (16-14), we obtain

$$\mathcal{A} = \frac{27}{70}(2L+1) \quad (16\text{-}20)$$

To compute the total contribution from each value L, $\mathcal{A}$ is replaced by its sum over L

$$\mathcal{A} = \frac{27}{70}[3+5+7+9+11] = \frac{27}{2} \quad (16\text{-}21)$$

in agreement with the results of Judd et al.[1]

This means that for a $5p \to 6p$ excitation the ratio R has the value determined by the radial integrals and appropriate energies,

$$R = \frac{27}{2}\frac{\langle 5p|r^2|6p\rangle\langle 6p|r^{-3}|5p\rangle \times \Delta_2}{\langle 4f|r^2|4f\rangle\langle 4f|r^{-3}|4f\rangle \times E(\mathcal{A})} \quad (16\text{-}22)$$

where it is assumed that the energy spread of the excited states with respect to L is small compared to $E(\mathcal{A})$. The negative value of $\langle 5p|r^2|6p\rangle$ leads to a negative value for R. For example, the results of the Hartree-Fock numerical calculations performed for the average energy of $4f^6$ configuration of Eu^{3+} are the following

$$\langle 4f|r^2|4f\rangle = 0.84$$
$$\langle 4f|r^{-3}|4f\rangle = 8.09$$
$$\langle 5p|r^2|6p\rangle = -1.66$$
$$\langle 6p|r^{-3}|5p\rangle = 21.88$$

In these calculations the one-electron state of $6p$ symmetry was generated for the frozen core orbitals of the ground configuration in order to avoid the problem of overlap of the orbitals occupied in both configurations. The latter would appear if the Hartree-Fock model is applied to each configuration separately, since the solutions obtained for each Fock operator do not need to be orthogonal.

With these values of the radial integrals the relative importance of the closed shell excitation from the occupied $5p^6$, when compared to the impact due to the interactions between the excited energy levels of the ground configuration, determined by R in (16-22), is reflected by the value

$$R = -72.54 \frac{\Delta_2}{E(A)}$$

It is interesting to analyze the results of a theoretical model extended by the perturbing impact of the excitations from the $4f^N$ shell of the optically active electrons that are responsible for the unusual spectroscopic properties of the lanthanides. The assumptions, the tensor operator tools and approximations are the same as those introduced in chapter 13 in the case of the analysis of the magnetic nuclear hyperfine interactions.

16.2   EXCITATIONS FROM THE $4f^N$ SHELL

In order to analyze the energy corrections that originate from the perturbing influence of singly excited configurations $4f^{N-1}n'\ell'$ taken into account via various physical interactions, the approach is formulated in the language of the perturbation theory applied for the Hamiltonian defined in (13-24). The electric multipole hyperfine interactions remain of primary interest, but now, because of new selection rules, they are not limited to the quadrupole part. The latter part is the only one that contributes to the energy defined within the single configuration approximation. Therefore, in the present discussion the general definition of V_{EM} in (16-6) is used, and the ranks of the spherical tensors k are allowed to be even and odd.

The second-order corrections to the energy are determined by terms with two possible arrangements of the operators V_{EM} and V_{cryst}, similarly as in (13-28),

$$\Gamma_{EM}^2 = \lambda \left[\langle \Psi | V_{EM} | \Psi^1 \rangle + \langle \Psi^1 | V_{EM} | \Psi \rangle \right] \qquad (16\text{-}23)$$

where Ψ^1 is the first-order correction to the wave function. As a first choice, in addition to V_{EM}, the perturbing influence of the crystal field potential is regarded here. In this case the second-order corrections are determined by the following matrix elements

$$\Gamma_{EM}^\Psi(cryst) = \sum_{n',\chi} \sum_{tp}^{all} \frac{B_p^t}{(E_\Psi - E_\chi)}$$

$$\times \left[\langle 4f^N \Psi | V_{EM} | 4f^{N-1} n'\ell' \chi \rangle \langle 4f^{N-1} n'\ell' \chi | r^t C_p^{(t)} | 4f^N \Psi \rangle \right.$$

$$\left. + \langle 4f^N \Psi | r^t C_p^{(t)} | 4f^{N-1} n'\ell' \chi \rangle \langle 4f^{N-1} n'\ell' \chi | V_{EM} | 4f^N \Psi \rangle \right] \qquad (16\text{-}24)$$

where the crystal field potential is defined by the spherical tensors $C_p^{(t)}$ and structural (crystal field) parameters B_p^t, and V_{EM} is defined in (16-6).

To analyze the global impact of the chosen excitations and to estimate the importance of various physical mechanisms, we have to apply the same approximations as those in chapter 13. Also here the zero order problem is solved for the average energy of the ground configuration, and the excited one-electron radial basis sets are generated for the frozen core. When it is recalled that the excited configurations are rather distant in an energy scale from the ground configuration, their energy levels look indeed like relatively degenerate. Therefore the average configuration model of calculations is justified. As a result, instead of the expression (16-24), which because of the summations over the intermediate states is difficult to evaluate in practical calculations, we have a new effective operator,

$$
\Gamma_{EM}^{\Psi}(cryst) = 2\sum_{t,p}\sum_{k,q}^{p(t)=p(k)} B_p^t \sum_{\ell'}^{all} \langle f\|C^{(t)}\|\ell'\rangle\langle\ell'\|C^{(k)}\|f\rangle R_{EM}^{tk}(\ell')
$$
$$
\times \sum_{\lambda,\mu}^{even}(-1)^\mu \begin{pmatrix} t & k & \lambda \\ p & q & -\mu \end{pmatrix} \begin{Bmatrix} k & \lambda & t \\ f & \ell' & f \end{Bmatrix} \langle 4f^N\Psi \mid U_\mu^{(\lambda)} \mid 4f^N\Psi\rangle
$$

$$(16\text{-}25)$$

where the radial integrals, for all the members of complete radial basis set of excited one-electron functions of ℓ' symmetry, are collected as a radial term,

$$
R_{EM}^{tk}(\ell') = \sum_{n'}^{exc.} \frac{\langle 4f \mid r^t \mid n'\ell'\rangle\,\langle n'\ell' \mid r^{-k-1} \mid 4f\rangle}{\Delta(4f, n'\ell')}
$$

$$(16\text{-}26)$$

The numerical procedure for its evaluation within the perturbed function approach[3] is presented in chapter 17.

The limitations of the summations in (16-25) result from adding both contributions of (16-23). Thus, the second correction to the energy is determined by the effective operators with the even ranks λ. The tensorial structure of the effective operators is very similar to the second-order effective operators that determine the amplitude of the $f \longleftrightarrow f$ transitions within the Judd-Ofelt theory (see chapter 17).

If $k = 2$ in (16-25), the quadrupole interactions are included, and these are the terms that are also included in the case of the closed shell excitations taken as intermediate states (discussed by Judd et al[1]). In fact, these are the standard terms that contribute to the energy when it is defined within the single configuration approximation. The requirement of the same parity of k and t indicates that the interactions are via the even part of the crystal field potential, and the even rank crystal field parameters are included in the expression for the energy corrections. In addition, if $t = 2$, there are three ranks of unit tensor operator allowed, $\lambda = 0, 2, 4$, while for $t = 4$, $\lambda = 2, 4, 6$; in both cases the ranks of unit tensor operators are even.

For $k = 1$ in (16-25), the electric dipole hyperfine interactions are included, and these terms are new. Again the parity conditions limit the crystal field potential, but this time to the odd part, for odd values of t. This means that the odd rank crystal field parameters determine the correction to the energy. This is almost a *revolutionary* conclusion! Indeed, as mentioned before, and as discussed in the following chapters devoted to the spectroscopic properties of the lanthanides, the lack of a reliable model for evaluation of the odd rank crystal field parameters that determine the amplitude of electric dipole $f \longleftrightarrow f$ transitions makes their direct evaluation impossible. At the same time, since only the even part of crystal field potential contributes to the energy, it is impossible to determine the odd parameters via the fitting procedure that reproduces the energy. However, as seen from the second-order contributions defined in (16-25) this difficulty no longer exists, since when including the electric dipole hyperfine interactions at the second order, it is possible to write that in general

$$E \equiv E\left(B_p^t(t = even) + B_p^t(t = odd)\right) \tag{16-27}$$

The first part of the energy is defined in a standard way, and the second one is a new correction, which is determined by (16-25). These contributions are associated with the effective operators with the ranks $\lambda = 0, 2$ when $t = 1$, and $\lambda = 2, 4, 6$ for $t = 3$.

The second-order corrections defined in (16-25) contain the perturbing influence of the single excited configurations $4f^{N-1}n'\ell'$ for all n' of excited one-electron states of ℓ' symmetry. The latter are the orbitals d, g, for odd values of k and t, and p, f orbitals for k and t even, as predicted by the triangular conditions for the non-vanishing matrix elements of spherical tensor.

Continuing the search for the energy corrections to the state $^7F_0$ of the electron configuration $4f^6$, it is important to note that there are two non-vanishing terms associated with the unit tensor operator in (16-25) with the rank $\lambda = 0$. In general it means that there are non-zero second-order contributions that originate from the electric dipole and electric quadrupole hyperfine interactions assisted by the interactions via the crystal field potential.

In particular, for the axial component of the electric dipole hyperfine interactions (for $q = 0$), the correction to the energy of $^7F_0$ is determined by the following term

$$\gamma_{ED}(cryst)B_0^1 = -\frac{4}{\sqrt{7}} B_0^1 R_{EM}^{11}(d) |\langle f||C^{(1)}||d\rangle|^2$$
$$+ R_{EM}^{11}(g) |\langle f||C^{(1)}||g\rangle|^2 \tag{16-28}$$

where the first contribution is due to the excitations $4f^{N-1}n'd$, and the second one includes the impact of $4f^{N-1}n'g$.

For the quadrupole interaction, the corrections are due to the excitations $4f^{N-1}n'p$ and $4f^{N-1}n'f$, and we have

$$\gamma_{EQ}(cryst)B_0^2 = \frac{12}{5\sqrt{7}} B_0^2 R_{EM}^{22}(f) |\langle f||C^{(1)}||f\rangle|^2$$
$$+ R_{EM}^{11}(p) |\langle f||C^{(1)}||p\rangle|^2 \tag{16-29}$$

Similarly as in the case of R in (16-22), it is now possible to evaluate the ratio of the new terms to the contributions that arise from the interactions between the $^7F_0$

and $^7F_2$ of the $4f^6$ configuration (the intra-shell interactions). Namely, we have now

$$R' = \frac{\gamma_{EQ}(cryst) \times \Delta_2}{\langle 4f|r^2|4f\rangle\langle 4f|r^{-3}|4f\rangle} \qquad (16\text{-}30)$$

where the crystal field parameter B_0^2 is canceled by the same in the denominator. Furthermore, the energy difference between the ground and excited configurations (see the definition of R in (16-22)) is included here in the radial term defined in (16-26), and Δ_2 is defined in (16-5b). With the values of the radial terms

$$R_{EM}^{22}(f) = 1.93 \qquad\qquad R_{EM}^{22}(p) = -0.0006$$

evaluated for the complete radial basis sets of f and p symmetry by means of the perturbed function approach[3] (discussed in chapter 17), the ratio R' has the value of $0.26\Delta_2$. As seen from the values of the radial terms, the major part of the contributions to $\gamma_{EQ}(cryst)$ is caused by excitations to the one-electron states of f symmetry, while those to the orbitals of p symmetry are relatively negligible.

In addition to the quadrupole term, there is a new correction, which is determined by (16-28), and which originates from the electric dipole hyperfine interactions; this is the case of the energy contributions that are determined by the odd crystal field parameters, as discussed above. In particular, we have

$$E_{ED}^2(cryst) = B_0^1\gamma_{ED}(cryst) \qquad (16\text{-}31)$$

Taking into account possible excitations from the $4f^N$ shell to the states of d and g symmetry, we have the following expression for the energy correction

$$E_{ED}^2(cryst) = -\frac{4}{\sqrt{7}}B_0^1\left(3R_{EM}^{11}(d) + 4R_{EM}^{11}(g)\right)$$

Since this is the only contribution to the energy, which is associated with the rank $k = 1$ of the expansion of the electric multipole hyperfine interactions, it is impossible to determine its relative importance. In particular the odd crystal field potential for $t = 1$ does not contribute to the energy of the interactions between the various excited states of the ground configuration, and therefore it is impossible to evaluate the analog of R and R', analyzed in the previous cases.

The perturbing influence of singly excited configurations in which an electron is promoted from the $4f$ shell to an excited one-electron state may be taken into account via different physical mechanisms, as for example, the spin-orbit interaction. This is exactly the same situation as in the case of the magnetic nuclear hyperfine interactions discussed in chapter 13. Also here the possible excited configurations as intermediate states in the product of the matrix elements contributing to the energy at the second order are limited to those of the same parity. In the general expression, which is analogous to (16-24), but includes V_{so} instead of V_{cryst}, we have

$$
\Gamma_{EM}^{\Psi}(so) = \sum_{n',\chi} \frac{1}{(E_{\Psi} - E_{\chi})}
$$

$$
\times \left[\langle 4f^N \Psi | V_{EM} | 4f^{N-1} n' f \chi \rangle \langle 4f^{N-1} n' f \chi | V_{so} | 4f^N \Psi \rangle \right.
$$

$$
\left. + \langle 4f^N \Psi | V_{so} | 4f^{N-1} n' f \chi \rangle \langle 4f^{N-1} n' \chi | V_{EM} | 4f^N \Psi \rangle \right]
\qquad (16\text{-}32)
$$

Since the spin-orbit interaction is expressed by the double tensor operators acting within the spin-orbital space, the spherical tensors of V_{EM} are also converted into such objects, and the rules of coupling and contraction of the double tensor operators in (13-43) have to be used. In the case of these inter-shell interactions via the V_{so}, instead of the spin-orbit constant ζ introduced for the intra-shell interactions (within the $4f^N$ configuration, for example), here the radial integral of r^{-3}, with the assumption of the spherically symmetric potential of the zero order problem, is treated in an explicit way.

The second-order energy corrections that are due to V_{EM} combined with the spin-orbit interactions are determined by the effective operators of the following structure

$$
\Gamma_{EM}^{\Psi}(so) = -2\sqrt{6} \sum_k R_{EM}^{-3k}(f) \langle f \| C^{(k)} \| f \rangle
$$

$$
\times \sum_{\lambda,\mu} \sum_x [\lambda]^{\frac{1}{2}} \frac{[x]}{[k]} \left\{ \begin{array}{ccc} k & x & 1 \\ f & f & f \end{array} \right\} \langle 4f^N \Psi | W_{\mu}^{(1x)\lambda} | 4f^N \Psi \rangle
$$

$$
(16\text{-}33)
$$

where the radial term contains the integrals together with the energy denominator, and it is defined by (16-26) with appropriate powers of the radial arguments (t in (16-26) has the value of -3 in the present application). From the reduced matrix element of the spherical tensor with rank k, which is present in (16-33), it follows that $k = even$. This means that these particular second-order energy corrections are caused by the electric quadrupole interactions, and therefore $k = 2$.

The double tensor operators of (16-33) for various ranks determine the energy, and therefore in each case the operators has to be hermitian. This condition requires that the parity of x is opposite to the parity of λ. As a consequence we have the following distinct terms:

$$
\begin{array}{ll}
x = 1 & \lambda = 0, 2 \\
x = 2 & \lambda = 1, 3 \\
x = 3 & \lambda = 2, 4
\end{array}
$$

For $\lambda = 0$, which possibly determines the correction to the energy of $^7F_0$, there is a term of the form

$$
\gamma_{EM}^{^7F_0}(so) = -\frac{2}{5}\sqrt{6} R_{EM}^{-32}(f) \langle f \| C^{(2)} \| f \rangle \left\{ \begin{array}{ccc} 2 & 1 & 1 \\ 3 & 3 & 3 \end{array} \right\} \langle 4f^6\ {}^7F_0 | W_0^{(11)0} | 4f^6\ {}^7F_0 \rangle
$$

$$
= 0.52 R_{EM}^{-32}(f) \langle 4f^6\ {}^7F_0 | W_0^{(11)0} | 4f^6\ {}^7F_0 \rangle
\qquad (16\text{-}34)
$$

In the direct calculations of the first order energy, which is due to the spin orbit interaction within the ground configuration, there is a contribution associated with the double tensor operator $W^{(11)0}$. Thus, in summary, taking into account the spin-orbit interaction up to the second order, and including its interplay with V_{EQ}, we have

$$\left(1^{st} order + \gamma_{EM}^{^7F_0}(so)\right) = \left[3\sqrt{2}\zeta_{4f} + 0.52 R_{EM}^{-32}(f)\right] \langle 4f^6 \, ^7F_0 \mid W_0^{(11)0} \mid 4f^6 \, ^7F_0 \rangle \tag{16-35}$$

The remaining term of this kind, which also is of the second-order and is associated with the double tensor operator with the same ranks in the spin and orbital spaces, but for $\lambda = 2$ (because of which it does not contribute to the energy of $^7F_0$), has the form

$$2.87 R_{EM}^{-32}(f) W_\mu^{(11)2}.$$

In order to analyze the combined effect of the inter-shell interactions taken into account via V_{so}, V_{cryst} and V_{EM}, the third-order terms determined by the triple products of the matrix elements of these operators have to be constructed. For example, the third-order contribution has the following form

$$\langle 4f^N \Psi | V_{so} | Aa \rangle \langle Aa | V_{cryst} | Bb \rangle \langle Bb | V_{EM} | 4f^N \Psi \rangle \tag{16-36}$$

where the energy denominators are omitted for the simplicity, and the excited configurations, together with their energy states, are represented in a symbolic way. Inspection of this general expression leads to the conclusion that at the third-order analysis one may expect an explosion of various terms. Only for this particular sequence of the operators do we have several possibilities of excited configurations. Namely, $A \equiv 4f^{N-1}n'f$, because of the structure of the spin-orbit interaction operator; $B \equiv 4f^{N-1}n'd$ and $4f^{N-1}n'g$ if V_{EM} contains the dipole interactions ($k = 1$) or $4f^{N-1}n'p$ and $4f^{N-1}n'f$ for the quadrupole part of V_{EM} (for $k = 2$). As a consequence, when the dipole interactions are taken into account, then the odd parts of the crystal field potential are involved, and for the quadrupole interactions, the even part of V_{cryst}.

In order to complete the analysis of important third-order corrections to the energy, all possible sequences of the operators in (16-36) have to be investigated. The situation is again the same as in the case of the previous discussion devoted to the hyperfine interactions. Indeed, the analysis of the tensorial structure of various effective operators provides only new selection rules for the non-vanishing contributions. In order to establish the importance of the distinct mechanism or a particular term, numerical calculations have to be performed. However, the clarity of our discussion would suffer, if the results of particular calculations would be presented here. Therefore only numerical examples are used to illustrate the points that are crucial for the detailed analysis.

16.3 EXERCISES

16-1. Verify that (16-14) follows from (16-11).

16-2. Show that

$$
\begin{aligned}
\left| \langle 4f^6 \ F(\ldots)S; \ F \| C^{(2)} \| 4f^6 \ F(\ldots)S; \ F \rangle \right|^2 \\
= \left| \langle 3 \| C^{(2)} \| 3 \rangle \langle 4f^6 \ ^7 F \| U^{(2)} \| 4f^6 \ ^7 F \rangle \right|^2 \\
= \langle 3 \| C^{(2)} \| 3 \rangle^2 \\
= \frac{28}{15}
\end{aligned}
\tag{16-37}
$$

REFERENCES

1. Judd B R, Lovejoy C A and Shirley D A, (1962) Anomalous Quadrupole Coupling in Europium Ethylsulfate *Phys. Rev.,* **128** 1733.
2. Elliott R J, (1957) Theory of Nuclear Magnetic Resonance in Eu^{3+} *Proc. Phys. Soc. (London)* **B70** 119.
3. Jankowski K, Smentek-Mielczarek L and Sokołowski A, (1986) Electron-correlation Third-order Contributions to the Electric Dipole Transition Amplitudes of Rare Earth Ions in Crystals I. Perturbed-function Approach *Mol. Phys.* **59** 1165.

17 Electric Dipole $f \longleftrightarrow f$ Transitions

> *If the experimental physicist has already done a great deal of work in this field, nevertheless the theoretical physicist has still hardly begun to evaluate the experimental material which may lead him to conclusions about the structure of the atom.*
>
> Johannes Stark

17.1 JUDD-OFELT THEORY OF $f \longleftrightarrow f$ INTENSITIES

The standard theory of $f \longleftrightarrow f$ electric dipole transitions, the so-called Judd-Ofelt theory[1,2] of intensities, forms a landmark in the theory of spectroscopic properties of lanthanides in crystals. The approach introduced in 1962 to describe the one-photon processes is based on the single configuration approximation adopted within the free ionic system approximation. Due to the parity selection rules, electric dipole transitions between the energy levels of the configuration of equivalent electrons are regarded as forced in origin. As a consequence of the lanthanide contraction the optically active $4f$ electrons are screened by closed shells of $5s$ and $5p$ symmetry from the perturbing influence of the environment of the ion in the crystal. Therefore, to break the selection rules and lower the spherical symmetry of the isolated ion, the perturbing influence of the crystal field potential produced by the surrounding ligands is taken into account. The standard Judd-Ofelt theory is based on the second-order perturbation theory applied for the Hamiltonian that describes the electronic structure of the lanthanide ion in the crystal (see chapter 7),

$$H = H_0 + \lambda(P V_{cryst} Q + Q V_{cryst} P) \qquad (17\text{-}1)$$

where the projection operators P and Q are the same as defined in (7-23). The perturbing operator in (17-1) represents the inter-shell interactions via the crystal field potential. This operator is responsible for mixing of the wave functions of $4f^N$ configuration with such components from the Q space (spanned by the states of excited configurations) which are of opposite parity. In this sense the crystal field potential is regarded as a forcing mechanism of one-photon electric dipole $f \longleftrightarrow f$ transitions. H_0 in (17-1) defines the zeroth order eigenvalue problem, which we expect to be able to solve in an exact way. This means that the zero order solutions Ψ_i^0 and E_i^0 of $4f^N$ are known. In practical realizations H_0 is usually a Hamiltonian of the central field approximation and describes the electronic structure of the $f-$ electron ion within the Hartree-Fock model. There are two remaining fragments of interactions via the crystal field potential in addition to those included in (17-1). Namely, $P V_{cryst} P$, that represents the interactions within the $4f^N$ configuration and contributes to the energy. This operator is included within H_0, if one evaluates the energy of a free ion modified by the impact due to the crystal environment. Finally, the part $QV_{cryst}Q$ describes the interactions within the excited configurations, and it does not contribute to the energy, if the approach is defined within the single configuration approximation. This part of the perturbing operator might modify the transition amplitude at higher orders of perturbation analysis than the second, which defines the Judd-Ofelt theory. Thus, as seen from this introduction, a formal partitioning of the space into two subspaces P and Q allows one to count and separate various interactions in order not to include them twice. For example if the zeroth order problem is solved via diagonalization of the whole Hamiltonian H_0 that contains $PV_{cryst}P$, the wave function obtained in this manner is appropriate for evaluation of the transition amplitude within the Judd-Ofelt theory, since in both cases two different parts of interactions are taken into account.

Following the standard procedure of Rayleigh-Schrödinger perturbation theory, the wave function defined up to the first order in perturbation V_{cryst} has the form

$$\Psi_i = \Psi_i^0 + \lambda \sum_{k \neq i} \frac{\langle \Psi_k^0 \mid QV_{cryst}P \mid \Psi_i^0 \rangle}{E_i^0 - E_k^0} \Psi_k^0 \qquad (17\text{-}2)$$

Since in (17-2) Ψ_k^0 belongs to Q and Ψ_i^0 belongs to P, and $PQ = 0$, the summation is over the states of excited configurations, and therefore the energy denominators are always non-zero (as a consequence the restriction on the summation may be removed).

The first non-vanishing terms contributing to the transition amplitude Γ are the terms of second order in the standard nomenclature of perturbation theory, namely

$$\Gamma = \lambda \{ \langle \Psi_f^0 \mid D_\rho^{(1)} \mid \Psi_i^1 \rangle + \langle \Psi_f^1 \mid D_\rho^{(1)} \mid \Psi_i^0 \rangle \} + \theta(\lambda^m), \quad m \geq 2 \qquad (17\text{-}3)$$

where the electric dipole transition is represented by a tensor operator

$$D_\rho^{(1)} = \sum_i^N r_i C_\rho^{(1)}(\vartheta_i, \phi_i) \qquad (17\text{-}4)$$

These matrix elements do not vanish only if the first order corrections to the wave functions represent the energy states of excited configurations of the opposite parity than the parity of $4f^N$. This means that the set of excited configurations in (17-2) (associated with Q) is limited to those of opposite parity,

$$X \equiv 4f^{N-1}n'\ell', \text{for } \ell' = even(\equiv d, g, \ldots) \qquad (17\text{-}5)$$

where due to the one-particle character of all operators in (17-3), only the perturbing influence of single excitations from the open $4f$ shell is taken into account. From a theoretical point of view, however, it is possible to regard also the excitations from the closed shells, but their influence upon the transition amplitude has been found to be numerically negligible [3]. In general, the second-order terms contributing to the transition amplitude have the form of double products of matrix elements that differ by the order of appropriate operators,

$$\Gamma = \sum_{Xx} \left\{ \langle \Psi_f^0 \mid D_\rho^{(1)} \mid Xx \rangle \langle Xx \mid QV_{CF}P \mid \Psi_i^0 \rangle / (E_i^0 - E_{Xx}^0) \right.$$
$$\left. + \langle \Psi_f^0 \mid PV_{CF}Q \mid Xx \rangle \langle Xx \mid D_\rho^{(1)} \mid \Psi_i^0 \rangle / (E_f^0 - E_{Xx}^0) \right\} \qquad (17\text{-}6)$$

where x denotes the states of the singly excited configuration X. The transition amplitude defined in (17-6) is suitable neither for numerical calculations nor for analysis of the selection rules for a given transition. The problem here is the same as discussed in chapter 7 and again is caused by a presence of energy denominators that prevent one to perform the closure procedure over the complete set of states of a given excited configuration. In order to derive the effective operator that reproduces the inter-shell interactions via the crystal field potential acting just within the $4f^N$ configuration, as mentioned in chapter 7, some additional approximations and assumptions have to be introduced. Since the same procedure is used in the following chapters when discussing higher order contributions to the transition amplitudes, here its key steps are outlined.

A great simplification of the general expression is obtained if the zero order problem is solved within the Hartree-Fock model for an average energy of a configuration (see the discussion in chapter 21). From a physical point of view it means that the states x of excited configurations X are seen as degenerate, since the actual energy differences between distinct states x are almost the same in relation to the energy difference between very distant excited configurations X and the ground configuration $4f^N$. The same assumption is made in the case of energy differences that involve various states of the ground configuration. In addition, when for particle one-electron states that are used for the construction of the states describing the excited configurations the zero order problem is solved within the frozen core orbital approach, the energy denominators of all expressions are reduced simply to a difference of the orbital energies of the one-electron states of appropriate symmetry. As a consequence the energy denominators are independent of all quantum numbers that identify the states, except the principal quantum number and orbital angular momentum of one-electron occupied states in bra and ket. The frozen core orbitals scheme not only simplifies the expression for the transition amplitude, but most of all it makes the calculations possible in practice. For example there are no overlap integrals in expression of (17-6), since all one-electron functions that build many electron functions, as solutions obtained for the same hamiltonian, are orthogonal to each other, if they belong to various eigenvalues.

Summarizing this analysis, both energy denominators in (17-6) have the same value, independent of the particular states of $4f^N$ and X, and therefore it is possible to perform the partial closure over the intermediate states Xx. It is convenient at this point to introduce the inter-shell unit tensor operators that are defined in an analogous way as those in (7-29), with the difference that their reduced matrix element here is equal to unity, namely

$$\langle n\ell \,||\, u^{(k)}(n_1\ell_1, n_2\ell_2) \,||\, n'\ell' \rangle = \delta(n\ell, n_1\ell_1)\delta(n_2\ell_2, n'\ell') \qquad (17\text{-}7)$$

Obviously, apart from this minor difference in the definition, both objects, v from (7-29) and u defined above, are interpreted as coupled products of annihilation and creation operators of one-electron states. In the terms of these unit tensor operators the perturbing operator has the following form

$$QV_{cryst}P \equiv \sum_{t,p} B_p^t \sum_j \langle n'\ell' \,|\, r_j^t \,|\, 4f \rangle \langle \ell' \,||\, C^{(t)} \,||\, f \rangle u_{j,p}^{(t)}(n'\ell', 4f) \qquad (17\text{-}8)$$

where the basic definition of the crystal field potential is taken from (9-2). The electric dipole transition operator is expressed as follows,

$$D_\rho^{(1)} \rightarrow P D_\rho^{(1)} Q \equiv \sum_i \langle 4f \mid r_i \mid n'\ell' \rangle \langle f \parallel C^{(1)} \parallel \ell' \rangle u_{i,\rho}^{(1)}(4f, n'\ell') \qquad (17\text{-}9)$$

The transition amplitude defined at the second order is determined by the product of the following operators, that are coupled together to an object with the rank k,

$$u_{i,\rho}^{(1)}(4f, n'\ell') u_{j,p}^{(t)}(n'\ell', 4f) = \sum_{k,q} (-1)^{1-t-q} [k]^{1/2} \begin{pmatrix} 1 & t & k \\ \rho & p & -q \end{pmatrix}$$
$$\times \left[u_i^{(1)}(4f, n'\ell') \times u_j^{(t)}(n'\ell', 4f) \right]_q^{(k)}. \qquad (17\text{-}10)$$

where $[k] = 2k + 1$, and the original summation over Xx is reduced now to the summation over $n'\ell'$ that defines the excited configurations specifying to which one-electron state an electron is promoted from the $4f$ shell. Using the commutation relation presented in (7-30) and the fact that the coordinates on which both operators act must be the same (otherwise the matrix element vanishes due to the orthogonality of one-electron states), the tensorial product in (17-10) is replaced by a single intra-shell unit tensor operator, namely

$$\left[u_i^{(1)}(4f, n'\ell') \times u_i^{(t)}(n'\ell', 4f) \right]_q^{(k)}$$
$$= (-1)^k [k]^{1/2} \begin{Bmatrix} t & k & 1 \\ f & \ell' & f \end{Bmatrix} u_{i,q}^{(k)}(4f, 4f) \qquad (17\text{-}11)$$

In this way, the transition amplitude defined up to the second order within the standard Judd-Ofelt theory is determined by the matrix element of one particle effective operators

$$\Gamma_{J-O} = 2 \sum_{t,p}^{odd} B_p^t \sum_{k,q}^{even} \sum_{\ell'}^{even} (-1)^q [k]^{1/2} \begin{pmatrix} t & 1 & k \\ p & \rho & -q \end{pmatrix} A_t^k(\ell') R_{JO}^t(\ell')$$
$$\times \langle 4f^N \Psi_f^0 \mid U_q^{(k)} \mid 4f^N \Psi_i^0 \rangle \qquad (17\text{-}12)$$

where the angular term is defined as

$$A_t^k(\ell') = [k]^{1/2} \begin{Bmatrix} t & k & 1 \\ f & \ell' & f \end{Bmatrix} \langle f \parallel C^{(1)} \parallel \ell' \rangle \langle \ell' \parallel C^{(t)} \parallel f \rangle \qquad (17\text{-}13)$$

All the terms that depend on the principal quantum number are collected in the radial term that consists of the product of two radial integrals and the appropriate energy denominator

$$R_{JO}^t(\ell') = \sum_{n'}^{exc.} \frac{\langle 4f \mid r \mid n'\ell' \rangle \langle n'\ell' \mid r^t \mid 4f \rangle}{\Delta(4f, n'\ell')} \qquad (17\text{-}14)$$

The factor 2 in (17-12) results from the fact that two terms in the perturbing expression (17-6) have the same value (the same energy denominators within the adopted approximation) that differ from each other by a sign $(-1)^k$, where k is the order of the effective operator. This condition is also a source of the limitation on the rank of the effective operators, and in the present formulation, only those with even ranks contribute to the transition amplitude defined by the Judd-Ofelt theory. This means that in the case of the $4f^N$ configuration of lanthanides there are only three effective operators with $k = 2, 4$ and 6 that determine all the transitions observed in a given system.

The small number of contributing terms is the greatest success of the Judd-Ofelt theory due to which it became a common tool used for the interpretation of lanthanide spectra. The simple expression for the transition amplitude and its usefulness in practical calculations compensates for all critical comments in the literature on the basic assumptions about the energy differences between various configurations and their relative degeneracy. As demonstrated above (and discussed also in chapter 21) without these approximations it would be impossible to perform practical calculations.

All the terms that depend on the principal quantum number of excited one-electron states are collected into a radial term defined in (17-14). Note that in our approach the energy denominator is simply a difference of the orbital energies of one-electron states. The summation is over all excited states of ℓ' symmetry, which means that all components of a discrete as well as the continuum part of the spectra have to be taken into account; the summation has to be performed over the complete radial basis set of one-electron states $n'\ell'$. This troublesome summation is the reason that in practice only a few discrete components of a set are taken into account. Due to the parity selection rules, ℓ' is limited to d and g symmetry. It is common practice to include the excitations just to $5d$ and $5g$, disregarding the remaining one-electron states for higher values of n'. It is possible to solve this problem in a rather straightforward way by applying the so called *perturbed function approach*. The concept of the perturbed functions was developed in the Many Body Perturbation Theory by I. Lingren [4], and applied to the lanthanides by J. Morrison [5,6]. Here the concept is used to evaluate the radial integrals in an exact way, including the impact caused by all possible excited states of a given symmetry. In this approach the difficulty of performing the summation over the complete radial basis set is replaced by the task of solving a hydrogen-like differential equation. The details of this procedure are presented in the proceeding subsection.

For a direct application of (17-12) one has to know the radial one-electron functions to evaluate the appropriate radial integrals. While the angular parts of the effective operators determining the transition amplitude are the same for all lanthanide ions (except the crystal field parameters, of course), the radial integrals carry the specific features of the electronic structure of each ion individually. In the *ab initio* -type calculations those integrals have to be evaluated in an exact way, within the adopted model (for a given choice of H_0). It is possible however to treat all these terms as parameters and adjust their values to reproduce experiment. Thus, introducing the intensity as a square of the transition amplitude defined within the Judd-Ofelt theory we have

$$\mathcal{I}_{f \leftarrow i} = \sum_{\lambda}^{2,4,6} \Omega_\lambda |\langle \Psi_f | U^{(\lambda)} | \Psi_i \rangle|^2 \qquad (17\text{-}15)$$

where Ω_λ, for $\lambda = 2, 4, 6$ are the so-called Judd-Ofelt parameters. From (17-15) it is seen again that the complete spectrum for a given system is defined in the terms of at most three intensity parameters. The actual number of parameters for a particular transition is determined by the selection rules for the non-vanishing matrix elements of unit tensor operators U.

When this one-particle parametrization scheme of $f-$spectra (the effective operators are one particle objects) expressed in the terms of the intensity parameters Ω_λ is introduced, the problem of evaluation of the radial integrals is removed. In addition, in such a fitting procedure another problem of direct calculations is excluded, which makes any numerical analysis not only simpler but possible. Namely, due to the parity requirements the summation in (17-12) is over the odd part of the crystal field potential that involves the odd rank crystal field parameters. Unfortunately there is no theoretical model for the direct evaluation of these structural parameters. The only reliable source of the values of the crystal field parameters is to perform semi-empirical calculations for the energy trying to reproduce the observed energy levels by diagonalizing the whole matrix of elements of the Hamiltonian. In the case of the model that is based on a single configuration approximation (this is the model we apply here choosing the Hartree Fock method to describe the zeroth order problem) it is possible to determine the values of even rank crystal field parameters, since the odd rank parameters do not contribute to the energy. Thus, the semi-empirical version of the Judd-Ofelt theory, based on (17-15), gives the only possibility of interpreting various experimental data reported over the years. Indeed, for more than forty years Judd-Ofelt theory in its semi-empirical version has served as the best, and in fact the only reliable tool for interpretation of the f-spectra of ions in various materials.

There are however electric dipole transitions observed that are forbidden by the selection rules of the Judd-Ofelt theory. The Wigner-Eckart theorem applied for the matrix element in (17-15) limits the transitions to the following ones,

$$|J - J'| \leq \lambda \leq J + J'$$

where J and J' denote the initial and final states between which the transition is observed. In particular, the one-photon electric dipole transitions $^7F_0 \longrightarrow {}^5D_0$ and $^7F_0 \longrightarrow {}^5D_1$ observed in Eu^{3+} in various hosts are not described by the standard Judd-Ofelt theory. Thus there is a demand for a theoretical model in which among the even rank effective operators also those with odd ranks are included. These particular transitions are a special challenge for all those involved in theoretical investigations devoted to the spectroscopy of the lanthanides. The problem of their description is addressed in the proceeding sections in various contexts, within extensions and modifications of the standard approach and including the relativistic theory of $f \longleftrightarrow f$ transitions.

While the Judd-Ofelt theory is successful as a parametrization scheme used in the fitting procedure, it fails when used in *ab initio*- type calculations. This means that its physical model is not rich enough to reproduce the subtleties of the electronic structure of f-electron systems. Therefore there is a search for various physical mechanisms which when taken into account might improve the description of $f \longleftrightarrow f$ transitions. This means, the question is which interactions, in addition to the influence of the crystal field potential taken into account in the derivation of the standard model of

(17-15), should be included. It is also important to extend the second-order analysis by new terms of higher orders, and to establish their relative importance.

17.2 DOUBLE PERTURBATION THEORY

In order to extend the physical model of the standard second-order approach of the $f \longleftrightarrow f$ transitions, double perturbation theory has to be applied[7]. Indeed, in addition to the perturbing influence of the crystal field potential, which is admixing the opposite parity components to the wave functions of the $4f^N$ configuration, a new physical model has to be introduced as a second perturbation. Thus, the Hamiltonian in (17-1) has to be extended by a second perturbing operator V,

$$H = H_0 + \lambda(P V_{cryst} Q + Q V_{cryst} P) + \mu(P V Q + Q V P) \qquad (17\text{-}16)$$

where there are two perturbing parameters and, as discussed above, only the inter-shell interactions via the perturbing operators are taken into account. V in this definition represents possibly an important physical mechanism that should be taken into account to improve the theory. This is a crucial step if model is used for direct *ab initio* calculations. It also provides the information about the possible importance of the chosen V and its impact on the intensity parameters evaluated from the fitting procedure. The first candidate for this investigation is an operator responsible for electron correlation effects between the $4f$ electrons of the lanthanide configuration. In this way, the limitations of the single configuration approximation of the standard Judd-Ofelt theory are broken, and configuration interaction is taken into account in a perturbative way via the non-central part of the Coulomb interaction taken for V.

In general, the eigenvalues and eigenfunctions are expanded in a double power series in both perturbing parameters,

$$\Psi_i = \sum_{m=0} \sum_{n=0} \lambda^m \mu^n \Psi_i^{(nm)}, \qquad (17\text{-}17)$$

$$E_i = \sum_{m=0} \sum_{n=0} \lambda^m \mu^n E_i^{(nm)}, \qquad (17\text{-}18)$$

where $E_i^{(00)} \equiv E_i^0$ and $\Psi_i^{00} \equiv \Psi_i^0$.

Similarly as in the case of the derivation of the Judd-Ofelt theory, the perturbing expansion of the wave functions is limited to the terms of the first order in both perturbations,

$$\Psi_k = \Psi_k^0 + \lambda \Psi_k^{(10)} + \mu \Psi^{(01)} + \theta(\lambda^n \mu^m), \qquad n + m \geq 2 \qquad (17\text{-}19)$$

$\Psi_k^{(10)}$ is the correction caused by the crystal field potential, and when used to evaluate the transition amplitude gives the Judd-Ofelt contributions defined in (17-6); this function has to be of opposite parity to Ψ_k^0, and it has the same form as before,

$$\Psi_i^{(10)} = \sum_{k \neq i}' \frac{\langle \Psi_k^0 \mid Q V_{cryst} P \mid \Psi_i^0 \rangle}{E_i^0 - E_k^0} \Psi_k^0 \qquad (17\text{-}20)$$

The first order correction $\Psi_k^{(01)}$ is caused by the second perturbation. If V represents the electron correlation effects or spin-orbit interaction, since both operators are scalar objects, this correction to the wave function has to be of the same parity as the parity of the states of $4f^N$. In general this function is defined in an analogous way to (17-20),

$$\Psi_i^{(01)} = \sum_{k \neq i}' \frac{\langle \Psi_k^0 \mid QVP \mid \Psi_i^0 \rangle}{E_i^0 - E_k^0} \Psi_k^0 \qquad (17\text{-}21)$$

The transition amplitude is now determined by the following contributions,

$$\Gamma = \lambda \{ \langle \Psi_f^0 \mid D_\rho^{(1)} \mid \Psi_i^{10} \rangle + \langle \Psi_f^{10} \mid D_\rho^{(1)} \mid \Psi_i^0 \rangle \}$$
$$+ \mu \{ \langle \Psi_f^0 \mid D_\rho^{(1)} \mid \Psi_i^{01} \rangle + \langle \Psi_f^{01} \mid D_\rho^{(1)} \mid \Psi_i^0 \rangle \}$$
$$+ \lambda \mu \{ \langle \Psi_f^{10} \mid D_\rho^{(1)} \mid \Psi_i^{01} \rangle + \langle \Psi_f^{01} \mid D_\rho^{(1)} \mid \Psi_i^{10} \rangle \} \qquad (17\text{-}22)$$

The terms associated with λ define the second-order contributions to the transition amplitude, which lead to the Judd-Ofelt theory, and their effective operator form is presented in (17-12). The terms proportional to the perturbing parameter μ vanish due to the parity requirements. The most interesting terms in (17-22) are those proportional to $\lambda \mu$. These terms are new contributions of the third order, and they represent the interplay between both mechanisms, crystal field potential and additional perturbation V. The third-order terms that are associated with λ^2 and μ^2, and which are evaluated with the second-order corrections to the wave functions Ψ^{20} and Ψ^{02}, are not included in (17-22). The former kind of contributions contain the perturbing influence of the odd (the same as in the Judd-Ofelt model) and also even parts of the crystal field potential; therefore these terms do not provide new insight into the physical nature of the $f \longleftrightarrow f$ electric dipole transitions. The terms of the latter kind, proportional to μ^2, due to the parity requirements, simply vanish.

    If the perturbing operators in the definition of the Hamiltonian in (17-16) are extended by the intra-shell interactions, $QV_{cryst}Q$ and QVQ, then there are additional, non-zero contributions of third order. They are expressed by the matrix elements with the function Ψ^{11}, which is the first order correction in both perturbations simultaneously,

$$\lambda \mu \left(\langle \Psi_f^0 \mid D^{(1)} \mid \Psi_i^{(11)} \rangle + \langle \Psi_f^{(11)} \mid D^{(1)} \mid \Psi_i^0 \rangle \right) \qquad (17\text{-}23)$$

where Ψ^{11} connects both physical mechanisms taken into account as perturbations through the following equation

$$\left(H_0 - E_i^0 \right) \Psi_i^{(11)} + V_{cryst} \Psi_i^{(01)} + V \Psi_i^{(10)} = 0 \qquad (17\text{-}24)$$

    When the perturbing influence of electron correlation effects upon the transition amplitude is taken into account, the third-order contributions that represent the inter-shell interactions via both perturbing operators have the following general form

$^3\Gamma_{corr}(inter)$

$$= \sum_{Xx} \sum_{Bb} \left\{ \frac{\langle \Psi_f^0 \mid PV_{corr}Q \mid Bb \rangle \langle Bb \mid D_\rho^{(1)} \mid Xx \rangle \langle Xx \mid QV_{cryst}P \mid \Psi_i^0 \rangle}{\left(E_i^0 - E_{Xx}^0\right)\left(E_f^0 - E_{Bb}^0\right)} \right.$$

$$\left. + \frac{\langle \Psi_f^0 \mid PV_{cryst}Q \mid Xx \rangle \langle Xx \mid D_\rho^{(1)} \mid Bb \rangle \langle Bb \mid QV_{corr}P \mid \Psi_i^0 \rangle}{\left(E_f^0 - E_{Xx}^0\right)\left(E_i^0 - E_{Bb}^0\right)} \right\} \qquad (17\text{-}25)$$

where V_{corr} is the non-central part of the Coulomb interaction

$$V_{corr} = \sum_{i<j} \frac{1}{r_{ij}} - u_{HF} \qquad (17\text{-}26)$$

and $|Bb\rangle$ and $|Xx\rangle$ are the states of the excited configurations of the same and opposite parities to the parity of $4f^N$ configuration, respectively. As a consequence of the two particle character of the Coulomb interaction in (17-26), the set of excited configurations taken into account in the Judd-Ofelt theory is extended at the third order by all the doubly excited configurations of the same parity as parity of $4f^N$.

The impact due to the interactions via V_{corr} and V_{cryst} within the Q space of excited configurations are represented by the third-order contributions

$^3\Gamma_{corr}(intra)$

$$= \sum_{Xx} \sum_{Yy} \left\{ \frac{\langle 4f^N \Psi_f^0 | D^{(1)} | Yy \rangle \langle Yy | QV_{corr}Q | Xx \rangle \langle Xx | V_{cryst} | 4f^N \Psi_i^0 \rangle}{\left(E_i^0 - E_{Yy}^0\right)\left(E_i^0 - E_{Xx}^0\right)} \right.$$

$$\left. + \frac{\langle 4f^N \Psi_f^0 | V_{cryst} | Xx \rangle \langle Xx | QV_{corr}Q | Yy \rangle \langle Yy | D^{(1)} | 4f^N \Psi_i^0 \rangle}{\left(E_f^0 - E_{Yy}^0\right)\left(E_f^0 - E_{Xx}^0\right)} \right\}$$

$$+ \sum_{Bb} \sum_{Yy} \left\{ \frac{\langle 4f^N \Psi_f^0 | D^{(1)} | Yy \rangle \langle Yy | QV_{cryst}Q | Bb \rangle \langle Bb | V_{corr} | 4f^N \Psi_i^0 \rangle}{\left(E_i^0 - E_{Yy}^0\right)\left(E_i^0 - E_{Bb}^0\right)} \right.$$

$$\left. + \frac{\langle 4f^N \Psi_f^0 | V_{corr} | Bb \rangle \langle Bb | QV_{cryst}Q | Yy \rangle \langle Yy | D^{(1)} | 4f^N \Psi_i^0 \rangle}{\left(E_f^0 - E_{Yy}^0\right)\left(E_f^0 - E_{Bb}^0\right)} \right\} \qquad (17\text{-}27)$$

where $|Xx\rangle$ and $|Yy\rangle$ are of the same parity, which is opposite to the parity of $4f^N$, and their set includes singly as well as doubly excited configurations.

The third-order contributions to the transition amplitude that are caused by the perturbing influence of the spin-orbit interaction taken as a perturbation V in (17-16) have an analogous form as (17-25) and (17-27) with V_{corr} replaced by V_{so}. All the derived perturbing expressions differ from each other by an order of three operators $D^{(1)}$, V_{so} and V_{cryst} in the triple products of matrix elements. Since the spin-orbit interaction operator is a one particle object, only the perturbing influence of the singly excited configurations are taken into account at the third order. In particular, there is a pair of excited configurations X and Y of the same parity, that is opposite to the parity of $4f^N$. There is a restriction however that these two configurations differ from each other by only the principal quantum number of excited one-electron states that have to be of the same symmetry. This limitation is caused by the fact that the matrix elements of V_{so} operator have to be diagonal in ℓ, but those off-diagonal in n do not vanish. In addition, the third-order spin-orbit terms involve the impact due to the excitation $B \equiv 4f^{N-1}n'f$ for $n' \geq 5$.

In connection with the nomenclature used in perturbation theory a comment is appropriate at this point of our discussion. It should be pointed out that the perturbation expansion is applied to the eigenvalue of the Hamiltonian, and to the eigenfunctions, as in (17-17) and (17-18). As a result, the energy consists of corrections of various orders. If the partitioning of the Hamiltonian is properly performed, the series of the corrections to the energy is convergent. The wave functions obtained in such a way, the best by means of the energy criterion, are very often subsequently used to evaluate the so-called properties of a system. This very procedure is used here (and it was used also in the discussion of the standard Judd-Ofelt theory) to evaluate the transition amplitude of certain transitions. Therefore, not the *corrections* but the *contributions* to the transition amplitude are evaluated and analyzed. Unfortunately, contrary to the case of energy, there is no *a priori* information about the relative magnitude of contributions of various orders, and it is possible that the higher order contributions may be greater than those of lower order (in fact this is a case for the third order electron correlation contributions). This means that in each case the relative importance of various contributions to the transition amplitude has to be verified numerically and analyzed separately.

17.3 THIRD-ORDER EFFECTIVE OPERATORS

Although the general expressions for the third-order contributions are very clear and illustrative, they are not easily applicable to numerical calculations. In addition, as seen above, at the third order analysis a large number of effective operators is expected. Since it is impossible to establish the relative importance of various terms by analyzing their tensorial structure, each one has to be numerically evaluated.

In order to obtain an effective operator form of all the perturbing expressions that contribute to the transition amplitude defined up to the third order, the same approximations that allowed one to perform the closure procedure in the case of the Judd-Ofelt theory are adopted. This means that we assume the relative degeneracy of the excited configurations, since this is the only way to define the third-order model of $f \longleftrightarrow f$ transitions within a reasonable size and simplicity.

Because of the practical applications of the formalism, similarly as in the case of the Judd-Ofelt approach, the calculations at the zero order (to generate the orbital basis sets of one-electron states of given symmetry) are performed for the average energy of the configuration of a free ion and, for the excited configurations, with the frozen core orbitals. As a consequence, all the energy denominators are expressed by the orbital energies, and this simplification allows one to perform a partial closure procedure. The rules of derivation of the effective operators, commutator relation and the interpretation of the inter-shell tensor operators are the same as in the section 17-1. The only difference now is that three unit tensor operators (taken from the triple product of matrix elements present in all perturbing expression of the third order) have to be coupled together and consequently *contracted* in the sense of the contraction of annihilation and creation operators to obtain the effective operator. Although the order of coupling of three tensor operators (or three angular momenta) is arbitrary, when once chosen, it has to be followed for all cases. Note, that in the case of coupling of three angular momenta the transition matrix between two possible schemes $|(1, 2)3, 4; 5\rangle - |1, (2, 4)3'; 5\rangle$ defines the $6 - j$ angular momentum coupling coefficient!

The form of the effective operators depends on the intermediate configurations and the order of operators in the triple products of matrix elements. In general the third-order electron correlation contributions to the transition amplitude has the general form

$$^3 D_{corr}^{eff} = \sum_k T_k^{corr} U^{(k)}(4f, 4f)$$

$$+ \sum_{xyk} \sum_t T_{xyk}^t \sum_{i<j} \left[u_i^{(x)}(4f4f) \times u_j^{(y)}(4f4f) \right]^{(k)} \qquad (17\text{-}28)$$

The first part of this expression is within the parametrization scheme of Judd-Ofelt theory as determined by single particle effective operators. These contributions to the transition amplitude represent the influence of the interactions via the Hartree-Fock potential (see the definition (17-26)). The second part expressed by the two-particle effective operators is beyond the standard parametrization, and it is caused by the perturbing influence of Coulomb interaction. However, when one of the ranks of operators in the tensorial product is zero, such a term is effectively one particle. This means that both parts of the electron correlation operator (17-26) contribute to the Judd-Ofelt intensity parameters when they are evaluated through the fitting procedure. At the same time this is evidence that the Judd-Ofelt theory, when applied in semi-empirical calculations, is more general than its derivation might suggest. Indeed, in spite of the fact that its standard formulation is based on single configuration approximation at the level of the Hartree-Fock model, the intensity parameters contain the impact due to the interactions that are beyond the central field approximation.

It should be mentioned that the two-particle effective operators that determine the transition amplitude might change the selection rules. Since for $k = 0$ in (17-28) the tensorial product is reduced to just a scalar one, still there might be a non-zero contribution to the amplitude of the unusual transitions $0 \longleftrightarrow 0$ that are highly forbidden by all selection rules of the standard one particle formulation of the intensity theory. Although the *ab initio*-type numerical calculations performed for the ions across the lanthanide series demonstrated that the purely two particle effective operators are negligible in relation to the one particle third-order contributions, when the latter vanish, as in the case of the $0 \longleftrightarrow 0$ transition, the former are the only ones that determine the transition amplitude.

All the two-particle effective operators have a form of a product of the angular terms and the radial integrals. Similarly as in the case of the Judd-Ofelt theory and its radial integral defined in (17-14), also at the third order the radial terms contain all the objects that depend on the principal quantum number of one-electron excited states, including their orbital energies. Therefore, it is expected that the radial terms of third-order effective operators contain the troublesome summations over the complete radial basis sets.

In the particular case of the perturbing influence of Coulomb interaction, when the single excitations are taken into account in (17-25), the two particle effective operator has the following form

$$^2T^t_{sck}(corr) = \sum_{\ell'}\sum_{\ell''} A^{scy}_{1t}(\ell'\ell'')^A R^s_{1t}(\ell'\ell''), \qquad (17\text{-}29)$$

with

$$A^{skc}_{1t}(\ell'\ell'') = [c]\begin{Bmatrix} s & c & k \\ f & \ell'' & f \end{Bmatrix}\begin{Bmatrix} 1 & k & t \\ f & \ell' & \ell'' \end{Bmatrix}\langle f \| C^{(t)} \| \ell' \rangle$$
$$\times \langle \ell' \| C^{(1)} \| \ell'' \rangle \langle \ell'' \| C^{(s)} \| f \rangle \langle f \| C^{(s)} \| f \rangle \sum_{i<j}\left[u^{(s)}_i \times u^{(c)}_j \right]^{(k)} \quad (17\text{-}30)$$

and the radial term is a product of three integrals,

$$^A R^s_{1t}(\ell'\ell'') = \sum_{n'}^{exc}\sum_{n''}^{exc} \frac{R^{(s)}(4f4f4fn''\ell''\ell'')\langle n''\ell'' \mid r^1 \mid n'\ell'\rangle \langle n'\ell' \mid r^t \mid 4f\rangle}{(\epsilon_{4f} - \epsilon_{n'\ell'})(\epsilon_{4f} - \epsilon_{n''\ell''})} \quad (17\text{-}31)$$

where t is a rank of tensor operator that defines crystal field potential in (9-2), and in the application here it is odd, while s is the rank of tensor operators that define the Coulomb interaction operator in (2-29), and it is even (see the last reduced matrix element above). Furthermore, in the radial term, in addition to one-electron radial integrals, there are also Slater integrals that contain the excited one-electron states. It should be also clarified that the power of 1 in the radial integral is written on purpose, since in the collection of all effective operators of third order there are objects defined as above, for the pairs of ranks $(1t)$, and also those with the positions of the ranks interchanged, $(t1)$, in appropriate places.

The angular part of each effective operator is a source of the selection rules for the non-vanishing matrix elements. Namely, the reduced matrix elements present in (17-30), in support of the conclusions derived from the analysis of the parity requirements for the general perturbing expression in (17-25), provide the limitations for the allowed one-electron states. Namely, one-electron excited states ℓ' and ℓ'' are of opposite parity (second reduced matrix element). At the same time, the parity of ℓ' is the same as the parity of $4f^N$, thus $\ell' \equiv f$. Thus, the effective operator defined by (17-29) represents the contributions due to the pairs of excited configurations $4f^{N-1}n'\ell'$ and $4f^{N-1}n''\ell''$ for all n' and $\ell' = d, g$, as in the case of the standard Judd-Ofelt theory, and all n'' and $\ell'' = p, f$.

When the perturbing influence of doubly excited configurations is taken into account in the pair $4f^{N-1}n'\ell' - 4f^{N-2}n''\ell''n'\ell'$, the two-particle effective operators have a different form, although still they are the product of the angular and radial parts, namely

$$^2T^t_{xyk}(corr) = \sum_{\ell'}\sum_{\ell''} B^{xky}_{1t}(\ell'\ell'')^B R^s_{1t}(\ell'\ell''),$$ (17-32)

where

$$B^{xky}_{1t}(\ell'\ell'') = (-1)^{k+t+x}[t]^{-1/2}[x,y]\sum_s^{odd}\langle\ell''\|C^{(s)}\|f\rangle\langle\ell'\|C^{(s)}\|f\rangle\langle\ell''\|C^{(t)}\|f\rangle$$

$$\times\langle\ell'\|C^{(1)}\|f\rangle\begin{Bmatrix}x & y & k\\ t & 1 & s\end{Bmatrix}\begin{Bmatrix}s & x & 1\\ f & \ell' & f\end{Bmatrix}$$

$$\times\begin{Bmatrix}s & y & t\\ f & \ell'' & f\end{Bmatrix}\sum_{i<j}[u^{(x)}_i \times u^{(y)}_j]^{(k)}$$ (17-33)

$$^B R^{(s)}_{1t}(\ell'\ell'') = \sum_{n'}^{exc.}\sum_{n''}^{exc.}\frac{\langle4f|r^1|n'\ell'\rangle R^{(s)}(n'\ell'4f4fn''\ell'')\langle n''\ell''|r^t|4f\rangle}{(\epsilon_{4f}-\epsilon_{n'\ell'})(\epsilon_{4f}-\epsilon_{n''\ell''})}$$ (17-34)

Inspection of the reduced matrix elements of the spherical tensors in this particular case shows that one-electron excited states ℓ' and ℓ'' are of the same parity, and they are both even. This means that $\ell', \ell'' = d, g$. Note at the same time that in the case of all radial terms the summations are always performed over all members of the radial basis sets of one-electron excited states of a given symmetry.

The third-order effective operators that are due to the Hartree-Fock potential are always one particle objects, and their tensorial structure is rather similar to those of the second-order Judd-Ofelt terms[8,9] (except for radial integrals, of course).

The third-order objects that originate from the perturbing influence of spin-orbit interaction are of a completely different character. In addition to new angular and radial factors, the effective operators act now within the spin-orbital space. Instead of unit tensor operators U in (17-30) and (17-33), which act only within the orbital space, new objects have to be introduced, namely the double tensor operators $w(s\ell s\ell)$ together with their inter-shell counterparts. In fact, the perturbing influence of the spin-orbit interactions is taken into account in an elegant way in a relativistic approach of the $f \longleftrightarrow f$ transitions, which is discussed in chapter 18. Indeed, it is rather straightforward to include the relativistic effects, among others also the spin orbit interactions, by applying the concept introduced by Sandras[10]. In this approach the relativistic analog of the standard (non-relativistic) Judd-Ofelt theory is constructed by using a relativistic form of all operators and replacing the radial integrals by the terms defined with the small and large components of the wave functions, while the matrix elements of effective operators are still evaluated in a non-relativistic way.

17.4 RADIAL INTEGRALS AND PERTURBED FUNCTION APPROACH

The radial terms of the effective operators contributing to the transition amplitude have the form of double (at the second order) and triple products (at the third order) of integrals that contain the radial parts of various operators. The radial terms of the standard Judd-Ofelt contributions are presented in (17-14), and the integrals of the third-order effective operators are defined in (17-31) and (17-34). In all cases the product of integrals contain appropriate energy denominators that prevent performing the closure procedure over the principal quantum numbers of one-electron excited states of a given symmetry. It is common practice to limit the members of the excited configurations and, in the case of the Judd-Ofelt contributions, to regard only the single excitations to the one-electron states $5d$. In such calculations the other states of d symmetry together with all the excitations to one-electron states of g symmetry are neglected. In more advanced numerical studies based on the direct calculations the first excited states of d and g are taken into account, and all the other members for $n' > 5$, not to mention the continuum states of each set, are neglected. Here, the problem of performing the summations in the radial terms in a direct way is replaced by solving differential equations for new functions, the so-called perturbed functions. The approach was originally introduced to the theory of $f-$ electron systems by Morrison in his configuration interaction calculations of the energy of lanthanides[5,6]. Morrison demonstrated that the single excitations from the core orbitals are described by single-particle functions that are the solutions of inhomogeneous differential equations that have form of the first order equation used by Sternheimer[11] to describe the distortion of the quadrupole moment of a nucleus upon the core electrons. Since 1986 the concept of single-particle perturbed functions has been used as a tool for evaluation of the radial terms of effective operators that contribute to the amplitude of $f \longleftrightarrow f$ transitions[3].

The simplest radial term of the intensity theory is that of the second order defined by (17-14), and formally rewritten below as explicit two integrals,

$$R^t_{JO}(\ell') = \int \left\{ \sum_{n'}^{exc} \frac{\langle 4f \mid r^t \mid n'\ell' \rangle}{(\varepsilon_{4f} - \varepsilon_{n'\ell'})} P_{n'\ell'}(r') \right\} r' P_{4f}(r') dr' \qquad (17-35)$$

Note that r and r^t are respectively the radial parts of the electric dipole radiation operator and crystal field potential, which plays the role of a perturbing operator. The term within the curly brackets depends on the principal quantum number of an excited state, and it is a linear combination of all one-electron states with ℓ' symmetry. This term is the most difficult to evaluate directly, and it defines the perturbed function associated with the perturbing influence of the crystal field potential (its radial part), namely

$$\varrho^t(4f \longrightarrow \ell'; r) = \sum_{n'}^{exc} \frac{\langle 4f \mid r^t \mid n'\ell' \rangle}{(\varepsilon_{4f} - \varepsilon_{n'\ell'})} P_{n'\ell'}(r) \qquad (17-36)$$

It is seen from this definition that a new function is a linear combination of all first order corrections caused by the perturbing influence of the excited configurations $4f^{N-1}n'\ell'$, for all n', which is taken into account via the odd part of crystal field potential. This function is a solution of the following differential equation

$$(\varepsilon_{4f} - h^{\ell'}_0)\varrho^t(4f \longrightarrow \ell'; r) = r^t P_{4f} - \sum_{n'}^{occ} P_{n'\ell'}(r)\langle n'\ell' \mid r^t \mid 4f \rangle \qquad (17-37)$$

where the operator has the form

$$h_0^{\ell'} = -\frac{1}{2}\frac{d^2}{dr^2} - \frac{Z}{r} + U(r) + \frac{\ell'(\ell'+1)}{2r^2} \tag{17-38}$$

The simplicity of this operator is the reason that the construction of the equation, as well as the task of its solving, is rather straightforward. At the right hand side of the equation there is the radial part of the perturbing operator, and the troublesome summation over the excited one particle states of ℓ' symmetry is eliminated by the equality

$$\overset{exc}{\underset{n'}{\sum}} \ldots = \overset{all}{\underset{n'}{\sum}} \ldots - \overset{occ}{\underset{n'}{\sum}} \ldots$$

where the first summation at the right hand side is over the complete radial basis set of one-electron states, occupied and excited (bound and continuum). Therefore for this term a complete closure is performed, and the only summation left is over the occupied one-electron states. In the case of excitations to one-electron states of d symmetry, the summation over the occupied orbitals contain terms with $3d$ and $4d$ states, while for $g-$ excitations there is no contribution, since g orbitals are not occupied in the ground configuration of the lanthanides.

Thus, in the terms of the perturbed functions, the Judd-Ofelt radial terms from (17-35) have the form of a single radial integral with a newly defined function,

$$R_{JO}^t(\ell') = \langle \varrho^t(4f \longrightarrow \ell') \mid r \mid 4f \rangle \tag{17-39}$$

The procedure of expressing the radial terms of the third-order effective operators in their perturbed function form is the same as demonstrated in the case of the Judd-Ofelt term. In particular the radial part of crystal field potential r^t at the right hand side of the equation in (17-37) should be replaced, for example, by the radial part of Coulomb interaction, if we want to find a perturbed function that is due to this perturbation. As a consequence, the complex form of the third-order radial terms defined in (17-31) and (17-34) is reduced to single integrals that involve two perturbed functions, namely

$$^A R_{k_1 k_2}^s(\ell'\ell'') = \langle \varrho^s(4f \longrightarrow \ell'') \mid r^{k_1} \mid \varrho^{k_2}(4f \longrightarrow \ell') \rangle \tag{17-40}$$

where $\varrho^s(4f \longrightarrow \ell'')$ is the solution of equation (17-37) with the radial part of Coulomb interaction at the right hand side, and $\varrho^{k_2}(4f \longrightarrow \ell')$ is the function of the Judd-Ofelt second-order approach (solution of (17-37)). The second example of the radial term has the following form

$$^B R_{k_1 k_2}^s(\ell'\ell'') = R^{(s)}(4f\varrho^{k_1}(4f \longrightarrow \ell'')\varrho^{k_2}(4f \longrightarrow \ell')4f) \tag{17-41}$$

In both cases, (17-40) and (17-41), $k_1 = 1$ and $k_2 = t$, or vice versa.

When the electron correlation third-order approach of $f \longleftrightarrow f$ transitions is analyzed, one has to expect a new kind of radial integral with the perturbed function caused by the Hartree-Fock potential. Namely, the radial term initially defined as (an analogue to (17-40))

$$^A R_{k_1 k_2}^{HF}(\ell'\ell'') = \sum_{n'}^{exc} \sum_{n''}^{exc} \frac{\langle 4f \mid u_{HF}(r) \mid n''\ell''\rangle \langle n''\ell'' \mid r^{k_1} \mid n'\ell'\rangle \langle n'\ell' \mid r^{k_2} \mid 4f\rangle}{(\varepsilon_{4f} - \varepsilon_{n''\ell''})(\varepsilon_{4f} - \varepsilon_{n'\ell'})}$$

(17-42)

in the terms of the perturbed functions has the form ($\ell'' \equiv f$),

$$^A R_{k_1 k_2}^{HF}(\ell' f) = \langle \varrho_{HF}(4f \longrightarrow f) \mid r^{k_1} \mid \varrho^{k_2}(4f \longrightarrow \ell')\rangle$$

(17-43)

where $\varrho_{HF}(4f \longrightarrow f)$ is the solution of the equation (17-34) with r^t replaced by the Hartree-Fock potential $u_{HF}(r)$. Thus the radial terms of all effective operators defined up to the third order, which include the perturbing influence of crystal field potential (forcing mechanism) and electron correlation effects, are determined by the integrals with three kinds of the perturbed functions,

$$\varrho^t(4f \longrightarrow \ell'), \varrho^s(4f \longrightarrow \ell''), \varrho_{HF}(4f \longrightarrow f)$$

for $t = 1, 3, 5$ (odd part of crystal field potential), $s = 0, 2, 4, 6$ (the rank of spherical tensors that define the Coulomb interaction), and the excitations: $\ell' = d, g$, and $\ell'' = p, f$. This means that all the radial terms are determined by at most 15 perturbed functions (one associated with the Hartree-Fock potential) and the $4f$ orbital, which is occupied in the ground configuration $4f^N$. The numerical values of various radial integrals contributing to the amplitude of the $f \longleftrightarrow f$ transitions are discussed in chapter 21.

17.5   OTHER CONTRIBUTIONS

So far we have looked at distinct contributions to the line strength of a forced electric dipole transition, as developed in the Judd-Ofelt theory and its extension by various third-order terms representing new physical mechanisms that might possibly play an important role in the understanding of spectroscopic properties of lanthanides in crystals. In order to complete the discussion in chapter 19 the magnetic dipole transitions are analyzed, since we have to realize that the square root of the line strength of a generic transition, the transition amplitude, could be written in fact as the sum of two terms,

$$S^{\frac{1}{2}}(\alpha\Gamma_i \leftrightarrow \alpha'\Gamma_j) = S^{\frac{1}{2}}(\alpha\Gamma_i \leftrightarrow \alpha'\Gamma_j)_{ed} + S^{\frac{1}{2}}(\alpha\Gamma_i \leftrightarrow \alpha'\Gamma_j)_{md}$$

(17-44)

The relevant quantity for the oscillator strength is the line strength, the square of (17-44). Hence interference terms may arise, if both electric dipole *and* magnetic dipole terms are simultaneously involved. The situation becomes more complicated if one considers other possible contributions to the transition amplitude. One such contribution is the so-called Wybourne-Downer mechanism[13-17]. The term Wybourne-Downer mechanism appears to have been so-named by Tanaka and Kushida[18] and arose in attempts to explain the origin of the observed $^5 D_0 \longleftrightarrow {}^7 F_0$ transitions of

Eu^{3+} and Sm^{2+} in low symmetry sites. Wybourne suggested[12,13] that these transitions could arise as a result of two mechanisms that involve linked terms of the form (apart from appropriate energy denominators)

$$\sum_{L,J,L',J'} \langle f^6 \, {}^7F_0|V_{cryst}^{odd}|f^5 d \, {}^7L_J\rangle\langle f^5 d \, {}^7L_J|\sum_i (\mathbf{s}\cdot\mathbf{l})_i|f^5 d \, {}^7L'_{J'}\rangle$$
$$\times \langle f^5 d \, {}^7L'_{J'}|e\mathbf{r}|f^6 \, {}^5D_0\rangle \qquad (17\text{-}45)$$

and

$$\sum_{J,L,J'} \langle f^6 \, {}^7F_0|V_{cryst}^{even}|f^6 \, {}^7F_J\rangle\langle f^6 \, {}^7F_J|V_{cryst}^{odd}|f^5 d \, {}^7LJ'\rangle$$
$$\times \langle f^5 d \, {}^7L'_J|e\mathbf{r}|f^6 \, {}^7F_0\rangle\langle f^6 \, {}^7F_0|\sum_i (\mathbf{s}\cdot\mathbf{l})_i|f^6 \, {}^5D_0\rangle \qquad (17\text{-}46)$$

The third-order contributions in (17-45) are the particular cases of intra-shell interactions via the spin-orbit interaction and are obtained from the first two terms of (17-27) with V_{corr} replaced by V_{so}. The second contributions are of the fourth order and they contain the perturbing influence of both parts of the crystal field potential. The interaction via the odd part of V_{cryst} is the same as in the case of the standard Judd-Ofelt theory, while the even part of crystal field mediates the interactions between various energy states of the ground configuration. In these expressions $PV_{cryst}^{even}P$ and $PV_{so}P$ are present, and at the same time these same operators contribute to the energy of the $4f^N$ configuration. Hence in order to regard them as perturbative contributions to the transition amplitude, one has to assume that the zero order egienvalue problem is defined at the level of the Hartree-Fock model of a free ionic system. Thus, it is seen from (17-46) that this second mechanism "borrows" intensity principally from the $0 \longleftrightarrow 2$ transitions. The first mechanism is the so-called Wybourne-Downer (WD) mechanism, suggested by Wybourne[12,13] and initially developed by Downer and his associates[14-16]. Note that in the above representation of the two mechanisms d-orbitals are indicated, whereas in realistic numerical calculations one must consider also g-orbitals, as discussed in section 17.4.

The WD mechanism involves the spin-orbit interaction among states of the excited configuration. As a consequence, if there is a breakdown of LS-coupling in the excited configuration, it is also reflected in the f^N configuration (even if there is not a breakdown in the f^N configuration). Thus the WD mechanism can lead to a violation of the spin selection rule and the J-selection rules. As a result *both* mechanisms can contribute to the strongly forbidden $^7F_0 \longleftrightarrow {}^5D_0$ transition. Furthermore, interference between the two mechanisms can occur[17] when computing line strengths by summing the square root of the line strength for each contribution and then square the resultant sum.

As mentioned the $^7F_0 \longleftrightarrow {}^5D_0$ transition is commonly associated with Eu^{3+} and Sm^{2+} ions in crystals of low symmetry; these are also the hosts that exhibit *hypersensitive* transitions.

Back in 1940 Spedding *et al*[18] noted that some transitions involving europium salts seem unusually sensitive to changes in their environment. These were considered in some detail by Jørgensen and Judd[19]. A noteworthy feature of hypersensitive transitions is that they involve the selection rule $\Delta J <= 2$. In the Judd-Ofelt theory of intensities they seem to be associated with the parameter Ω_2 (see (17-15)) and the matrix elements of unit tensor operator $U^{(2)}$. We note that the Judd-Ofelt and Wybourne-Downer mechanisms both involve *odd* rank t crystal field interactions. Generally, t is limited to the values $t = 3, 5, 7$ ($t = 3, 5$ for d−orbitals, $t = 3, 5, 7$ for g−orbitals). Only in the low symmetry fields C_1, C_2, C_s, C_{2v} is it possible for the rank of tensor operator of crystal field potential $t = 1$ to arise. An odd rank $t = 1$ crystal field component can only contribute to the matrix elements of the tensor operator $U^{(2)}$ and its associated Judd-Ofelt parameter Ω_2. Such hypersensitivity is well known for the $^7F_2 \longleftrightarrow {}^5D_0$ and $^7F_0 \longleftrightarrow {}^5D_2$ transitions in Eu^{3+} and Sm^{2+} ions in low symmetry sites as opposed to the corresponding transitions in higher symmetry sites. The enhancement of these transitions leads, by the "borrowing" and Wybourne-Downer mechanisms, to an enhancement of the $^7F_0 \longleftrightarrow {}^5D_0$ transition. This strict requirement of the Judd-Ofelt theory of $\lambda = 2$ for the hypersensitive transitions is the reason for the intensive search for additional physical mechanisms, which may introduce new selection rules. This is especially important for all the cases for which hypersensitivity is observed, in spite of the fact that the expansion of the crystal field potential does not contain the term for $t = 1$. In order to introduce such a modification of the theoretical model, the hyperfine induced $f \longleftrightarrow f$ electric dipole transitions are analyzed in chapter 20.

It should be mentioned that relativistic effects also lead to additional contributions that go beyond the limitations of the Judd-Ofelt formalism. It is clear at the same time that the multitude of contributions make any meaningful parametric fitting to the observed spectral intensities impossible. The problem of estimating the relative size of the various contributions is considerable. In the relativistic approach many of the relevant radial integrals can be calculated directly using modern implementations of relativistic Hartree-Fock-Dirac theory. However, this has still not given a truly *ab initio* computation of the crystal field parameters, and our knowledge of the *odd* crystal field parameters, essential in understanding intensities, remains very limited.

REFERENCES

1. Judd B R, (1962) Optical Absorption Intensities of Rare-Earth Ions *Phys. Rev.*,**127** 750.

2. Ofelt G S, (1962) Intensities of Crystal Spectra of Rare-Earth Ions *J. Chem. Phys.*, **37** 511.

3. Jankowski K, Smentek-Mielczarek L and Sokołowski A, (1986) Electron-correlation Third-order Contributions to the Electric Dipole Transition Amplitudes of Rare Earth Ions in Crystals I. Perturbed-function Approach *Mol. Phys.* **59** 1165.

4. Lindgren I and Morrison J, (1982) *Atomic Many-Body Theory* Springer-Verlag Berlin, Heidelberg, New York.

5. Morrison J C, (1972) Effect of Core Polarization upon the f-f Interactions of Rare-earth and Actinide Ions *Phys. Rev.* **A6** 643.

6. Morrison J C, (1973) Many-body Calculations for the Heavy Atoms. III. Pair Correlations *J. Phys. B,* **6** 2205.
7. Hirschfelder J O, Byers-Brown W and Epstein S T (1964) Recent Developments in Perturbation Theory *Adv. Quant. Chem.* **1** 255.
8. Smentek L, (1998) Theoretical Description of the Spectroscopic Properties of Rare Earth Ions in Crystals, *Phys. Rep.,* **297** 155.
9. Smentek L, (2004) Effective Operators and Spectroscopic Properties, *J. Alloys Compd.*, **380** 89.
10. Sandars P G H and Beck J, (1965) Relativistic Effects in Many Electron Hyperfine Structure I. Theory *Proc. R. Soc.,* **A287** 97.
11. Sternheimer R M, (1950) On Nuclear Quadrupole Moments *Phys. Rev.,* **80** 102.
12. Wybourne, B G, (1967) Coupled Products of Annihilation and Creation Operators in Crystal Energy Problems, pp35-52 in *Optical Properties of Ions in Crystals* Wiley - Interscience Publication, John Wiley and Sons Inc., New York.
13. Wybourne, B G, (1968) Effective Operators and Spectroscopic Properties, *J. Chem. Phys.,* **48** 2596.
14. Downer M C (1988) The Puzzle of Two-Photon Rare Earth Spectra in Solids in *Laser Spectroscopy of Solids II*, edited by W M Yen ,Springer, Heidelberg, pp29-75.
15. Downer M C, Burdick G W and Sardar D K, (1988) A New Contribution to Spin-forbidden Rare Earth Optical Transition Intensities: Gd^{3+} and Eu^{3+} *J. Chem. Phys.,* **89** 1787.
16. Burdick G W, Downer M C and Sardar D K, (1989) A New Contribution to Spin-forbidden Rare Earth Optical Transition Intensities: Analysis of all Trivalent Lanthanides *J. Chem. Phys.,* **91** 1511.
17. Tanaka M and Kushida T, (1996) Interference between Judd-Ofelt and Wybourne-Downer Mechanisms in the $^5D_0 - {}^7F_J (J = 2, 4)$ Transitions of Sm^{2+} in Solids *Phys. Rev.,* **B 53** 588.
18. Spedding F H, Moss C C and Waller R C, (1940) The Absorption Spectra of Europium Ion in Some Hydrated Salts *J. Chem. Phys.,* **8** 908.
19. Jørgensen C K and Judd B R, (1964) Hypersensitive Pseudoquadrupole Transitions in Lanthanides *Mol. Phys.* **8** 281.

18 Relativistic Effects

The fundamental laws necessary for the mathematical treatment of a large part of physics and the whole of chemistry are thus completely known, and the difficulty lies only in the fact that application of these laws leads to equations that are too complex to be solved.

Paul Dirac

In this chapter we discuss the effects that arise in crystal field theory when one considers relativity. Relativistic crystal field theory[1] started with attempts to calculate the electric quadrupole moment of the ground state $4f^7 6s^2 [^8 S_{7/2}]$ of neutral europium in an atomic beam[2,3], a problem very analogous to the crystal field splitting of the ground state $4f^7 [^8 S_{7/2}]$ of trivalent gadolinium[1,4]. If we solve Dirac's equation for an electron $n\ell j$ in a central field, we obtain two radial functions, F and G, which are associated with the small and large components of the Dirac wave function, respectively, and which depend on the total angular momentum j of the electron. Here we consider the simple case of a single electron in an f−orbital and then remark upon some of the consequences in the Judd-Ofelt theory of $f \longleftrightarrow f$ transitions intensities when looking for effects that go beyond the standard non-relativistic approach.

18.1 RELATIVISTIC CRYSTAL FIELD THEORY

To give a specific example, assume a crystal field potential

$$V = B_0^2 r^2 C_0^{(2)} + B_0^4 r^4 C_0^{(4)} + B_0^6 r^6 C_0^{(6)} + B_6^6 r^6 \left(C_6^{(6)} + C_{-6}^{(6)} \right) \qquad (18\text{-}1)$$

For a single f−electron we have the fourteen states

$$|\tfrac{7}{2} \pm \tfrac{1}{2}\rangle, \ |\tfrac{7}{2} \pm \tfrac{3}{4}\rangle, \ |\tfrac{7}{2} \pm \tfrac{5}{2}\rangle, \ |\tfrac{7}{2} \pm \tfrac{7}{2}\rangle, \ |\tfrac{5}{2} \pm \tfrac{1}{2}\rangle, \ |\tfrac{5}{2} \pm \tfrac{3}{4}\rangle, \ |\tfrac{5}{2} \pm \tfrac{5}{2}\rangle \qquad (18\text{-}2)$$

Noting that the reduced matrix element of spherical tensor for the functions expressed in the coupled scheme (s coupled with ℓ to the final value j) has the form[2]

$$\langle s\ell j || C^{(k)} || s\ell j' \rangle = \langle \ell || C^{(k)} || \ell \rangle (-1)^{s+\ell+j+k} [j, j']^{\frac{1}{2}} \begin{Bmatrix} j & k & j' \\ \ell & s & \ell \end{Bmatrix} \qquad (18\text{-}3)$$

the matrix element of the crystal field potential operator (the structural factors are omitted) is expressed as follows

$$\langle s\ell jm|r^k C_q^{(k)}|s\ell j'm'\rangle$$

$$= \langle \ell||C^{(k)}||\ell\rangle R_{jj'}(-1)^{j-m} \begin{pmatrix} j & k & j' \\ -m & q & m' \end{pmatrix}$$

$$\times (-1)^{s+\ell+j+k}[j, j']^{\frac{1}{2}} \begin{Bmatrix} j & k & j' \\ \ell & s & \ell \end{Bmatrix} \qquad (18\text{-}4)$$

where each non-relativistic radial integral from previous investigations

$$R_{\ell\ell}^k \equiv \langle n\ell|r^k|n\ell\rangle = \int_0^\infty R_{n\ell}(r) r^k R_{n\ell} dr, \qquad (18\text{-}5)$$

is now replaced by three relativistic radial integrals $R_{jj'}^k$ such that

$$R_{++}^k = \int_0^\infty r^k (F_+^2 + G_+^2) dr,$$

$$R_{+-}^k = \int_0^\infty r^k (F_+ F_- + G_+ G_-) dr,$$

$$R_{--}^k = \int_0^\infty r^k (F_-^2 + G_-^2) dr, \qquad (18\text{-}6)$$

with the $+$ referring to $j = \ell + \frac{1}{2}$ and the $-$ to $j = \ell - \frac{1}{2}$.

For fourteen states of a single f−electron we obtain the crystal field matrices

$\mu = \pm\frac{1}{2}$ $\qquad\qquad |\frac{7}{2} \pm \frac{1}{2}\rangle$ $\qquad\qquad\qquad |\frac{5}{2} \pm \frac{1}{2}\rangle$

$$
\langle\frac{7}{2} \pm \frac{1}{2}|
\begin{pmatrix}
\frac{5}{21} B_0^2 R_{++}^2 + \frac{9}{77} B_0^4 R_{++}^4 + \frac{25}{429} B_0^6 R_{++}^6 & \sqrt{3}(\pm\frac{2}{105} B_0^2 R_{+-}^2 \pm \frac{10}{231} B_0^4 R_{+-}^4 \\
& \pm \frac{50}{429} B_0^6 R_{+-}^6) \\
\langle\frac{5}{2} \pm \frac{1}{2}| \quad \sqrt{3}(\pm\frac{2}{105} B_0^2 R_{+-}^2 \pm \frac{10}{231} B_0^4 R_{+-}^4 & \frac{8}{35} B_0^2 R_{--}^2 + \frac{2}{21} B_0^4 R_{--}^4 \\
\pm \frac{50}{429} B_0^6 R_{+-}^6)
\end{pmatrix}
$$

$$(18\text{-}7a)$$

$\mu = \pm\frac{3}{2}$ $\qquad\qquad |\frac{7}{2} \pm \frac{3}{2}\rangle$ $\qquad\qquad\qquad |\frac{5}{2} \pm \frac{3}{2}\rangle$

$$
\langle\frac{7}{2} \pm \frac{3}{2}|
\begin{pmatrix}
\frac{1}{7} B_0^2 R_{++}^2 - \frac{3}{77} B_0^4 R_{++}^4 & \sqrt{10}(\pm\frac{1}{35} B_0^2 R_{+-}^2 \pm \frac{8}{231} B_0^4 R_{+-}^4 \\
- \frac{15}{143} B_0^6 R_{++}^6 & \mp \frac{5}{143} B_0^6 R_{+-}^6) \\
\langle\frac{5}{2} \pm \frac{3}{2}| \quad \sqrt{10}(\pm\frac{1}{35} B_0^2 R_{+-}^2 \pm \frac{8}{231} B_0^4 R_{+-}^4 & \frac{2}{35} B_0^2 R_{--}^2 - \frac{1}{7} B_0^4 R_{--}^4 \\
\mp \frac{5}{143} B_0^6 R_{+-}^6)
\end{pmatrix}
$$

$$(18\text{-}7b)$$

$$\mu = \pm\tfrac{5}{2} \qquad |\tfrac{7}{2} \pm \tfrac{7}{2}\rangle \qquad\qquad |\tfrac{7}{2} \mp \tfrac{5}{2}\rangle \qquad\qquad |\tfrac{5}{2} \mp \tfrac{5}{2}\rangle$$

$$
\begin{array}{c}
\langle \tfrac{7}{2} \pm \tfrac{7}{2}| \\[2em]
\langle \tfrac{7}{2} \mp \tfrac{5}{2}| \\[2em]
\langle \tfrac{5}{2} \mp \tfrac{5}{2}|
\end{array}
\begin{pmatrix}
-\tfrac{1}{3}B_0^2 R_{++}^2 + \tfrac{1}{11}B_0^4 R_{++}^4 & -\tfrac{10\sqrt{33}}{429}B_6^6 R_{++}^6 & \mp\tfrac{10\sqrt{22}}{143}B_6^6 R_{+-}^6 \\
\quad -\tfrac{5}{429}B_0^6 R_{++}^6 & & \\[1em]
-\tfrac{10\sqrt{33}}{429}B_6^6 R_{++}^6 & -\tfrac{1}{21}B_0^2 R_{++}^2 - \tfrac{13}{77}B_0^4 R_{++}^4 & \sqrt{6}(\mp\tfrac{1}{21}B_0^2 R_{+-}^2 \\
& \quad +\tfrac{25}{429}B_0^6 R_{++}^6 & \quad \pm\tfrac{10}{231}B_0^4 R_{+-}^4 \mp \tfrac{5}{429}B_0^6 R_{+-}^6) \\[1em]
\mp\tfrac{10\sqrt{22}}{143}B_6^6 R_{+-}^6 & \sqrt{6}(\mp\tfrac{1}{21}B_0^2 R_{+-}^2 & -\tfrac{2}{7}B_0^2 R_{--}^2 + \tfrac{1}{21}B_0^4 R_{--}^4 \\
& \quad \pm\tfrac{10}{231}B_0^4 R_{+-}^4 \mp \tfrac{5}{429}B_6^6 R_{+-}^6) &
\end{pmatrix}
$$

$$(18.7c)$$

If in the above matrices we were to make all the radial integrals $R_{jj'}^k$ of the same rank k equal, we would obtain the standard non-relativistic crystal field matrices. The next problem is to extend this formulation to many-electron configurations. Two ways are open:

(1). Do the entire calculation in a jj−coupling basis; or

(2). Follow Sandars and Beck[2] and continue to use the traditional LS−coupling basis by making the operator replacements

$$r^k \mathbf{C}^{(k)} \rightarrow \sum_{\kappa,\kappa'} b_k(\kappa\kappa')\mathbf{w}^{(\kappa\kappa')k}, \qquad (18\text{-}8)$$

where the $\mathbf{w}^{(\kappa\kappa')k}$ are single-particle double tensor operators[5], which act within the spin-orbital space (as tensor operator of rank κ in spin part, and tensor operator with rank k in the orbital part). The coefficients $b_k(\kappa\kappa')$ involve the relativistic radial integrals. For the spherical tensors that define the crystal field potential (the even part since only such contributes to the energy) one finds that

$$r^2 \mathbf{C}^{(2)} \rightarrow b_2(11)\mathbf{w}^{(11)2} + b_2(13)\mathbf{w}^{(13)2} + b_2(02)\mathbf{w}^{(02)2}, \qquad (18\text{-}1)$$

$$r^4 \mathbf{C}^{(4)} \rightarrow b_4(13)\mathbf{w}^{(13)4} + b_4(15)\mathbf{w}^{(15)4} + b_4(04)\mathbf{w}^{(04)4}, \qquad (18\text{-}2)$$

$$r^6 \mathbf{C}^{(6)} \rightarrow b_6(15)\mathbf{w}^{(15)6} + b_6(06)\mathbf{w}^{(06)6}, \qquad (18\text{-}3)$$

where the coefficients of double tensor operators are linear combinations of appropriate relativistic radial integrals, namely

$$b_2(11) = 4\sqrt{21}\left[-5R_{++}^2 + 3R_{+-}^2 + 2R_{--}^2\right]/245$$

$$b_2(13) = 4\sqrt{7}\left[5R_{++}^2 + 4R_{+-}^2 - 9R_{--}^2\right]/245$$

$$b_2(02) = -2\sqrt{42}\left[25R_{++}^2 + 6R_{+-}^2 + 18R_{--}^2\right]/735 \qquad (18\text{-}10a)$$

$$b_4(13) = 4\sqrt{21}\left[6R_{++}^4 - 5R_{+-}^4 - R_{--}^4\right]/441$$

$$b_4(15) = 2\sqrt{2310}\left[-3R_{++}^4 - 8R_{+-}^4 + 11R_{--}^4\right]/4851$$

$$b_4(04) = 2\sqrt{77}\left[18R_{++}^4 + 20R_{+-}^4 + 11R_{--}^4\right]/1617 \qquad (18\text{-}10b)$$

$$b_6(15) = 20\sqrt{77}\left[-R_{++}^6 + R_{+-}^6\right]/1001$$

$$b_6(06) = -10\sqrt{462}\left[R_{++}^6 + 6R_{+-}^6\right]/3003 \qquad (18\text{-}10c)$$

Not surprisingly, the calculation with the replacement operators yields exactly the same results for a single electron, as found in (18-7a-c)[6]. For states involving N equivalent electrons the $\mathbf{w}^{(\kappa\kappa')}$ are simply replaced by N-electron operator

$$\mathbf{W}^{(\kappa k)} = \sum_{i=1}^{N} \mathbf{w}_i^{(\kappa k)} \qquad (18\text{-}11)$$

and the matrix elements may be evaluated in the usual non-relativistic $LS-$coupling basis but with the associated radial integrals being taken from appropriate relativistic Dirac-Hartree-Fock wave functions.

The important point to notice is that the replacement operators are *double tensor operators* that act in *both* the spin and orbital spaces, whereas the non-relativistic crystal field operators act only in the orbital space. It is this property that leads to a second-order contribution to the ground state splitting for rare earth and actinide ions having a half-filled $f-$shell[1,4,7]. Smentek *et al*[8] have given detailed calculations of a relativistic crystal field for $S-$state f electron ions. Free ion non-relativistic calculations were performed using Froese-Fischer's MCHF programme. The relativistic radial integrals were evaluated using the GRASP[9] package.

18.2   RELATIVISTIC $f \longleftrightarrow f$ TRANSITIONS IN CRYSTAL FIELDS

The relativistic effective operator theory of $f \longleftrightarrow f$ transitions is based on the extension of the original proposal of Sandras and Beck introduced for calculating relativistic effects in many electron hyperfine structure. Similarly as in the case of the crystal field energy, we also apply here the concept of $LS-$coupled relativistic states. Also in the case of the intensity theory the problem of evaluation of the relativistic impact upon the transition amplitude becomes a non-relativistic task, which might be treated more easily.

In the second order intensity theory we have products of the electric dipole and crystal field matrix elements coupling the $4f^N$ configuration to those of opposite parity (see (17-6)). Again, whereas in the non-relativistic Judd-Ofelt theory, transitions depend on the tensor operators $\mathbf{U}^{(k)}$, the relativistic treatment leads to double-tensor operators $\mathbf{W}^{(\kappa t)k}$. In the Judd-Ofelt theory the single particle unit tensor operators $\mathbf{u}(n\ell, n'\ell')^{(t_{odd})}$ link a ground configuration orbital $n\ell$ to an orbital $n'\ell'$ in an excited configuration and closure results in single particle tensor operators $\mathbf{u}(n\ell, n\ell)^{(\lambda_{even})}$. In the relativistic extension the double tensor operators $\mathbf{w}(n\ell, n'\ell')^{(\kappa t)k}$ couple the orbitals and closure results in single particle double tensor operators $\mathbf{w}(n\ell, n\ell)^{(\kappa K)\lambda}$ where even if λ is *even*, K is *even* or *odd* as κ is 0 or 1 respectively.

The derivation of the standard, non-relativistic Judd-Ofelt theory presented in chapter 17 is the scheme to be repeated in the language of the effectively relativistic approach. All the perturbing operators that define the Hamiltonian in (17-1) have to be converted into a relativistic form. For this purpose, having the inter-shell tensor operators that contribute to the transition amplitude in accordance with the perturbing expressions, the concept of Sandras and Beck is extended by the following substitution,

$$\langle n\ell|r^x|n'\ell'\rangle\langle\ell||C^{(x)}||\ell'\rangle u_\varrho^{(x)}(\ell,\ell') \Rightarrow$$

$$\sum_{\kappa_1 k_1}\sum_{i_1,i_2}^2 \beta_{\kappa_1 k_1}^{x\ell'}(j_{i_1}j_{i_2}')R^x(j_{i_1}j_{i_2}')\langle j_{i_1}||C^{(x)}||j_{i_2}'\rangle w_\varrho^{(\kappa_1 k_1)x}(s\ell, s\ell'), \qquad (18\text{-}12)$$

where $j_\pm \equiv \ell \pm 1/2$ is numbered by $i_1 = 1, 2$, and in the general case, $j_\pm' \equiv \ell' \pm 1/2$ is numbered independently by $i_2 = 1, 2$; the ranks κ_1 and k_1 are limited by the triangular conditions for the non-vanishing $9j$–symbol of the angular factor

$$\beta_{\kappa_1 k_1}^{x\ell'\ell}(j_{i_1}j_{i_2}') = (-1)^{\kappa_1+k_1+x}[j_{i_1}, j_{i_2}']^{1/2}\begin{Bmatrix} \ell' & \ell & k_1 \\ s & s & \kappa_1 \\ j_{i_2}' & j_{i_1} & x \end{Bmatrix}, \qquad (18\text{-}13)$$

The radial integrals contain the *large* and *small* components, and they are defined as follows

$$R^x(j_{i_1}, j_{i_2}') = \langle P^{j_{i_1}}|r^x|P^{j_{i_2}'}\rangle + \langle Q^{j_{i_1}}|r^x|Q^{j_{i_2}'}\rangle, \qquad (18\text{-}14)$$

The reduced matrix element of spherical tensor is now generalized for the inter-shell case, and it has the form

$$\langle j_{i_1}||C^{(x)}||j_{i_2}'\rangle = (-1)^{j_{i_1}+1/2}[j_{i_1}, j_{i_2}']^{1/2}\varepsilon(\ell+x+\ell')\begin{pmatrix} j_{i_1} & x & j_{i_2}' \\ \tfrac{1}{2} & 0 & \tfrac{1}{2} \end{pmatrix} \qquad (18\text{-}15)$$

which vanishes unless $\ell + x + \ell'$ is even ($\varepsilon(\ell + x + \ell')$ is 1 for even sum of arguments, 0 otherwise), as concluded from the triangular condition for the non-vanishing $3j$–symbol in the reduced matrix element at the left hand side of (18-12).

The double inter-shell operator $w_\varrho^{(\kappa_1 k_1)x}(s\ell, s\ell')$ is a unit tensor operator, and it is defined by its reduced matrix element in an analogous way to the inter shell unit tensor operator u introduced in (17-7), namely

$$\langle s\ell''||w^{\kappa_1 k_1}(s\ell, s\ell')||s\ell'''\rangle = \delta(\ell'', \ell)\delta(\ell', \ell''') \qquad (18\text{-}16)$$

For the evaluation of any matrix element of double tensor operator the Wigner-Eckart theorem has to be used twice, separately for spin and orbit parts. Also the commutation relation of double tensor operators that is applicable for the derivation of the effective operators (see (17-11)) has two parts. Indeed, appropriate ranks of distinct operators have to be coupled together to give the final ranks for spin and orbital parts, and subsequently they both are coupled again to define the rank of the resultant operator.

This procedure is very clearly described by the coupling rule of the inter-shell objects, namely

$$w_{\varrho}^{(\kappa_1 k_1)x}(s\ell, s\ell')w_{\eta}^{(\kappa_2 k_2)y}(s\ell', s\ell)$$

$$= \sum_{z\zeta}\sum_{\kappa_3 k_3}(-1)^{x-y-\zeta+\kappa_3+k_3+2s}\,[z, x, y]^{1/2}[\kappa_3, k_3]\begin{pmatrix} x & y & z \\ \varrho & \eta & -\zeta \end{pmatrix}$$

$$\times \begin{Bmatrix} \kappa_2 & \kappa_3 & \kappa_1 \\ \frac{1}{2} & \frac{1}{2} & \frac{1}{2} \end{Bmatrix}\begin{Bmatrix} k_2 & k_3 & k_1 \\ \ell & \ell' & \ell \end{Bmatrix}\begin{Bmatrix} \kappa_1 & k_1 & x \\ \kappa_2 & k_2 & y \\ \kappa_3 & k_3 & z \end{Bmatrix} w_{\zeta}^{(\kappa_3 k_3)z}(\ell\ell) \quad (18\text{-}17)$$

where as a result of contraction of two inter-shell double tensor operators, an effective operator acting within the ground configuration is obtained. The coupling, uncoupling and re-coupling of various ranks of the operators are all recorded in the $9j$–symbol of (18-17). Indeed, in each row the coupling scheme for a distinct operator is displayed. Namely $(\kappa_1 k_1)x$ and $(\kappa_2 k_2)y$ are the ranks of the initial operators, which are at the left hand side of (18-17), and which have to be at first coupled together into a tensorial product. Then, if they act on the same coordinate in the spin-orbital space, both objects have to be contracted to give an effective operator. Finally the third row of the $9j$–symbol shows the ranks of the resultant operators, namely $(\kappa_3 k_3)z$. The columns have also a very clear interpretation, since they show the coupling within each subspace separately $(\kappa_1\kappa_2)\kappa_3$, $(k_1 k_2)k_3$ and $(xy)z$. It is seen also in (18-17) that the information collected in the columns of the $9j$–symbol is extended in the remaining angular momentum coefficients which add details about one particle states involved in the contraction ($6j$–symbols) and about the spatial orientation of all objects ($3j$–symbol).

18.3   EFFECTIVE OPERATORS OF RELATIVISTIC $f \longleftrightarrow f$ THEORY

The effective operators that contribute to the transition amplitude[10,11] defined in an effectively relativistic way generalize the standard Judd-Ofelt theory. The second-order Judd-Ofelt contributions defined in (17-12) by a single particle effective operator now have a more complex structure, namely

$$T_2^{rel} = \sqrt{3}\sum_{tp} B_p^t[t]^{1/2}\sum_{\kappa_1=0,1}\sum_{k_1\leq|\kappa_1-1|}^{\kappa_1+1}\sum_{\kappa_2=0,1}\sum_{k_2\leq|\kappa_2-t|}^{\kappa_2+t}$$

$$\times \sum_{\ell'}\varepsilon(f+1+\ell')\varepsilon(f+t+\ell')A_{k_1 k_2}^{\kappa_1\kappa_2}(t\ell')$$

$$\times \sum_{\kappa_3=0,1}\sum_{k_3\leq|t-1|}^{t+1}a\sum_{\lambda\leq|\kappa_3-k_3|}^{\kappa_3+k_3}[\lambda]^{1/2}\sum_{q}(-1)^{\kappa_3+k_3+t-q}[\kappa_3, k_3]\begin{pmatrix} 1 & t & \lambda \\ \varrho & p & -q \end{pmatrix}$$

$$\times \begin{Bmatrix} \kappa_2 & \kappa_3 & \kappa_1 \\ \frac{1}{2} & \frac{1}{2} & \frac{1}{2} \end{Bmatrix}\begin{Bmatrix} k_2 & k_3 & k_1 \\ f & \ell' & f \end{Bmatrix}\begin{Bmatrix} \kappa_1 & k_1 & 1 \\ \kappa_2 & k_2 & t \\ \kappa_3 & k_3 & \lambda \end{Bmatrix} W_q^{(\kappa_3 k_3)\lambda}(ff) \quad (18\text{-}18)$$

where the factor a is 2, if the parity of $\kappa_1 + k_1 + \kappa_2 + k_2$ is the same as parity of $\kappa_3 + k_3$, otherwise $a = 0$, and the whole expression vanishes. The coefficients $A_{k_1 k_2}^{\kappa_1 \kappa_2}$ contain the radial integrals (18-14), factors that result from the substitution in (18-12) and which are defined in (18-13), and the appropriate reduced matrix elements of spherical tensors from (18-15),

$$A_{k_1 k_2}^{\kappa_1 \kappa_2}(t\ell') = \sum_{i_1, i_2}^{2} \beta_{\kappa_1 k_1}^{1\ell' f}(j_{i_1} j_{i_2}') \beta_{\kappa_2 k_2}^{t f \ell'}(j_{i_2}' j_{i_1})$$

$$\times R^1(j_{i_1} j_{i_2}') R^t(j_{i_2}' j_{i_1}) \langle j_{i_1} || C^{(1)} || j_{i_2}' \rangle \langle j_{i_2}' || C^{(t)} || j_{i_1} \rangle \quad (18\text{-}19)$$

The effective operators defined in (18-18) represent the relativistic effects taken into account in an effective way. At the same time these are the effective double tensor operators, since they act within the ground configuration $4f^N$. This means that the objects that determine the transition amplitude within relativistic approach are *doubly effective double tensor operators*. In spite of the fact that their structure is complex, still they are one particle objects as in the Judd-Ofelt theory. However here they act within the spin-orbital space. The contributions associated with the double operator $W^{(11)\lambda}$ represent the perturbing influence of the spin orbit interaction upon the transition amplitude, and here it is automatically included at the second order, while in the non-relativistic theory it is possible to take it into account at the third order analysis. In addition, if all $\kappa's$ in (18-18) are made zero, the expression is reduced to the non-relativistic Judd-Ofelt standard effective operators.

In order to break the limitations of the single configuration approximation of the second-order approach, electron correlation effects have to be included as a perturbation. Thus, the transition amplitude defined by (18-18) has to be modified by new effectively relativistic operators of the third order. In the case where the Coulomb interaction is taken into account, the effective operator, an analogue to (17-29), is also a two-particle object but expressed as a tensorial product of double tensor operators,

$$T_3^{rel}(corr) = \sqrt{3} \sum_{tp} B_p^t[t]^{1/2} \sum_s \sum_{\ell', \ell''} \varepsilon_s \varepsilon_i^* \varepsilon_{\ell'} \varepsilon_{\ell''}^* \sum_{\kappa_1, k_1} \sum_{\kappa_2, k_2} \sum_{\kappa_3, k_3} \sum_{\kappa_4, k_4}$$

$$\times A_{k_1 k_2 k_3 k_4}^{\kappa_1 \kappa_2 \kappa_3 \kappa_4}(st\ell'' \ell') \sum_{\lambda, q} (-1)^{t-q} [s, \lambda]^{\frac{1}{2}} \begin{pmatrix} 1 & t & \lambda \\ \varrho & p & -q \end{pmatrix} \sum_{\kappa_5 k_5} [\kappa_5, k_5]$$

$$\times \begin{Bmatrix} \kappa_2 & \kappa_5 & \kappa_1 \\ \frac{1}{2} & \frac{1}{2} & \frac{1}{2} \end{Bmatrix} \begin{Bmatrix} k_2 & k_5 & k_1 \\ \ell'' & \ell' & f \end{Bmatrix} \begin{Bmatrix} \kappa_1 & k_1 & 1 \\ \kappa_2 & k_2 & t \\ \kappa_5 & k_5 & \lambda \end{Bmatrix}$$

$$\times \sum_x (-1)^{x+\lambda} [x]^{\frac{1}{2}} \sum_{\kappa_6 k_6} (-1)^{\kappa_6 + k_6 + 1} [\kappa_6, k_6]$$

$$\times \begin{Bmatrix} \kappa_5 & \kappa_6 & \kappa_4 \\ \frac{1}{2} & \frac{1}{2} & \frac{1}{2} \end{Bmatrix} \begin{Bmatrix} k_5 & k_6 & k_4 \\ f & \ell'' & f \end{Bmatrix} \begin{Bmatrix} \kappa_4 & k_4 & s \\ \kappa_5 & k_5 & \lambda \\ \kappa_6 & k_6 & x \end{Bmatrix}$$

$$\times \sum_{i<j} \left[w_i^{(\kappa_3 k_3)s}(ff) \times w_j^{(\kappa_6 k_6)x}(ff) \right]^{(\lambda)} \quad (18\text{-}20)$$

where $\varepsilon_s \varepsilon_t^* \varepsilon_{\ell'} \varepsilon_{\ell''}^*$ shows a required parity of each summation index, and ε denotes its even parity, while ε^* odd-parity. An inspection of the third-order perturbing expression in (17-25) shows that there are four tensor operators to be converted into their relativistic form. Therefore the product of four transformation factors β from (18-12) and four reduced matrix elements are expected in the final result. Also the number and the structure of relativistic integrals are as expected, since as defined below, $A_{k_1 k_2 k_3 k_4}^{\kappa_1 \kappa_2 \kappa_3 \kappa_4}$ contains the following terms,

$$
A_{k_1 k_2 k_3 k_4}^{\kappa_1 \kappa_2 \kappa_3 \kappa_4}(st\ell''\ell') = \sum_{i_1, i_2}^{2} \sum_{i_3, i_4}^{2} \beta_{\kappa_3 k_3}^{sff}(j_{i_3} j_{i_3}) \beta_{\kappa_4 k_4}^{s\ell''f}(j_{i_1} j_{i_4}'') \beta_{\kappa_1 k_1}^{1\ell'\ell''}(j_{i_4}'' j_{i_2}') \beta_{\kappa_2 k_2}^{tf\ell'}(j_{i_2}' j_{i_1})
$$

$$
\times R^s(j_{i_3} j_{i_1} j_{i_3} j_{i_2}'') R^1(j_{i_4}'' j_{i_2}') R^t(j_{i_2}' j_{i_1}) \langle j_{i_3} || C^{(s)} || j_{i_3} \rangle
$$

$$
\times \langle j_{i_1} || C^{(s)} || j_{i_4}'' \rangle \langle j_{i_4}'' || C^{(1)} || j_{i_2}' \rangle \langle j_{i_2}' || C^{(t)} || j_{i_1} \rangle \qquad (18\text{-}21)
$$

When one of the ranks of the operators in the tensorial product is equal to 0, for example $x = 0$ in (18-20), then the whole expression is reduced and instead of a two-particle effective operator, we have a one-particle object, but still defined by double tensor operator,

$$
T_{corr}^{rel}(x = 0) \Longrightarrow \frac{\sqrt{3}}{\sqrt{14}} \sum_{tp} B_p^t[t]^{1/2} \sum_{\lambda, q}^{even} [\lambda]^{1/2} (-1)^{t-q} \sum_{\ell', \ell''} \varepsilon_t^* \varepsilon_{\ell'} \varepsilon_{\ell''}^*
$$

$$
\times \sum_{\kappa_1, k_1} \sum_{\kappa_2, k_2} \sum_{\kappa_3, k_3} \sum_{\kappa_4, k_4} A_{k_1 k_2 k_3 k_4}^{\kappa_1 \kappa_2 \kappa_3 \kappa_4}(\lambda t\ell''\ell') \begin{pmatrix} 1 & t & \lambda \\ \varrho & p & -q \end{pmatrix}
$$

$$
\times \begin{Bmatrix} \kappa_2 & \kappa_4 & \kappa_1 \\ \frac{1}{2} & \frac{1}{2} & \frac{1}{2} \end{Bmatrix} \begin{Bmatrix} k_2 & k_4 & k_1 \\ \ell'' & \ell' & f \end{Bmatrix} \begin{Bmatrix} \kappa_1 & k_1 & 1 \\ \kappa_2 & k_2 & t \\ \kappa_4 & k_4 & \lambda \end{Bmatrix} W^{(\kappa_3 k_3)\lambda}(ff). \qquad (18\text{-}22)
$$

There are also one particle effective operators that originate from the Hartree-Fock potential taken as a perturbation. In this particular case the relativistic contributions are represented by the effective operators of the form

$$
T_{HF}^{rel} = \sqrt{3} \sum_{tp} B_p^t[t]^{1/2} \sum_{\ell', \ell''} \delta(\ell'', f) \varepsilon_t^* \varepsilon_{\ell'} \sum_{\kappa_1, k_1} \sum_{\kappa_2, k_2} \sum_{\kappa_3, k_3} \delta(\kappa_1, k_1)
$$

$$
\times A_{k_1 k_2 k_3}^{\kappa_1 \kappa_2 \kappa_3}(0t\ell''\ell') \sum_{\lambda, q}^{all} (-1)^{t-q} [\lambda]^{\frac{1}{2}} [k_1]^{-\frac{1}{2}} \begin{pmatrix} 1 & t & \lambda \\ \varrho & p & -q \end{pmatrix} \sum_{\kappa_5 k_5} [\kappa_5, k_5]
$$

$$
\times \begin{Bmatrix} \kappa_2 & \kappa_5 & \kappa_1 \\ \frac{1}{2} & \frac{1}{2} & \frac{1}{2} \end{Bmatrix} \begin{Bmatrix} k_2 & k_5 & k_1 \\ f & \ell'' & \ell' \end{Bmatrix} \begin{Bmatrix} \kappa_2 & k_2 & 1 \\ k_5 & \kappa_5 & k_1 \end{Bmatrix}
$$

$$
\times \sum_{\kappa_6 k_6} (-1)^{\kappa_6 + k_6 + k_5} [\kappa_6, k_6] \begin{Bmatrix} \kappa_3 & \kappa_6 & \kappa_5 \\ \frac{1}{2} & \frac{1}{2} & \frac{1}{2} \end{Bmatrix} \begin{Bmatrix} k_3 & k_6 & k_5 \\ f & \ell' & f \end{Bmatrix}
$$

$$
\times \begin{Bmatrix} \kappa_5 & k_5 & 1 \\ \kappa_3 & k_3 & t \\ \kappa_6 & k_6 & \lambda \end{Bmatrix} W^{(\kappa_6 k_6)\lambda}(ff), \qquad (18\text{-}23)
$$

with

$$
\begin{aligned}
\mathcal{A}_{k_1 k_2 k_3}^{\kappa_1 \kappa_2 \kappa_3}(xt\ell''\ell') = \sum_{i_1, i_2, i_3}^{2} & \beta_{\kappa_3 k_3}^{x\ell''f}(j_{i_1} j_{i_3}'') \beta_{\kappa_1 k_1}^{1\ell'\ell''}(j_{i_3}'' j_{i_2}') \beta_{\kappa_2 k_2}^{tf\ell'}(j_{i_2}' j_{i_1}) \\
& \times R^{HF}(j_{i_1} j_{i_3}'') R^1(j_{i_3}'' j_{i_2}') R^t(j_{i_2}' j_{i_1}) \langle j_{i_1} || C^{(x)} || j_{i_3}'' \rangle \\
& \times \langle j_{i_3}'' || C^{(1)} || j_{i_2}' \rangle \langle j_{i_2}' || C^{(t)} || j_{i_1} \rangle .
\end{aligned}
\tag{18-24}
$$

At this point of the presentation of the relativistic approach formulated for the description of $f \longleftrightarrow f$ transitions it is becoming apparent that the analysis of possible contributions becomes rather complex, since already at the third order there is an out-of-control explosion of various terms. Thus, in order to summarize this discussion, it is easier to describe the f−spectra in the terms of two possible parametrization schemes, for which the angular and radial parts of effective operators are regarded simply as intensity parameters. However, as concluding remarks derived from the discussion presented in this chapter it is worthwhile to notice that in terms of the *angular* parts, which reflect the symmetry properties of f−orbitals, there is a remarkable (and understandable) similarity between crystal field theory and the theory of $f \longleftrightarrow f$ transitions. In the non-relativistic approach both involve, to second-order, the matrix elements of the unit tensor operators $\mathbf{U}^{(\lambda)}$ with $\lambda = 0, 2, 4, 6$. In the extension to the relativistic theory both involve the $\mathbf{U}^{(\lambda)}$ operators replaced by the double tensor operators $\mathbf{W}^{(\kappa k)\lambda}$. The fundamental difference comes in the radial integrals which are involved. It is only relatively recently that it has become possible to give serious consideration to the detailed calculation of such integrals. Future work will undoubtedly be more directed to such calculations and to estimates of the significance of relativistic effects both for crystal field interactions and transition intensities. One expects these effects to become increasingly important as the calculations, and hopefully experiments, are made on the heavy actinides. It may well be that future studies will be directed towards calculations in the jj−coupling basis, which is the natural basis to use when relativistic effects become significant.

18.4   PARAMETRIZATION SCHEMES OF f-SPECTRA

Are the higher-order contributions to the amplitude of electric dipole transitions important? This question is valid in both non-relativistic and relativistic theoretical models of $f \longleftrightarrow f$ transitions. The results of numerical *ab initio* calculations performed for the ions across the lanthanide series and based on the third-order non-relativistic model demonstrated that the major modification of the transition amplitude is provided by one particle effective operators originating from the electron correlation perturbing influence. In fact the third-order contributions of this origin are dominant in comparison to the standard ones that are defined within the second-order Judd-Ofelt model. This result fulfills the expectations that electron correlation effects are very important for the proper description of the electronic structure of many electron systems as is evident in all investigations which show the impact of these effects on energy. At the same time, the result that the third-order contributions to the transition amplitude are greater than those of the second order is not in contradiction with the properties of a properly designed perturbation approach, since not the *corrections* of certain order (as in the case of energy) but the *contributions* are compared to each other.

The analysis of the tensorial structure of non-relativistic effective operators may suggest that the inclusion of the electron correlation effects is a difficult task. One gets just the opposite impression from the analysis of the expression that defines the transition amplitude up to the third order and includes the most important one particle effective operators. Indeed, the transition amplitude defined in such a way has a very simple form,

$$^1\Gamma \approx \sum_\lambda^{even} \sum_t^{odd} B_p^t T_t^\lambda U^{(\lambda)}(ff) \tag{18-25}$$

where B_p^t is the structural parameter, and the factor T_t^λ is a product of angular part of the Judd-Ofelt effective operator $A_t^\lambda(\ell')$ defined in (17-13) and the radial terms, that are modified by the third-order integrals,

$$T_t^\lambda = \sum_{\ell'}^{even} \left[R_{JO}^t(\ell') - R_{HF}^t(\ell', f) + \frac{1}{2}(N-1)R_t^0(\ell', f) \right] A_t^\lambda(\ell') \tag{18-26}$$

This means that in order to include the major part of electron correlation contributions, it is enough to modify the Judd-Ofelt radial integral by the values of the appropriate integrals of the third-order one-particle effective operators. This is a straightforward procedure, if the direct calculations are performed and the perturbed functions of a given kind are generated. At the same time, however, the structure of the transition amplitude in (18-25) indicates that also at the third order analysis the one-particle parametrization scheme defined in (17-15) is valid,

$$S_{f\leftarrow i} = \sum_{\lambda=2,4,6} \Omega_\lambda \mid \langle \Psi_f \mid\mid U^{(\lambda)} \mid\mid \Psi_i \rangle \mid^2 \tag{18-27}$$

but the intensity parameters Ω_λ now represent the perturbing influence of the crystal field potential (second-order Judd-Ofelt theory) and in addition also the major part of electron correlation third-order contributions. This means, as discussed in chapter 17, that the Judd-Ofelt intensity parameters are indeed more general than the original assumptions of the model may suggest. In fact a detailed analysis performed for various lanthanide ions, and based on *ab initio* numerical calculations, demonstrated that the one-particle parametrization scheme represents an extensive list of physical mechanisms, and the intensity parameters Ω_λ have a rich morphology[12].

Judd-Ofelt theory and its third-order electron correlation implementation are based on the so-called static model in which it is assumed that the surrounding ligands perturb the central ion stimulating its spectroscopic properties. This means that in a given system the environment of the ion is treated as a source of a static potential that perturbs the wave function of the lanthanide. The *dynamic* model, known also in the literature as the inhomogeneous dielectric or ligand polarization model, is formulated when the mutual interactions between the ion and the ligands are taken into account[13-16]. Thus, the Judd-Ofelt theory, in its standard formulation based on the static model, is now complemented by the interactions between the multipoles on a central ion and the dipole localized on the ligand. From these interactions originate new contributions to the transition amplitude, and they are a part of the intensity parameters in (18-27). Subsequently, breaking again the single configuration approximation, electron correlation effects are included within the dynamic model, again giving rise to a new part of the intensity parameters of one-particle scheme. The list of physical mechanisms that are represented by the fitted parameters in (18-27) is not complete. For example, rather exotic for the problems discussed here, the direct perturbing influence by the specific mass shift effect is also included in Ω_λ[12]. At the same time, however, the intensity parameters associated with the unit tensor operators U that act only within the orbital space do not include all relativistic effects, including the perturbing influence of the spin-orbit interaction.

This short presentation of the structure of the intensity parameters of one-particle (non-relativistic) scheme defines the size of the task when *ab initio* calculations are performed in order to understand the mechanisms that are responsible for the $f \longleftrightarrow f$ transitions. At the same time it shows the elegance and applicability of the semi-empirical approach in which at most three intensity parameters describe all (or almost all) transitions for a given system. However, since the selection rules of the effective third-order model are not changed, and they are the same as at the second order standard formulation, still certain transitions are forbidden from a theoretical point of view.

The selection rules are changed in the effectively relativistic description of $f \longleftrightarrow f$ transitions. As presented above, the second-order relativistic terms, the analogues of the Judd-Ofelt contributions, as well as the majority of the third-order electron correlation terms are all defined by one particle operators. This means that the relativistic scheme of parametrization is also one-particle, however the intensity parameters now are associated with the double tensor operators acting within the spin-orbital space. The line strength of a transition is determined by the reduced matrix element of the W operator weighted by newly defined parameters,

$$S^{rel}_{f \leftarrow i} = \sum_\lambda^{all} \left| \sum_{\kappa k} \omega_{(\kappa k)\lambda} \langle \Psi_f \, || \, W^{(\kappa k)\lambda} \, || \, \Psi_i \rangle \right|^2 \qquad (18\text{-}28)$$

where λ is *even* or *odd*. This possibility gives the chance for a theoretical description of the transition $0 \longleftrightarrow 1$, which is highly forbidden in the non-relativistic scheme; in this formulation its intensity is determined by the following terms

$$\omega_{(01)1} \langle 1 \, || \, W^{(01)1} \, || \, 0 \rangle + \omega_{(10)1} \langle 1 \, || \, W^{(10)1} \, || \, 0 \rangle$$
$$+ \omega_{(11)1} \langle 1 \, || \, W^{(11)1} \, || \, 0 \rangle \qquad (18\text{-}29)$$

The first two operators that determine the transition amplitude define the transformation properties of $L^{(1)}$ and $S^{(1)}$ that define in turn the magnetic dipole radiation operator. Is it the origin of the borrowing of intensity mechanism introduced in (17-44) of the previous chapter? This possibility is discussed together with the theory of the magnetic dipole transition in the next chapter. It should be also mentioned that the last double operator in (18-29) with the odd sum of its ranks as non-hermitian represents here the contribution to the transition amplitude, which is not in fact an observable, while in the definition of the line strength the requirement for the even sum of all ranks should be satisfied.

It is also possible to have in (18-28) $\lambda = 0$ that means only that $\kappa = k = 1$, and the matrix element of the double unit tensor operator does not vanish (as it does in the case of operator U in (18-27)). This gives, for the first time, a non-zero contribution to the intensity of the $0 \longleftrightarrow 0$ transition, which is proportional to

$$\omega_{(11)0} \langle 0 \parallel W^{(11)0} \parallel 0 \rangle \tag{18-30}$$

where the double unit tensor operator with ranks $(11)0$ represents the spin orbit interaction. It should be noted in addition, that in principle, the amplitude of this transition does not vanish (if evaluated directly) only if the crystal field potential contains in its expansion a term with $t = 1$. This is the particular case of C_{2v} symmetry, for which in fact the most interesting $0 \longleftrightarrow 0$ transitions are observed in Eu^{+3}.

The structure of the expression in (18-28) is similar to the Judd-Ofelt expression. The intensity parameters in both cases represent second-order crystal field terms, and third-order electron correlation contributions. However the number of the intensity parameters now is increased.

In order to relate the newly introduced intensity parameters with those of the non-relativistic theory it is convenient to define the ratio of the reduced matrix elements

$$X^{(\kappa t)\lambda}(\Psi, \Psi') = \langle \ell^N[\alpha SL]J \parallel W^{(\kappa t)\lambda} \parallel \ell^N[\alpha'S'L']J' \rangle /$$
$$\langle \ell^N[\alpha SL]J \parallel U^{(\lambda)} \parallel \ell^N[\alpha'S'L']J' \rangle \tag{18-31}$$

The line strength defined in (18-28) now has the following form

$$S_{f \leftarrow i}^{rel} = \sum_\lambda \Omega_\lambda^{rel}(\Psi_f \Psi_i) \mid \langle \Psi_f \parallel U^\lambda \parallel \Psi_i \rangle \mid^2 \tag{18-32}$$

where the new intensity parameters that are dependent on the states involved in the transition are defined as follows

$$\Omega_\lambda^{rel}(\Psi_f \Psi_i) = \left(\sum_{\kappa k} \omega_{(\kappa k)\lambda} X^{(\kappa k)\lambda}(\Psi_f, \Psi_i) \right)^2 \tag{18-33}$$

It is seen that in the relativistic parametrization scheme the simplicity, universality and elegance of the non-relativistic scheme is lost. At the same time, however, the applicability of the relativistic model is broader, and the electric dipole transitions that are forbidden in the standard theory now have a theoretical interpretation. As mentioned, the number of newly introduced parameters is increased, and quite often there are more unknown parameters than the experimental data available for the fitting procedure. Therefore to make the relativistic model applicable in practice, all attempts should be directed to such investigations that allow one to reduce the number of the parameters to a realistic size.

It is a formal step to rewrite the line strength in (18-28) in the following way

$$S_{f \leftarrow i} = \sum_\lambda \mid \sum_{\kappa k} \omega_\lambda^{\kappa k} \langle \Psi_f \mid W^{(\kappa k)\lambda} \mid \Psi_i \rangle \mid^2 \tag{18-34}$$

with $(\omega_\lambda^{\kappa k})^2 = \Omega_\lambda^{\kappa k}$ being the analogs of the standard intensity parameters of the Judd and Ofelt scheme. Thus, reconstructing the standard expression it is possible to write

$$S_{f \leftarrow i} = \sum_\lambda \sum_{\kappa k} \Omega_\lambda^{\kappa k} \mid \langle \Psi_f \mid W^{(\kappa k)\lambda} \mid \Psi_i \rangle \mid^2 + cross\ terms \qquad (18\text{-}35)$$

In fact, when the semi-empirical procedure is applied for calculations it is believed that the proposed scheme is powerful enough to compensate, through the set of newly introduced parameters, $\Omega_\lambda^{\kappa k}$, for the truncation of the *cross terms* and as a consequence use for the fitting the simplified expression

$$S_{f \leftarrow i} = \sum_\lambda \sum_{\kappa k} \Omega_\lambda^{\kappa k} \mid \langle \Psi_f \mid W^{(\kappa k)\lambda} \mid \Psi_i \rangle \mid^2 \qquad (18\text{-}36)$$

In this way the standard Judd-Ofelt intensity parameters are replaced formally by new objects, namely

$$\Omega_2 \Rightarrow \Omega_2^{02}, \Omega_2^{11}, \Omega_2^{12}, \Omega_2^{13}$$
$$\Omega_4 \Rightarrow \Omega_4^{04}, \Omega_4^{13}, \Omega_4^{14}, \Omega_4^{15}$$
$$\Omega_6 \Rightarrow \Omega_6^{06}, \Omega_6^{15}, \Omega_6^{16}, \Omega_6^{17}.$$

However only those for which the hermiticity condition is satisfied contribute to the line strength, and therefore in practice the number of the intensity parameters is reduced to nine. In addition, since there is no limitation on the values of the final rank λ of the double operator, the parameters for odd ranks are also present in this scheme of parametrization. This means that in practice very often the number of parameters is larger than the measured intensities available for the purpose of performing the fitting procedure. It is crucial then to establish the hierarchy of those terms that play the most important role. This issue is addressed in a recent paper [17], where the results of the pioneering calculations performed for the Eu^{3+} ion are presented. Among the transitions analyzed there all those forbidden within the standard scheme are included, and for the first time the oscillator strengths of $0 \longrightarrow 0$ and $0 \longleftrightarrow 1$ transitions are reproduced. The final conclusions derived from the numerical analysis showed much better accuracy of the reproduction of all transitions, and in particular of the hypersensitive ones.

REFERENCES

1. Wybourne B G, (1965) Use of Relativistic Wave Functions in Crystal Field Theory *J. Chem. Phys.*, **43** 4506.
2. Sandars P G H and Beck J, (1965) Relativistic Effects in Many Electron Hyperfine Structure I. Theory *Proc. R. Soc.*, **A 287** 97.
3. Bordarier Y, Judd B R and Klapisch M, (1965) Hyperfine Structure of EuI *Proc. R. Soc.*, **A 289** 81.
4. Wybourne B G, (1966) Energy Levels of Trivalent Gadolinium and Ionic Contributions to the Ground-State Splitting, *Phys. Rev.,* **148** 317.

5. Judd B R, (1998) *Operator Techniques in Atomic Spectroscopy* Princeton: Princeton University Press.

6. Kędziorski A, Smentek L and Wybourne B G, (2004) Net-value of the Relativistic Crystal Field Effect *J. Alloys Compd.,* **380** (2004) 151.

7. Judd B R and Lindgren I, (1961) Theory of the Zeeman Effect in the Ground Multiplets of Rare-Earth Atoms *Phys. Rev.,* **122** 1802.

8. Smentek L, Wybourne B G and Kobus J, (2001) Relativistic Crystal Field for S-state f-electron Ions *J. Phys. B,* **34** 1513.

9. Parpia F A, Froese Fischer C and Grant I P, (1996) GRASP92: A Package for Large-scale Relativistic Atomic Structure Calculations *Comput. Phys. Commun.,* **94** 249.

10. Smentek L and Wybourne B G, (2000) A Relativistic Model of $f \leftrightarrow f$ Transitions *J. Phys. B,* **33** 3647.

11. Smentek L and Wybourne B G, (2001) Relativistic $f \leftrightarrow f$ Transitions in Crystal Fields II: Beyond the Single Configuration Approximation *J. Phys. B,* **34** 625.

12. Smentek L, (2000) Morphology of Intensity Parameters *Mol. Phys.,* **98** 1233.

13. Jørgensen C K and Judd B R, (1964) Hypersensitive Pseudoquadrupole Transitions in Lanthanides *Mol. Phys.,* **8** 281.

14. Peacock R D, (1975) The Intensities of Lanthanide f-f Transitions, *Structure and Bonding,* **22** 83.

15. Mason S F, Peacock R D and Steward B, (1975) Ligand-polarization Contributions to the Intensity of Hypersensitive Trivalent Lanthanide Transitions *Mol Phys.,* **30** 1829.

16. Peacock R D, (1977) The Charge-transfer Contribution to the Intensity of Hypersensitive Trivalent Lanthanide Transitions *Mol. Phys.,* **33** 1239.

17. Smentek L and Kędziorski A, (2006) New Parametrization of f-spectra (submitted for publication).

19 Magnetic Dipole Transitions in Crystals

A fact is a simple statement that everyone believes. It is innocent, unless found guilty. A hypothesis is a novel suggestion that no one wants to believe. It is guilty, until found effective.

Edward Teller

Here we return to the question of the role of magnetic dipole transitions in crystals. So much attention has been given, and rightly so, to forced electric dipole transitions in crystals that the possibility of magnetic dipole transitions is often overlooked. In the following we outline the principal features of magnetic dipole transitions and those characteristics that distinguish them from electric dipole transitions.

19.1 POLARIZATION OF LIGHT AND TRANSITIONS

We already noted in Table 7-1 the selection rules for magnetic dipole transitions in free atoms or ions. They involve transitions between states of the *same* parity within a given electron configuration, and in the absence of spin-orbit coupling, with the same LS multiplet. The basic theory was worked out by Pasternack[1] and Shortley[2]. The magnetic dipole moment is given by

$$\mu = -\frac{e}{2mc}(\mathbf{L} + g_s \mathbf{S}) \tag{19-1}$$

We note that the selection rules are essentially the same as in the Zeeman effect.

The oscillator strength, f_σ, of a magnetic dipole transition may be written as

$$f_\sigma = \frac{8\pi^2 mc}{3he^2}\sigma \left| \frac{-e}{2mc} \langle \alpha SLJM_J|(\mathbf{L} + g_s\mathbf{S})^{(1)}_\rho|\alpha SLJ'M'_J\rangle \right|^2 \eta$$

$$= 4.028 \times 10^{-11}\sigma \left| \langle \alpha SLJM_J|(\mathbf{L} + g_s\mathbf{S})^{(1)}_\rho|\alpha SLJ'M'_J\rangle \right|^2 \eta \tag{19-2}$$

where σ is in units of cm^{-1} and η is the refractive index of the medium. Using the Wigner-Eckart theorem, the matrix elements of $(\mathbf{L} + g_s\mathbf{S})^{(1)}_\rho$ are given by

$$\langle \alpha SLJM_J|(\mathbf{L} + g_s\mathbf{S})^{(1)}_\rho|\alpha SLJ'M'_J\rangle = (-1)^{J-M_J} \begin{pmatrix} J & 1 & J' \\ -M_J & \rho & M'_J \end{pmatrix}$$

$$\times \langle \alpha SLJ\|(L + g_s S)^{(1)}\|\alpha SLJ'\rangle \tag{19-3}$$

where $\rho = 0$ gives the z component of the magnetic vector, which corresponds to the absorption or emission of $\sigma-$polarized light and $\rho = \pm 1$ gives the $x \pm iy$ components corresponding to π −polarized light. The calculation of the matrix elements in (19-3) is similar to that for the Zeeman effect discussed in chapter 2.

Polarization of light provides an important experimental tool for distinguishing between electric dipole and magnetic dipole transitions. In electric dipole transitions it is the electric vector of the light that is active, whereas in magnetic dipole transitions the magnetic vector is active. Recall that in light the directions of the electric vector, magnetic vector and direction of propagation are mutually perpendicular to one another. Consider a uniaxial crystal; the light path *parallel* to the crystallographic axis is termed the *"axial"* spectrum, while for a light path *perpendicular* to the axis is termed the *"transverse"* spectrum. The *transverse* spectrum is split into π or σ spectra depending on whether the axis of polarization is such that the direction of the electric vector is parallel to or perpendicular to the crystal axis. If a line in the axial spectrum and the σ spectra coincide, the transition is an electric dipole, while if the axial and π spectra coincide, the transition is either a magnetic dipole or electric quadrupole. Electric quadrupole transitions are usually a couple of orders of magnitude weaker than magnetic dipole transitions, and henceforth they are ignored in the present discussion. Practically, in the absorption of light in a uniaxial crystal the incident light may be polarized so that the axial or transverse spectra are observed, and conversely in flourescence the axial and transverse spectra are determined by examining the polarization of the emitted radiation.

19.2   SELECTION RULES FOR TRANSITIONS IN CRYSTALS

In many cases a given transition between a pair of crystal field levels of an ion located at a site whose symmetry is characterized by a particular crystallographic point group G may be assigned to be of electric dipole or magnetic dipole origin by determining the polarization state of the photons emitted or absorbed by the transition. In the case of electric dipole transitions the transformation of the components (x, y, z) with respect to the operations of the group G leads directly to the symmetry selection rules. In the case of magnetic dipole transitions, one needs to know the transformation properties of the rotation components (R_x, R_y, R_z), or equivalently those of (L_x, L_y, L_z). If components (x, y, z) transform according to different representations of G from those of (L_x, L_y, L_z), then the electric dipole transitions may be unequivocally distinguished from those of magnetic dipole transitions by experimentally determining the polarization of the transitions.

As an example the point symmetry group D_{3h} that is associated with a wide range of uniaxial crystals, such as is found in the lanthanide trichlorides and ethylsulphates, is analyzed. The group characters are given in Table 19-1 below[3] with the spin (or projective) representations being separated from the ordinary irreducible representations by an empty line. The ordinary irreducible representations $(\Gamma_i, i = 1, \ldots, 6)$ are associated with an *even* number of electrons, and the spin irreducible representations $(\Gamma_i, i = 7, \ldots, 9)$ with an *odd* number of electrons.

For electric dipole transitions z transforms as the Γ_4 representation with (x, y) spanning the Γ_6 representation. For magnetic dipole transitions L_z transforms as the Γ_2 representation with (L_x, L_y) spanning the Γ_5 representation. Since the irreducible

TABLE 19-1
The Characters of D_{3h}

Γ_i	E	$\bar{E}$	σ_h	$\bar{\sigma}_h$	$2C_3$	$2\bar{C}_3$	$2S_3$	$2\bar{S}_3$	$3C'_2$	$3\bar{C}'_2$	$3\sigma_v$	$3\bar{\sigma}_v$
Γ_1	1	1	1	1	1	1	1	1	1	1	1	1
Γ_2	1	1	1	1	1	1	1	1	-1	-1	-1	-1
Γ_3	1	1	-1	-1	1	1	-1	-1	1	1	-1	-1
Γ_4	1	1	-1	-1	1	1	-1	-1	-1	-1	1	1
Γ_5	2	2	-2	-2	-1	-1	1	1	0	0	0	0
Γ_6	2	2	2	2	-1	-1	-1	-1	0	0	0	0
Γ_7	2	-2	0	0	1	-1	$\sqrt{3}$	$-\sqrt{3}$	0	0	0	0
Γ_8	2	-2	0	0	1	-1	$-\sqrt{3}$	$\sqrt{3}$	0	0	0	0
Γ_9	2	-2	0	0	-2	2	0	0	0	0	0	0

representations of D_{3h} are all real, we may succinctly write the magnetic dipole (Md) and electric dipole (Ed) selection rules as

$$\Gamma_i \times \Gamma_j \supset \begin{cases} \Gamma_2 & \sigma \text{ Md} \\ \Gamma_5 & \pi \text{ Md} \\ \Gamma_4 & \pi \text{ Ed} \\ \Gamma_6 & \sigma \text{ Ed} \end{cases} \tag{19-4}$$

The Kronecker products $\Gamma_i \times \Gamma_j$ for the ordinary irreducible representations may be readily evaluated[3] to give

$$\begin{array}{c c c c c c c} & \Gamma_1 & \Gamma_2 & \Gamma_3 & \Gamma_4 & \Gamma_5 & \Gamma_6 \\ \Gamma_1 & \Gamma_1 & \Gamma_2 & \Gamma_3 & \Gamma_4 & \Gamma_5 & \Gamma_6 \\ \Gamma_2 & \Gamma_2 & \Gamma_1 & \Gamma_4 & \Gamma_3 & \Gamma_5 & \Gamma_6 \\ \Gamma_3 & \Gamma_3 & \Gamma_4 & \Gamma_1 & \Gamma_2 & \Gamma_6 & \Gamma_5 \\ \Gamma_4 & \Gamma_4 & \Gamma_3 & \Gamma_2 & \Gamma_1 & \Gamma_6 & \Gamma_5 \\ \Gamma_5 & \Gamma_5 & \Gamma_5 & \Gamma_6 & \Gamma_6 & \Gamma_1+\Gamma_2+\Gamma_6 & \Gamma_3+\Gamma_4+\Gamma_5 \\ \Gamma_6 & \Gamma_6 & \Gamma_6 & \Gamma_5 & \Gamma_5 & \Gamma_3+\Gamma_4+\Gamma_5 & \Gamma_1+\Gamma_2+\Gamma_6 \end{array} \tag{19-5}$$

and for the spin irreducible representations

$$\begin{array}{c c c c} & \Gamma_7 & \Gamma_8 & \Gamma_9 \\ \Gamma_7 & \Gamma_1+\Gamma_2+\Gamma_5 & \Gamma_3+\Gamma_4+\Gamma_6 & \Gamma_5+\Gamma_6 \\ \Gamma_8 & \Gamma_3+\Gamma_4+\Gamma_6 & \Gamma_1+\Gamma_2+\Gamma_5 & \Gamma_5+\Gamma_6 \\ \Gamma_9 & \Gamma_5+\Gamma_6 & \Gamma_5+\Gamma_6 & \Gamma_1+\Gamma_2+\Gamma_3+\Gamma_4 \end{array} \tag{19-6}$$

Noting (19-4) and (19-5) it is a simple matter to deduce that for an *even* number of electrons the D_{3h} symmetry electric dipole selection rules are

$$
\begin{array}{c|cccccc}
 & \Gamma_1 & \Gamma_2 & \Gamma_3 & \Gamma_4 & \Gamma_5 & \Gamma_6 \\
\hline
\Gamma_1 & - & - & - & \pi & - & \sigma \\
\Gamma_2 & - & - & \pi & - & - & \sigma \\
\Gamma_3 & - & \pi & - & - & \sigma & - \\
\Gamma_4 & \pi & - & - & - & \sigma & - \\
\Gamma_5 & - & - & \sigma & \sigma & \sigma & \pi \\
\Gamma_6 & \sigma & \sigma & - & - & \pi & \sigma
\end{array}
\tag{19-7}
$$

and for an *odd* number of electrons

$$
\begin{array}{c|ccc}
 & \Gamma_7 & \Gamma_8 & \Gamma_9 \\
\hline
\Gamma_7 & - & \sigma\pi & \sigma \\
\Gamma_8 & \sigma\pi & - & \sigma \\
\Gamma_9 & \sigma & \sigma & \pi
\end{array}
\tag{19-8}
$$

Note the appearance of $\sigma\pi$ transitions that effectively correspond to elliptical polarization.

In exactly the same manner we can deduce the corresponding selection rules for magnetic dipole transitions for an *even* number of electrons in D_{3h} symmetry as

$$
\begin{array}{c|cccccc}
 & \Gamma_1 & \Gamma_2 & \Gamma_3 & \Gamma_4 & \Gamma_5 & \Gamma_6 \\
\hline
\Gamma_1 & - & \sigma & - & - & \pi & - \\
\Gamma_2 & \sigma & - & - & - & \pi & - \\
\Gamma_3 & - & - & - & \sigma & - & \pi \\
\Gamma_4 & - & - & \sigma & - & - & \pi \\
\Gamma_5 & \pi & \pi & - & - & \sigma & \pi \\
\Gamma_6 & - & - & \pi & \pi & \pi & \sigma
\end{array}
\tag{19-9}
$$

and for an *odd* number of electrons

$$
\begin{array}{c|ccc}
 & \Gamma_7 & \Gamma_8 & \Gamma_9 \\
\hline
\Gamma_7 & \sigma\pi & - & \pi \\
\Gamma_8 & - & \sigma\pi & \pi \\
\Gamma_9 & \pi & \pi & \sigma
\end{array}
\tag{19-10}
$$

Comparison of the selection rules for electric dipole transitions with those for magnetic dipole transitions shows significant differences and allows us to often distinguish experimentally between the two types of transitions.

19.3   THE OSCILLATOR STRENGTHS FOR THE $^7F_{00} \longleftrightarrow {}^7F_{1M}$ TRANSITIONS

Magnetic dipole transitions are well-known in trivalent Europium and divalent Samarium. In both cases $4f^6$ is the ground configuration with the ground state ^7F. In the free ion the spin-orbit interaction results in seven levels ^7F_J with $J = 0, 1, \ldots, 6$. The ground state has $J = 0$ and in the absence of hyperfine interactions is non-degenerate. The first excited level is $^7F_1$. In a crystal with D_{3h} point symmetry the $^7F_1$ level splits into two sub-levels. In terms of the crystal quantum numbers introduced in section 10.2 the sublevels may be labeled as $\mu = 0$ and $\mu = \pm1$, or in the terms of irreducible representations of D_{3h} group, since they transform as Γ_2 and Γ_5, with the latter being two-fold degenerate. The ground state corresponds to a Γ_1 level. Absorption from the ground state to the two sub-levels involves two distinct transitions, $\Gamma_1 \rightarrow \Gamma_2$ and $\Gamma_1 \rightarrow \Gamma_5$. Inspection of (19-4) shows that these two transitions are electric dipole forbidden, whereas inspection of (19-6) shows that both transitions are allowed magnetic dipole; the $\Gamma_1 \rightarrow \Gamma_2$ transition being σ−polarized and the $\Gamma_1 \rightarrow \Gamma_5$ transition being π−polarized.

To calculate the oscillator strengths of these transitions we first note that

$$\langle ^7F0\|(L + g_s S)^{(1)}\|^7F1\rangle = -2\sqrt{3}(g_s - 1) \tag{19-11}$$

Evaluation of the $3j$−symbol in (19-3) then yields

$$\langle ^7F00|(L + g_s S)^{(1)}_0|^7F10\rangle = -2(g_s - 1) \tag{19-12a}$$

$$\langle ^7F00|(L + g_s S)^{(1)}_{\pm1}|^7F10 \mp 1\rangle = 2(g_s - 1) \tag{19-12b}$$

The two matrix elements are, to within a sign, equal. To a good approximation we can take $g_s = 2$, and hence the squares of both matrix elements become the integer 4. Using these values in (19-2), and remembering that the Γ_5 is two-fold degenerate we find for the two transitions the oscillator strengths as

$$f(\Gamma_1 \rightarrow \Gamma_2) = 1.6 \times 10^{-10}\sigma\eta \tag{19-13a}$$

$$f(\Gamma_1 \rightarrow \Gamma_5) = 3.2 \times 10^{-10}\sigma\eta \tag{19-13b}$$

The ratio of the two oscillator strengths is $1 : 2$ and independent of σ. Thus we would expect this ratio to be the same for both Sm^{2+} and Eu^{3+} when in the same medium.

19.4   INTERMEDIATE COUPLING AND $^5D_1 \longleftrightarrow {}^7F_0$ TRANSITIONS

The selection rules for magnetic dipole transitions, given in Table 7.1, are highly restrictive and at first sight seem to preclude the possibility of magnetic dipole transitions between states belonging to different multiplets, for example,$^5D_1 \longleftrightarrow {}^7F_0$ transitions. Nevertheless polarization studies show that these transitions are definitely of magnetic dipole origin. This possibility arises as a result of intermediate coupling via the spin-orbit interaction. Let us consider the case of Eu^{3+} in a $LaCl_3$ lattice where the Eu^{3+} is located at a site of D_{3h} symmetry. Ofelt[4] has diagonalized the complete Coulomb and spin-orbit energy matrices for a set of parameters appropriate to this case and has given the eigenfunctions for many levels. In particular he gives for the two levels being considered here the principal components of their eigenvectors as linear combination of the following terms

$$|^7\mathcal{F}_0\rangle = 0.9680|^7 F_0\rangle + 0.0016|^5 D_0\rangle + 0.1659|^5 D_0'\rangle - 0.1815|^5 D_0''\rangle \quad (19\text{-}14a)$$
$$|^5\mathcal{D}_1\rangle = -0.2096|^7 F_1\rangle - 0.2066|^5 D_1\rangle + 0.7162|^5 D_1'\rangle - 0.5561|^5 D_1''\rangle \quad (19\text{-}14b)$$

For brevity let us write

$$H_{Md} = (L + g_s S)^{(1)} \qquad (19\text{-}15)$$

Using the eigenvectors we can write

$$\langle^7\mathcal{F}_0\|H_{Md}\|^5\mathcal{D}_1\rangle = -0.2028\langle^7 F_0\|H_{Md}\|^7 F_1\rangle + 0.2201\langle^5 D_1\|H_{Md}\|^5 D_1\rangle \quad (19\text{-}16)$$

The reduced matrix elements evaluate as

$$\langle^7 F_0\|H_{Md}\|^7 F_1\rangle = -2\sqrt{3} \qquad (19\text{-}17a)$$
$$\langle^5 D_1\|H_{Md}\|^5 D_1\rangle = -\sqrt{6} \qquad (19\text{-}17b)$$

leading to the numerical value, again putting $g_s = 2$,

$$\langle^7\mathcal{F}_0\|H_{Md}\|^5\mathcal{D}_1\rangle = 0.1634 \qquad (19\text{-}18)$$

Use of the Wigner-Eckart theorem in (19-3), with evaluation of the $3j-$symbols, leads to the oscillator strengths of the two transitions as

$$f(\Gamma_1 \rightarrow \Gamma_2) = 3.58 \times 10^{-13}\sigma\eta \qquad (19\text{-}19a)$$
$$f(\Gamma_1 \rightarrow \Gamma_5) = 7.17 \times 10^{-13}\sigma\eta \qquad (19\text{-}19b)$$

At first sight the above oscillator strengths appear much smaller than those found in (19-13). However the factor σ is $\sim 300 cm^{-1}$ in (19-13), but $\sim 19,000 cm^{-1}$ in (19-19), leading to the two sets of oscillator strengths being comparable.

19.5   OSCILLATOR STRENGTHS FOR THE $^5 D_1 \longleftrightarrow {}^7 F_1$ MAGNETIC DIPOLE TRANSITIONS

Ofelt[4] gives the principal components of the eigenvector of the $|^7\mathcal{F}_1\rangle$ state as

$$|^7\mathcal{F}_1\rangle = 0.9742|^7 F_1\rangle + 0.0052|^5 D_1\rangle + 0.1472|^5 D_1'\rangle - 0.1645|^5 D_1''\rangle \quad (19\text{-}20)$$

Making use of the corresponding expansion for the $|^5\mathcal{D}_1\rangle$ state given in (19-14b) we obtain the reduced matrix element as

$$\langle^7\mathcal{F}_1\|H_{Md}\|^5\mathcal{D}_1\rangle = -0.2042\langle^7 F_1\|H_{Md}\|^7 F_1\rangle + 0.1959\langle^5 D_1\|H_{Md}\|^5 D_1\rangle \quad (19\text{-}21)$$

In this case the two reduced matrix elements on the right-hand-side of (19-21) are both equal to

$$\langle^7 F_1\|H_{Md}\|^7 F_1\rangle = \langle^5 D_1\|H_{Md}\|^5 D_1\rangle = \frac{\sqrt{6}}{2}(g_s + 1) \qquad (19\text{-}22)$$

As a result there is strong cancelation of the two terms in (19-21) and as noted by Judd[5] the $^5D_1 \longleftrightarrow {}^7F_1$ are of electric dipole character rather than magnetic dipole.

19.6 J−MIXING AND "INTENSITY BORROWING"

In the preceding sections we have seen how intermediate coupling, via the spin-orbit interaction can overcome the usual selection rules on the spin, S, and orbital, L, quantum numbers while still leaving J as a "good" quantum number. The presence of a crystal field can destroy J and M_J as "good" quantum numbers leading to $J-$ and $M-$mixing. This can make possible some transitions, which at first sight seem to be highly forbidden. A case in point would be the transition $^7F_0 \leftrightarrow {}^5D_0$. Evidence that such a transition can come from $J-$mixing via a crystal field comes from the observation that the ground state of Eu^{3+} in crystals exhibits an electric quadrupole hyperfine splitting originating from the crystal field coupling via the axial crystal field term $A_0^2 C_0^{(2)}$. This results in the ground state having a small admixture of $|^7F20\rangle$ character. We can estimate the amount of mixing by calculating the matrix element

$$\langle {}^7F00|C_0^{(2)}|^7F20\rangle = (-1)^{0-0} \begin{pmatrix} 0 & 2 & 2 \\ 0 & 0 & 0 \end{pmatrix} \langle 3\|C^{(2)}\|3\rangle \langle {}^7F0\|U^{(2)}\|^7F2\rangle$$

$$= -\frac{2}{15}\sqrt{21}\langle {}^7F0\|U^{(2)}\|^7F2\rangle$$

$$= -\frac{2}{15}\sqrt{21}(-1)^{3+3+0+2}\sqrt{1\times 5}\begin{Bmatrix} 3 & 2 & 3 \\ 2 & 3 & 0 \end{Bmatrix}\langle {}^7F\|U^{(2)}\|^7F\rangle$$

$$= -\frac{2}{15}\sqrt{3}\langle {}^7F\|U^{(2)}\|^7F\rangle$$

$$= +\frac{2}{15}\sqrt{3} \tag{19-23}$$

where in the last step we have used

$$\langle {}^7F\|U^{(2)}\|^7F\rangle = -1 \tag{19-24}$$

Let

$$\Delta_{20} = E(^7F20) - E(^7F00) \tag{19-25}$$

Then recalling section 15.2 and a note on the rank two matrices presented there, we can express the ground state energy as

$$E(^7\mathcal{F}00) \sim E(^7F00) - \frac{4}{75}\frac{\left(A_0^2\right)^2}{\Delta_{20}} \tag{19-26}$$

and the corresponding ground state eigenvector as

$$|^7\mathcal{F}00\rangle \sim \frac{1}{1+\frac{\sqrt{3}A_0^2}{15\Delta_{20}}}\left(|^7F00\rangle - \frac{2\sqrt{3}}{15}\frac{A_0^2}{\Delta_{20}}|^7F20\rangle\right) \tag{19-27}$$

We can estimate the value of the crystal field parameter from the observed crystal field splitting of the $^7F_1$ level in D_{3h} symmetry. Calculating in a similar manner to (19-23) we readily find that

$$\langle ^7F10|A_0^2C_0^{(2)}|^7F10\rangle = +\frac{1}{5}A_0^2 \tag{19-28a}$$

$$\langle ^7F1\pm 1|A_0^2C_0^{(2)}|^7F1\pm 1\rangle = -\frac{1}{10}A_0^2 \tag{19-28b}$$

and hence the splitting, Δ_1, of the $^7F_1$ level by such an axial field is approximately

$$\Delta_1 = \frac{3}{10}A_0^2 \tag{19-29}$$

In europium ethyl sulphate the observed splitting is $\sim 42cm^{-1}$, which gives the value of the crystal field parameter

$$A_0^2 \sim 145cm^{-1} \tag{19-30}$$

and the matrix element

$$\langle ^7F00|A_0^2C_0^{(2)}|^7F20\rangle =\sim 33.5cm^{-1} \tag{19-31}$$

Given that experimentally the $^7F_1$ level is about $1100cm^{-1}$ above the ground state we can deduce that the ground state, to a first approximation, has an admixture of the state $^7F_{20}$,

$$|^7\mathcal{F}00\rangle = 0.9995|^7F00\rangle - 0.03041|^7F20\rangle \tag{19-32}$$

While the mixing coefficient is very small, there is nevertheless some probability, even at this simple level of numerical calculations, to expect that the forbidden transition $^7F_0 \longleftrightarrow {}^5D_0$ is "borrowing intensity" from the electric dipole transition $^7F_2 \longleftrightarrow {}^5D_0$, which is allowed via the Judd-Ofelt mechanism. The square of the coefficient of $|^7F20\rangle$ gives a direct measure of the "borrowing of intensity". Clearly our calculation is overly simplistic. Ideally we should consider a complete crystal field $J-$mixing calculation that includes $J-$mixing in at least the ^5D multiplet. The whole subject of $J-$mixing as it affects the $^7F_0 \longleftrightarrow {}^5D_0$ transition has been extensively studied by Tanaka and Kushida and their associates[6-13], both theoretically and experimentally. Are there other contributing physical mechanisms to the intensity of $^7F_0 \longleftrightarrow {}^5D_0$? This subject is discussed in a separate section.

19.7 PERTURBATION APPROACH AND HIGHER-ORDER CONTRIBUTIONS

The amplitude of the magnetic dipole transition is evaluated above taking into account the spin-orbit interactions represented by the coefficients of linear combination that defines the wave function within the intermediate coupling scheme. These coefficients of various components mixed together via the spin-orbit interaction operator are evaluated through the diagonalization of the appropriate matrix that contain the corrections to the energy caused by this particular mechanism. In order to break the free ionic system approximation of this scheme, the J–mixing via the crystal field potential is included, and finally the transition amplitude discussed above contains the impact due to both interactions. It is possible however, and in fact it is clearly illustrative, to regard directly the crystal field potential and spin-orbit interaction in a perturbative way, similarly as it is done in the case of electric dipole transitions in chapter 17. While it is standard procedure to evaluate the amplitude of magnetic dipole transition in the way presented above (which, in the nomenclature of the perturbation approach is based on the first order term), a new formulation which includes higher order contributions would eventually be adequate for theoretical model of electric dipole transitions. This is required not only to improve the accuracy of the description, but also to be able to verify the importance of the "borrowing mechanism" and its role in the interpretation of the unusual transitions by means of (17-44).

Thus, the present task is to evaluate the following matrix element in the best way possible for a lanthanide ion in a crystal

$$\langle 4f^N \alpha \Psi \mid (L + g_s S)_\rho^{(1)} \mid 4f^N \alpha \Psi' \rangle \tag{19-33}$$

where Ψ and Ψ' are the wave functions expressed in the terms of the corrections evaluated for the following hamiltonian (see for comparison (17-16)),

$$H = H_0 + \lambda V_{cryst} + \mu V_{so} \tag{19-34}$$

where H_0 is the operator that describes the system within the central field approximation of Hartree-Fock model, V_{cryst} is the crystal field potential, and V_{so} represents the operator of the spin-orbit interaction of the form,

$$V_{so} = \sum_i r_i^{-3} \left(s_i^{(1)} \cdot l_i^{(1)} \right) \tag{19-35}$$

The matrix elements of D_{Md} are non-zero only between the functions of the same configuration. Note that there is no radial term associated with the magnetic dipole radiation operator, and therefore even the configurations that differ just by one-electron states with various principal quantum numbers give zero contributions. Therefore, the perturbing operators in (19-34) are not limited to those that connect various configurations as in the case of description of electric dipole transitions in (17-16). In general they represent the impact of a given interaction in the whole space. This means that the intra part associated with the projection operator P, and also the inter part, which is within the orthogonal space Q, are possibly taken into account via V_{cryst} and V_{so}. As a consequence, in order not to include certain interactions twice, the zero order problem has to be based on the single configuration approximation of the Hartree-Fock model for a free ionic system. This means that the energy states involved in the matrix element in (19-33) are expressed in a pure LS-coupling. Thus instead of the coefficients of linear combinations in the intermediate coupling scheme (spin-orbit impact), and instead of identification of the states by the crystal field quantum numbers, now we have the contributions to the transition amplitude defined at various orders of the perturbation expansion. Again, since we are discussing the so-called *properties* of many electron system rather than its energy, instead of *corrections*, which contribute only to the energy, various *contributions* to the transition amplitude are presented here.

Taking into account the corrections to the wave functions up to the second order in the crystal field perturbation, and the first order with respect to the spin orbit interaction, the transition amplitude consists of the following terms

$$\langle 4f^N \Psi \mid D_{Md} \mid 4f^N \Psi' \rangle = \langle 4f^N \psi \mid D_{Md} \mid 4f^N \psi' \rangle$$

$$+ \lambda \sum_x \langle 4f^N \psi \mid D_{Md} \mid 4f^N x \rangle \langle 4f^N x \mid V_{cryst} \mid 4f^N \psi' \rangle$$

$$+ \mu \sum_y \langle 4f^N \psi \mid D_{Md} \mid 4f^N y \rangle \langle 4f^N y \mid V_{so} \mid 4f^N \psi' \rangle$$

$$+ \lambda\mu \sum_x \sum_{Yy} \langle 4f^N \psi | D_{Md} | 4f^N x \rangle \langle 4f^N x | V_{cryst} | Yy \rangle \langle Yy | V_{so} | 4f^N \psi' \rangle$$

$$+ \lambda^2 \mu \sum_x \sum_{Yy} \sum_{Zz} \langle 4f^N \psi | D_{Md} | 4f^N x \rangle \langle 4f^N x | V_{cryst} | Yy \rangle \langle Yy | V_{so} | Zz \rangle \langle Zz | V_{cryst} | 4f^N \psi' \rangle$$

$$+ \lambda^2 \mu \sum_x \sum_{Yy} \sum_{Zz} \langle 4f^N \psi | D_{Md} | 4f^N x \rangle \langle 4f^N x | V_{cryst} | Yy \rangle \langle Yy | V_{cryst} | Zz \rangle \langle Zz | V_{so} | 4f^N \psi' \rangle$$

$$+ \langle 4f^N \psi | D_{Md} \longleftrightarrow V_{cryst} \longleftrightarrow V_{so} | 4f^N \psi' \rangle \tag{19-36}$$

where the energy denominators are omitted for simplicity. However their presence is crucial, since they prevent performing the complete closure over the intermediate states of these perturbing expressions.

The first matrix element at the right-hand-side of (19-36) represents the first order contribution, the same that was analyzed in the previous section with various functions used for its evaluation. All the terms higher than the first order matrix element are beyond the standard formulation, and they improve the theoretical description of magnetic dipole transitions. The terms that are proportional to the perturbing parameters λ or μ are of second-order, and they represent the direct impact of the crystal field potential and spin-orbit interaction, respectively. Note that these terms are already in the effective operator forms, since they are products of matrix elements between the functions describing the energy levels of the ground configuration. Thus, the summations are only performed over the excited states x of $4f^N$. Furthermore, in these terms there are no contributions caused by the interactions with the excited configurations, since $V_{cryst} \equiv P V_{cryst} P$ and $V_{so} \equiv P V_{so} P$.

The terms associated with $\lambda\mu$ are of the third order, and they represent the interplay of both mechanisms via the excited configurations Y that connect the last two matrix elements proportional to $\lambda\mu$. The parity selection rules for the non-vanishing matrix elements of the spin orbit interaction operator require that Y is a singly excited configuration of the same parity as the parity of $4f^N$. This parity condition also allows the inclusion of the impact of this particular sequence of interactions that takes place between the excited states of the ground configuration. Thus, among the third-order terms that originate from the perturbing influence of the excited configurations, there are also terms of the following structure,

$$D_{Md}(\psi, x)V_{cryst}(x, y)V_{so}(y, \psi') \qquad (19\text{-}37)$$

$$V_{cryst}(\psi, x)D_{Md}(x, y)V_{so}(y, \psi') \qquad (19\text{-}38)$$

where

$$D_{Md}(\psi, x) = \langle 4f^N \psi \mid L^{(1)} + g_s S^{(1)} \mid 4f^N x \rangle \qquad (19\text{-}39)$$

$$V_{cryst}(x, y) = \sum_{tp} B_p^t \langle 4f|r^t|4f\rangle\langle f||C^{(t)}||f\rangle\langle 4f^N x|U_p^{(t)}(ff)|4f^N y\rangle \qquad (19\text{-}40)$$

and all matrix elements are evaluated between the functions of the $4f^N$ configuration.

However the most interesting contributions are those which represent the inter shell interactions via V_{cryst} and V_{so}, which are linked by singly excited intermediate configurations $Y \equiv 4f^{N-1}n'f$,

$$D_{Md}(\psi, x)V_{cryst}(x, 4f^{N-1}n'fy)V_{so}(4f^{N-1}n'fy, \psi') \qquad (19\text{-}41)$$

and the summation is performed over the set of the excited configurations for $n' >= 5$. This is the case if the radial basis sets of one-electron functions are generated for the average energy of the configurations and with the frozen core orbitals for the excited orbitals. The singly excited configurations in (19-41) are limited to those with one electron promoted from the $4f$ shell to the other one-electron excited state of f symmetry, since these are the selection rules for the non-vanishing matrix elements of the spin orbit interaction operator. This implies, at the same time (due to the parity requirements), that in the expression (19-41) the even part of the crystal field potential contributes to the transition amplitude.

Adopting the same approximations as in the Judd-Ofelt model of $f \longleftrightarrow f$ electric dipole transitions about the excited configurations, which are seen as degenerate in relation to the energy distance from the ground configuration, it is possible to perform the partial closure to derive from the product of the last two matrix elements in (19-41) a new effective operator F_2 that contributes to the transition amplitude defined at the third order in the following way

$$D_{Md}(\psi, x)F_2(x, \psi') \qquad (19\text{-}42)$$

where the new effective operator has the following tensorial structure

$$F_2(x, \psi') = \frac{2}{\sqrt{3}} \sum_{tp}^{even}(-1)^{t+1}[t]^{-1/2}B_p^t\langle 4f|r^{-3}|\varrho^t(4f \longrightarrow f)\rangle a_{so}(f)\langle f||C^{(t)}||f\rangle$$

$$\times \sum_k^{even}[k]\begin{Bmatrix} t & k & 1 \\ f & f & f \end{Bmatrix} \langle 4f^N x|W_p^{(1k)t}(ff)|4f^N \psi'\rangle \qquad (19\text{-}43)$$

The effective double tensor operator in (19-43) originates from the spin-orbit interaction

$$V_{so}(x, y) = \langle 4f | r^{-3} | 4f \rangle a_{so}(f) \langle 4f^N x | W^{(11)0}(ff) | 4f^N y \rangle \tag{19-44}$$

where $a_{so}(\ell)$ denotes the reduced matrix elements of angular momentum operators $\ell^{(1)}$ and $s^{(1)}$ and it has the value $\frac{\sqrt{3}}{\sqrt{2}}[\ell(\ell+1)(2\ell+1)]^{1/2}$; $\varrho^t(4f \longrightarrow f)$ is the perturbed function, which in fact is the same as defined by (17-36) of the Judd-Ofelt theory of electric dipole $f \longleftrightarrow f$ transitions.

The effective operator F_2 also contributes to the amplitude of magnetic dipole transition at the fourth order. These are the terms in (19-36) that are proportional to $\lambda^2 \mu$, and for a particular choice of the intermediate configurations we have,

$$D_{Md}(\psi, x) V_{cryst}(x, y) F_2(y, \psi') \tag{19-45}$$

$$D_{Md}(\psi, x) F_2(x, y) V_{cryst}(y, \psi') \tag{19-46}$$

The analysis of the higher order terms contributing to the magnetic dipole transition amplitude is extended here to the fourth order as seen in (19-36). These are the terms associated with $\lambda^2 \mu$, which means that the perturbing influence of the crystal field potential is included twice, while the spin orbit interaction is taken into account only once. These are the very terms suggested by Wybourne in 1968 [14,15] as important in the description of the electric dipole transitions (see (17-46)). Thus, it is interesting to verify whether the so-called "borrowing intensity" mechanisms, extended here by the impact due to the excited configurations, is also important in the case of magnetic dipole transitions [16,17]. Inspection of the last two products of the matrix elements in (19-36) shows that at the fourth order the two effective operators, which are connected via the excited configurations, have to be introduced. They represent two possible sequences of operators, namely $V_{cryst} V_{so} V_{cryst}$ and $V_{cryst} V_{cryst} V_{so}$. From a physical point of view they both represent intra-, within P, and inter-shell, between $P - Q$, interactions via V_{cryst}. The difference is in the spin-orbit interaction part, since in the first term the spin orbit interaction within the Q space is included, and in the case of the second one the inter-shell interactions $P - Q$ are taken into account. Thus, two different effective operators are derived from the contraction of the inter-shell operators. Performing the partial closure we get

$$F_4(x, \psi') = \frac{2}{\sqrt{3}} \sum_{t_1 p_1} \sum_{t_2 p_2} \sum^{p(t_1)=p(t_2)} B_{p_1}^{t_1} B_{p_2}^{t_2} \sum_{\ell'} \langle \varrho^{t_1}(4f \longrightarrow \ell') | r^{-3} | \varrho^{t_2}(4f \longrightarrow \ell') \rangle$$

$$\times \langle f || C^{(t_1)} || \ell' \rangle \langle \ell' || C^{(t_2)} || f \rangle a_{so}(\ell') \sum_k^{odd} [k] \sum_{\lambda\mu} (-1)^{t_1 - \mu + \lambda} [\lambda]^{1/2}$$

$$\times \begin{pmatrix} t_1 & t_2 & \lambda \\ p_1 & p_2 & -\mu \end{pmatrix} \begin{Bmatrix} 1 & k & \lambda \\ \ell' & f & t_2 \\ \ell' & f & t_1 \end{Bmatrix} \langle 4f^N x | W_\mu^{(1k)\lambda}(ff) | 4f^N \psi' \rangle$$

$$\tag{19-47}$$

and

$$F_5(x, \psi') = \frac{2}{\sqrt{3}} \sum_{t_1 p_1} \sum_{t_2 p_2}^{p(t_1) = p(t_2)} B_{p_1}^{t_1} B_{p_2}^{t_2} \sum_{\ell'} \langle \varrho^{t_1}(4f \longrightarrow \ell') | r^{t_2} | \varrho^{-3}(4f \longrightarrow f) \rangle$$

$$\times \langle f || C^{(t_1)} || \ell' \rangle \langle \ell' || C^{(t_2)} || f \rangle a_{so}(f) \sum_k \overset{even}{\sum_{\lambda\mu}} (-1)^{k-\mu} [\lambda]^{1/2}$$

$$\times \begin{pmatrix} t_1 & t_2 & \lambda \\ p_1 & p_2 & -\mu \end{pmatrix} \begin{Bmatrix} t_2 & \lambda & t_1 \\ f & \ell' & f \end{Bmatrix} \begin{Bmatrix} 1 & k & \lambda \\ f & f & f \end{Bmatrix}$$

$$\times \langle 4f^N x | W_\mu^{(1k)\lambda}(ff) | 4f^N \psi' \rangle \tag{19-48}$$

where a_{so} is the same as defined in (19-44). These new effective operators, F_4 and F_5, determine the fourth order contributions to the transition amplitude,

$$D_{Md}(\psi, x) F_4(x, \psi') \tag{19-49}$$

$$D_{Md}(\psi, x) F_5(x, \psi'). \tag{19-50}$$

In (19-47) and (19-48) the following radial integrals defined in the terms of the perturbed functions are used

$$R_{F_2}^{t_1 t_2}(\ell') = \langle \varrho^{t_1}(4f \longrightarrow \ell') | r^{-3} | \varrho^{t_2}(4f \longrightarrow \ell') \rangle \tag{19-51}$$

$$R_{F_5}^{t_1 t_2}(\ell') = \langle \varrho^{t_1}(4f \longrightarrow \ell') | r^{t_2} | \varrho^{-3}(4f \longrightarrow f) \rangle \tag{19-52}$$

where only the function $\varrho^{-3}(4f \longrightarrow f)$ is new, and it is a solution of the differential equation (17-37) with the the radial part of the spin orbit interaction at the right-hand-side. The remaining integrals of the effective operators are the same as defined in the case of theoretical description of $f \longleftrightarrow f$ electric dipole transitions.

The amplitude of the magnetic dipole transition presented in (19-36) contains the perturbing influence of the crystal field potential and the spin orbit interaction. Since the zero order eigenvalue problem is defined at the level of Hartree-Fock, the approach is based on the central field approximation and all the correlation effects are not included. As discussed previously the impact due to these effects is very important in the theoretical description of electric dipole transitions. Thus, in order to formulate an adequate model of magnetic dipole transitions, the Hamiltonian in (19-34) should be implemented by a non-central part of the Coulomb interaction as defined in (17-26). It is easily deduced that already at the second order very important terms arise,

$$\sum_x \langle 4f^N \psi | D_{Md} | 4f^N x \rangle \langle 4f^N x | V_{corr} | 4f^N \psi' \rangle \tag{19-53}$$

and they represent the perturbing influence of all the excited states of $4f^N$ configuration taken into account via the electrostatic interaction. Here, due to the selection rules for the non-vanishing matrix elements of D_{Md}, the intermediate states are limited to the excited energy states of ground configuration. In the terms of effective operators, (19-53) leads to the two particle objects, which originate from the Coulomb interaction taken as a perturbation. There are also one particle operators derived from (19-53) for the Hartree-Fock potential. Only at the third order is the perturbing influence of the excited configurations included. Indeed, in the case of the following triple product of the matrix elements,

$$\sum_{x}\sum_{Yy}\langle 4f^{N}\psi|D_{Md}|4f^{N}x\rangle\langle 4f^{N}x|V_{cryst}|Yy\rangle\langle Yy|V_{corr}|4f^{N}\psi'\rangle \qquad (19\text{-}54)$$

Y denotes a singly excited configuration. A similar expression should be constructed with the V_{so} replacing V_{cryst}. Also all the other terms that differ from each other by a sequence of operators in the product of the matrix elements have to be taken into account in order to have the complete picture of this physical model. It is evident that already at the third order there is an explosion of various effective operators. Their angular parts and radial integrals depend on the kind of the mechanism taken into account and also on the intermediate states.

Similarly as in the discussion of the electric dipole transitions a crucial question arises here: *are the higher order contributions important?* To answer this question and to establish a hierarchy of important effective operators, detailed numerical calculations have to be performed for each effective operator. Application of the perturbed functions in all radial integrals eliminates the inaccuracy caused by the incompleteness of the radial basis sets. It guarantees at the same time that the results of performed calculations are exact within the model applied, and therefore they are the source of reliable conclusions.

At this point of the discussion, repeating the pattern established in the case of the electric dipole transitions in chapters 17 and 18, a relativistic version of the approach should be presented. This would be however beyond the scope of this presentation (not to mention the patience of the reader), and therefore this part of the formalism is preserved for another occasion. However, as seen in (19-43), (19-47) and (19-48), the effective operators contributing to the magnetic transition amplitude at various orders are associated with double tensor operators that act within the spin orbital space. Returning to the discussion of the borrowing of intensity, especially in the case of the unusual electric dipole transitions, one may conclude that part of the amplitude of the magnetic transition is defined within the same parametrization scheme as formulated for the electric dipole transitions in (18-36). This means that when the intensity parameters $\Omega_{\lambda}^{\kappa k}$ are obtained from the fitting procedure they include also the contributions from the magnetic dipole transitions; this conclusion gives another dimension to our understanding of the borrowing intensity mechanism in (17-44).

The theory of magnetic dipole transitions presented so far compensates for the lack of interest noted in the literature, and obviously their importance in the description of the spectroscopic properties of lanthanides in crystals is not overlooked by us.

19.8 EXERCISES

19-1. Use the character table for the point group C_{3v} (Table 19-2) to construct two tables of Kronecker products, one for the ordinary irreducible representations and one for the spin irreducible representations of C_{3v}.

19-2. Given that z transforms as Γ_1, and (x, y) as Γ_3 deduce the selection rules for electric dipole transitions for even and odd numbers of electrons for C_{3v} symmetry.

19-3. Given that L_z transforms as Γ_2 and (L_x, L_y) as Γ_3 deduce the selection rules for magnetic dipole transitions for even and odd numbers of electrons for C_{3v} symmetry.

TABLE 19-2
The Character Table of the Point Group
C_{3v}

	E	$\bar{E}$	$2C_3$	$2\bar{C}_3$	$3\sigma_v$	$3\bar{\sigma}$
Γ_1	1	1	1	1	1	1
Γ_2	1	1	1	1	-1	-1
Γ_3	2	2	-1	-1	0	0
Γ_4	2	-2	1	-1	0	0
Γ_5	1	-1	-1	1	i	$-i$
Γ_6	1	-1	-1	1	$-i$	i

REFERENCES

1. Pasternack S, (1940) Transition Probabilities of Forbidden Lines. *Astrophys. J.* **92** 129.
2. Shortley G H, (1940) The Computation of Quadrupole and Magnetic-dipole Transition Probabilities *Phys. Rev.,* **57** 225.
3. Koster G F, Dimock J O, Wheeler R G, and Statz H, (1963) *Properties of the Thirty-Two Point Groups* Cambridge, Mass.: MIT Press.
4. Ofelt G S, (1963) Structure of the f^6 Configuration with Application to Rare-Earth Ions *J. Chem. Phys.,* **38** 2171.
5. Judd B R, (1959) An Analysis of the Fluorescence Spectrum of Europium Ethyl Sulphate *Mol. Phys.,* **3** 407.
6. Nishimura G and Kushida T, (1991) Luminescence Studies in $Ca(PO_3)_2$: Eu^{3+} Glass by Laser-Induced Fluorescence Line-Narrowing Technique. I. Optical Transition Mechanism of the $^5D_0 - ^7F_0$ Line *J. Phys. Soc. Jap.,* **60** 683.
7. Nishimura G and Kushida T, (1991) Luminescence Studies in $Ca(PO_3)_2$: Eu^{3+} Glass by Laser-Induced Fluorescence Line-Narrowing Technique. II. *J. Phys. Soc. Jap.,* **60** 695.
8. Tanaka M, Nishimura G and Kushida T, (1994) Contribution of J Mixing to the $^5D_0 - ^7F_0$ Transition of Eu^{3+} Ions in Several Host Matrices *Phys. Rev.,* **B 49** 16917.
9. Tanaka M and Kushida T, (1995) Effects of Static Crystal Field on the Homogeneous Width of the $^5D_0 - ^7F_0$ Line of Eu^{3+} and Sm^{2+} in Solids *Phys. Rev.,* **B 52** 4171.
10. Tanaka M and Kushida T, (1999) Cooperative Vibronic Transitions in Eu^{3+} -doped Oxide Glasses *Phys. Rev.,* **B 60** 14752.
11. Kushida T and Tanaka M, (2002) Transition Mechanisms and Spectral Shapes of the $^5D_0 - ^7F_0$ Line of Eu^{3+} and Sm^{2+} in Solids *Phys. Rev.,* **B 65** 195118.
12. Kushida T, Kurita A and Tanaka M, (2003) Spectral Shape of the $^5D_0 - ^7F_0$ Line of Eu^{3+} and Sm^{2+} in Glass *J. Lumin.,* **102-103** 301.

13. Tanaka M, Nishisu Y, Kobayashi M, Kurita A, Hanzawa H and Kanematsu Y, (2003) Optical Characterization of Spherical Fine Particles of Glassy Eu^{3+} Doped Yttrium Basic Carbonates *J. Non-cryst. Solids* **318** 175.
14. Wybourne, B G, (1967) Coupled Products of Annihilation and Creation Operators in Crystal Energy Problems, pp35-52 in *Optical Properties of Ions in Crystals*, Wiley - Interscience Publication, John Wiley and Sons Inc., New York.
15. Wybourne, B G, (1968) Effective Operators and Spectroscopic Properties, *J. Chem. Phys.,* **48** 2596.
16. Wybourne, B G, Smentek, L, Kędziorski, A, (2004) Magnetic Dipole Transitions in Crystals, *Mol. Phys.* **102**, 1105.
17. Kędziorski, A, (2006) Magnetic Dipole f-f Transitions in Crystal Field; Fourth-order Perturbation Approach, (submitted for publication).

20 Hyperfine Induced Transitions

His advisor, Harold Davenport, was an eminent number theorist. "He gave me a very difficult problem - proving a conjecture that said every integer can be written as the sum of thirty seven numbers, each raised to the fifth power. When I told him that I had solved it, he didn't believe me." But the proof was correct.

Charles Seife, Interview with John Horton Conway[1]

In this chapter the hyperfine induced transitions are discussed, since this issue must be addressed in a monograph concerned with hyperfine interactions in lanthanides in crystalline environments. In 1962 Wybourne suggested[2] that under certain circumstances highly forbidden transitions might become allowed in a crystalline environment by a nuclear magnetic moment mixing nearby crystal field levels, particularly in the case of holmium salts. Baker and Bleaney[3] reported an apparent violation of selection rules in their paramagnetic resonance studies and suggested a mechanism based on the Jahn-Teller effect. It was only in the late 1980's that high resolution optical studies of hyperfine structure by Popova and her associates[4] revealed that indeed the interaction of nuclear magnetic dipole and electric quadrupole moments with $4f-$electrons leads to the observation of the "forbidden transitions". Remarkably, the possibility of hyperfine induced transitions started with the spectroscopic observation of gaseous nebulae. In this chapter we first define what is meant by *forbidden transitions* and after a brief sketch of the early history of hyperfine induced transitions we outline the calculation of such processes in atoms and ions focussing particularly upon the recently observed highly forbidden transition[5] $(2s2p) \, ^3P_0^o \to (2s^2) \, ^1S_0^e$ in $N \, IV$. This discussion is finalized by the analysis of possible importance of the hyperfine interactions in the description of the $f \longleftrightarrow f$ transitions in lanthanides in crystals.

In 1930 Huff and Houston[6] appended to their paper the following:-

Note added August 15: Dr. Bowen has called our attention to the fact that the line $\lambda 2270$ in Hg is probably due to the coupling of the nuclear spin with the electronic angular momentum. An estimate based on the relative separation of the multiplets and the hyperfine structure gives the right order of magnitude for the intensity. The statements made above with respect to J really refer strictly to the total angular momentum.

That note appears to be the first time that the possibility of hyperfine induced transitions was considered. Much of the early history is covered in an important article by Garstang[7] who also outlines details of the calculation of line intensities for hyperfine induced transitions. Here some of the methodology is illustrated in the particular case of the highly forbidden transition $(2s2p) \, ^3P_0^o \to (2s^2) \, ^1S_0^e$ in $N \, IV$.

20.1   THE ELECTRON CONFIGURATIONS $(2s2p)$ AND $(2s^2)$ IN N IV IONS

The lowest energy levels of triply ionized nitrogen are associated with the $(2s^2)$ and $(2s2p)$ electron configurations, which are of opposite parity. Their energy levels are given in cm^{-1} in Table 20-1. The transition $(2s2p)\,^3P_0^o \rightarrow (2s^2)\,^1S_0^e$ occurs at 1487Å, and is thus expected to appear in the vacuum ultraviolet and to be extraordinarily weak. It violates the $\Delta J = 0$ $(0 \nleftrightarrow 0)$ selection rule as well as the spin selection rule $\Delta S = 0$. To overcome the spin selection rule one might consider the effect of spin-orbit mixing, but there is only one $J = 0$ state in $(2s2p)$. To overcome the J selection rule would require an interaction that couples different J values. This would suggest that a combination of nuclear magnetic hyperfine interactions and spin-orbit interaction might be required. To that end we shall first consider the calculation of the energy levels of the $(2s2p)$ and $(2s^2)$ electron configurations in N IV.

The Coulomb matrix elements for the two terms, ^3P and ^1P of the $(2s2p)$ configuration may be calculated by standard tensor operator methods to give

$$E(^3P) = F_0(2s, 2p) - G_1(2s, 2p) \qquad (20\text{-}1a)$$

$$E(^1P) = F_0(2s, 2p) + G_1(2s, 2p) \qquad (20\text{-}1b)$$

The spin-orbit interaction matrices may likewise be computed in terms of the spin-orbit coupling constant, ζ_{2p}, to give

$$
\begin{array}{cc}
J = 2 \quad |^3P_2\rangle \\
\langle^3P_2| \;\; \left(\;\; \tfrac{1}{2} \;\; \right)
\end{array}
\qquad
\begin{array}{cc}
& J = 1 \quad |^3P_1\rangle \quad |^1P_1\rangle \\
\langle^3P_1| & \begin{pmatrix} -\tfrac{3}{2} & \tfrac{\sqrt{2}}{2} \\ \tfrac{\sqrt{2}}{2} & 0 \end{pmatrix} \\
\langle^1P_1| &
\end{array}
\qquad
\begin{array}{cc}
J = 0 \quad |^3P_0\rangle \\
\langle^3P_0| \;\; (\; -1 \;)
\end{array}
\qquad (20\text{-}2)
$$

Instead of calculating a radial integral which defines the spin-orbit coupling constant, we estimate its size using the expressions for the energy levels in (20-1a) and (20-1b) together with the data collected in Table 20-1 and the matrix elements in (20-2) to get in cm^{-1},

$$E\left(^3P_2\right) - E\left(^3P_0\right) = \tfrac{3}{2}\zeta_{2p} = 67416.3 - 67209.2 = 207.1 \qquad (20\text{-}3)$$

and hence

$$\zeta_{2p} = 138.1 cm^{-1}$$

TABLE 20-1
$2s^2$ and $2s2p$ Energy Levels of N IV

Configuration	Term	J	Level
$2s^2$	$^1S^e$	0	0.0
$2s2p$	$^3P^o$	0	67209.2
		1	67272.3
		2	67416.3
$2s2p$	$^1P^o$	1	130693.9

We can now estimate the Coulomb integral, $G_1(2s, 2p)$, by noting that

$$E(^1P_1) - E(^3P_1) \sim 2G_1(2s, 2p) + \tfrac{1}{2}\zeta_{2p} = 63421.6 \tag{20-4}$$

and therefore

$$G_1(2s, 2p) = 36726 cm^{-1} \tag{20-5}$$

Returning to (20-1a) we are led to the estimated value of

$$F_0(2s, 2p) = 104073 cm^{-1} \tag{20-6}$$

We are now in a position to estimate the spin-orbit mixing between the $^3P_1$ and $^1P_1$ states of the $(2s2p)$. We have the energy matrix for the two $J = 1$ states as

$$
\begin{array}{cc}
J = 1 & \qquad |^3P_1\rangle \qquad\qquad\qquad |^1P_1\rangle \\
\begin{array}{c} \langle ^3P_1| \\ \langle ^1P_1| \end{array} &
\left(
\begin{array}{cc}
F_0(2s, 2p) - G_1(2s, 2p) - \tfrac{3}{2}\zeta_{2p} & \frac{\sqrt{2}}{2}\zeta_{2p} \\
\frac{\sqrt{2}}{2}\zeta_{2p} & F_0(2s, 2p) + G_1(2s, 2p)
\end{array}
\right)
\end{array}
\tag{20-7}
$$

Putting in the numerical values for the integrals and diagonalizing the matrix to obtain the eigenvalues and eigenvectors leads to

$$E(^3P_1) = 67,278 \text{ cm}^{-1} \tag{20-8a}$$

$$|E(^3P_1)\rangle = 0.9999988|^3P_1\rangle - 0.00154214|^1P_1\rangle \tag{20-8b}$$

$$E(^1P_1) = 130600 \text{ cm}^{-1} \tag{20-9a}$$

$$|E(^1P_1)\rangle = 0.00154214|^3P_1\rangle + 0.9999988|^1P_1\rangle \tag{20-9b}$$

There is thus a very small amount of spin-mixing of the two $J = 1$ states that could lead to a very weak breakdown of the spin selection rule. However, that, by itself, cannot explain the observed transition unless there is also some J−mixing. In order to verify this possibility we need to consider the role of the nucleus. Note that the small spin admixture in (20-8b) is to a very good approximation

$$-\left(\frac{\sqrt{2}}{2}\zeta_{2p}\right) \Big/ \left(2G_1(2s, 2p) + \frac{3}{2}\zeta_{2p}\right) \tag{20-10}$$

20.2   NUCLEAR MAGNETIC DIPOLE HYPERFINE MATRIX ELEMENTS IN $(2s2p)$

It is possible to obtain information on the nuclear properties of the stable isotopes of nitrogen from the Internet[8] and the data are given in Table 20-2.

Note that indeed both isotopes of nitrogen have nuclear magnetic moments. It is significant that while the 1487Å spectral line occurs in triply ionized nitrogen, there is no analogous spectral line seen in ^{12}C, which has no nuclear spin and hence no nuclear magnetic moment.

TABLE 20-2

Nuclear Properties of the Stable Isotopes of Nitrogen

Isotope	Atomic Mass	Abundance	Nuclear Spin(I)	Magnetic Moment
^{14}N	14.003074002	99.632	1	0.4037607
^{15}N	15.00010897	0.368	$\frac{1}{2}$	-0.2831892

The calculation of the matrix elements of the magnetic dipole hyperfine interactions among the various states of the $(2s2p)$ electron configuration is fairly straightforward, though a somewhat tedious, tensor operator task. One is essentially calculating the matrix elements of the scalar product $(\mathbf{N}^{(1)} \cdot \mathbf{I}^{(1)})$ where

$$N^{(1)} = \sum_i \left[\ell_i^{(1)} - \sqrt{10}(s^{(1)}C^{(2)})_i^{(1)} + (8\pi/3)\delta(r_i)s_i^{(1)} \right] \tag{20-11}$$

where the last term in (20-11) is the Fermi contact term that arises when $s-$ orbitals are involved. Basically the matrix elements involve the hyperfine constants a_p and a_s. Using the results of chapter 2 it is left as an exercise to derive the relevant expressions for the matrix elements. Below we give the results for $I = \frac{1}{2}$ and $I = 1$, which are the nuclear spins of the stable isotopes of nitrogen. The results are given for LS-basis states.

For $I = \frac{1}{2}$ we obtain

$$
\begin{array}{cc}
F = \frac{5}{2} & |^3P_2\rangle \\
\langle ^3P_2| & \left(\frac{1}{20}(8a_p + 5a_s) \right)
\end{array}
\tag{20-12a}
$$

$$
\begin{array}{cccc}
F = \frac{3}{2} & |^3P_2\rangle & |^3P_1\rangle & |^1P_1\rangle \\
\langle ^3P_2| & \left(-\frac{3}{40}(8a_p + 5a_s) \right. & \frac{\sqrt{5}}{40}(6a_p - 5a_s) & \frac{\sqrt{10}}{40}(a_p - 5a_s) \\
\langle ^3P_1| & \frac{\sqrt{5}}{40}(6a_p - 5a_s) & \frac{1}{8}(4a_p + a_s) & -\frac{\sqrt{2}}{8}(a_p + a_s) \\
\langle ^1P_1| & \frac{\sqrt{10}}{40}(a_p - 5a_s) & -\frac{\sqrt{2}}{8}(a_p + a_s) & \left. \frac{1}{2}a_p \right)
\end{array}
\tag{20-12b}
$$

$$
\begin{array}{cccc}
F = \frac{1}{2} & |^3P_1\rangle & |^3P_0\rangle & |^1P_1\rangle \\
\langle ^3P_1| & \left(-\frac{1}{4}(4a_p + a_s) \right. & \frac{\sqrt{2}}{4}(3a_p - a_s) & \frac{\sqrt{2}}{4}(3a_p + a_s) \\
\langle ^3P_0| & \frac{\sqrt{2}}{4}(3a_p - a_s) & 0 & -\frac{1}{4}(2a_p - a_s) \\
\langle ^1P_1| & \frac{\sqrt{2}}{4}(3a_p - a_s) & -\frac{1}{4}(2a_p - a_s) & \left. -a_p \right)
\end{array}
\tag{20-12c}
$$

and for $I = 1$ we obtain

$$
\begin{array}{cc}
F = 3 & |^3P_2\rangle \\
\langle ^3P_2| & \left(\frac{1}{10}(8a_p + 5a_s) \right)
\end{array}
\tag{20-13a}
$$

$$
\begin{array}{cccc}
F = 2 & |^3P_2\rangle & |^3P_1\rangle & |^1P_1\rangle \\
\langle ^3P_2| & \left(-\frac{1}{20}(8a_p + 5a_s) \right. & \frac{\sqrt{3}}{20}(6a_p - 5a_s) & \frac{\sqrt{6}}{20}(a_p - 5a_s) \\
\langle ^3P_1| & \frac{\sqrt{3}}{20}(6a_p - 5a_s) & \frac{1}{4}(4a_p + a_s) & -\frac{\sqrt{2}}{4}(a_p + a_s) \\
\langle ^1P_1| & \frac{\sqrt{6}}{20}(a_p - 5a_s) & -\frac{\sqrt{2}}{4}(a_p + a_s) & \left. a_p \right)
\end{array}
\tag{20-13b}
$$

$$
\begin{array}{c}
\begin{array}{ccccc}
F = 1 & |^3P_2\rangle & |^3P_1\rangle & |^1P_1\rangle & |^3P_0\rangle
\end{array}\\[4pt]
\begin{array}{c}
\langle^3P_2| \\ \langle^3P_1| \\ \langle^1P_1| \\ \langle^3P_0|
\end{array}
\left(
\begin{array}{cccc}
-\frac{3}{20}(8a_p + 5a_s) & \frac{\sqrt{15}}{60}(6a_p - 5a_s) & \frac{\sqrt{30}}{60}(a_p - 5a_s) & 0 \\[4pt]
\frac{\sqrt{15}}{60}(6a_p - 5a_s) & -\frac{1}{4}(4a_p + a_s) & \frac{\sqrt{2}}{4}(a_p + a_s) & \frac{\sqrt{3}}{3}(3a_p - a_s) \\[4pt]
\frac{\sqrt{30}}{60}(a_p - 5a_s) & \frac{\sqrt{2}}{4}(a_p + a_s) & -a_p & -\frac{\sqrt{6}}{6}(2a_p - a_s) \\[4pt]
0 & \frac{\sqrt{3}}{3}(3a_p - a_s) & -\frac{\sqrt{6}}{6}(2a_p - a_s) & 0
\end{array}
\right)
\end{array}
$$

(20-13c)

$$
\begin{array}{c}
\begin{array}{ccc}
F = 0 & |^3P_1\rangle & |^1P_1\rangle
\end{array}\\[4pt]
\begin{array}{c}
\langle^3P_1| \\ \langle^1P_1|
\end{array}
\left(
\begin{array}{cc}
-\frac{1}{2}(4a_p + a_s) & \frac{\sqrt{2}}{2}(a_p + a_s) \\[4pt]
\frac{\sqrt{2}}{2}(a_p + a_s) & -2a_p
\end{array}
\right)
\end{array}
$$

(20-13d)

In the above results we omit in the bra and ket vectors the quantum numbers I, F which may be readily inferred.

It is not difficult to derive the corresponding results for a jj-coupling basis for $I = 1$, which we give below. Note again that the quantum numbers I, F have been omitted as well as the principal quantum numbers, those of the fixed $2s_{\frac{1}{2}}$ state.

$$
\begin{array}{c}
\begin{array}{cc}
F = 3 & |p_{\frac{3}{2}}, 2\rangle
\end{array}\\[4pt]
\langle p_{\frac{3}{2}}, 2|
\left(
\frac{1}{10}(8a_p + 5a_s)
\right)
\end{array}
$$

(20-14a)

$$
\begin{array}{c}
\begin{array}{cccc}
F = 2 & |p_{\frac{3}{2}}, 2\rangle & |p_{\frac{3}{2}}, 1\rangle & |p_{\frac{1}{2}}, 1\rangle
\end{array}\\[4pt]
\begin{array}{c}
\langle p_{\frac{3}{2}}, 2| \\ \langle p_{\frac{3}{2}}, 1| \\ \langle p_{\frac{1}{2}}, 1|
\end{array}
\left(
\begin{array}{ccc}
-\frac{1}{20}(8a_p + 5a_s) & \frac{1}{20}(8a_p - 15a_s) & \frac{\sqrt{2}}{4}a_p \\[4pt]
\frac{1}{20}(8a_p - 15a_s) & \frac{1}{12}(8a_p - 3a_s) & -\frac{\sqrt{2}}{12}a_p \\[4pt]
\frac{\sqrt{2}}{4}a_p & -\frac{\sqrt{2}}{12}a_p & \frac{1}{6}(8a_p + 3a_s)
\end{array}
\right)
\end{array}
$$

(20-14b)

$$
\begin{array}{c}
\begin{array}{ccccc}
F = 1 & |p_{\frac{3}{2}}, 2\rangle & |p_{\frac{3}{2}}, 1\rangle & |p_{\frac{1}{2}}, 1\rangle & |p_{\frac{1}{2}}, 0\rangle
\end{array}\\[4pt]
\begin{array}{c}
\langle p_{\frac{3}{2}}, 2| \\ \langle p_{\frac{3}{2}}, 1| \\ \langle p_{\frac{1}{2}}, 1| \\ \langle p_{\frac{1}{2}}, 0|
\end{array}
\left(
\begin{array}{cccc}
-\frac{3}{20}(8a_p + 5a_s) & \frac{\sqrt{5}}{60}(8a_p - 15a_s) & \frac{\sqrt{10}}{12}a_p & 0 \\[4pt]
\frac{\sqrt{5}}{60}(8a_p - 15a_s) & -\frac{1}{12}(8a_p - 3a_s) & \frac{\sqrt{2}}{12}a_p & \frac{1}{3}a_p \\[4pt]
\frac{\sqrt{10}}{12}a_p & \frac{\sqrt{2}}{12}a_p & -\frac{1}{6}(8a_p + 3a_s) & \frac{\sqrt{2}}{6}(8a_p - 3a_s) \\[4pt]
0 & \frac{1}{3}a_p & \frac{\sqrt{2}}{6}(8a_p - 3a_s) & 0
\end{array}
\right)
\end{array}
$$

(20-14c)

$$
\begin{array}{c}
\begin{array}{ccc}
F = 0 & |p_{\frac{3}{2}}, 1\rangle & |p_{\frac{1}{2}}, 1\rangle
\end{array}\\[4pt]
\begin{array}{c}
\langle p_{\frac{3}{2}}, 1| \\ \langle p_{\frac{1}{2}}, 1|
\end{array}
\left(
\begin{array}{cc}
-\frac{1}{12}(8a_p - 3a_s) & \frac{\sqrt{2}}{12}a_p \\[4pt]
\frac{\sqrt{2}}{12}a_p & \frac{1}{6}(8a_p + 3a_s)
\end{array}
\right)
\end{array}
$$

(20-14d)

20.3   THE MAPLE PROCEDURES USED TO CALCULATE THE HYPERFINE MATRIX ELEMENTS

The preceding hyperfine matrix elements were calculated using MAPLE. The procedure "njsym" is a collection of MAPLE routines for calculating the various $3nj$ symbols. Note that the nuclear spin is designated by the lower case "i", since MAPLE reserves "I" for $\sqrt{-1}$. To run the programme you need to have in your MAPLE directory the two files njsym and hsp.map. Once MAPLE is running you need to issue the command <read"hsp.map";>. To compute the matrix element $\langle(2s2p)^3 P1, I = 1, F = 1|H_{hfs}|(2s2p)^3 P0, I = 1, F = 1\rangle$ you issue the command <hsp(1,1,1,0,1,1);> and MAPLE should respond with

$$3^{\frac{1}{2}}ap - 1/3\ 3^{\frac{1}{2}}as$$

We list below the actual MAPLE procedures.

```
# These procedures compute the matrix elements of the magnetic and contact#
# hyperfine interactions in the sp electron configuration in the LS basis #
# (hsp) or in the jj basis (hjj)#
read"njsym";
hss:=proc(S1,J1,S2,J2,i,F)
local result,X,Y,Z,as;
X:=(-1)^ (S2)*sqrt(6*(2*S1+1)*(2*S2+1))
*sixj(S1,1,S2,1/2,1/2,1/2)/2;
Y:=(-1)^ (S1+J2)*sqrt((2*J1+1)
*(2*J2+1))*sixj(J1,1,J2,S2,1,S1)*X;
Z:=(-1)^ (J2+i+F)*sqrt(i*(i+1)*(2*i+1))
*sixj(J2,i,F,i,J1,1)*Y*as;
result:=combine(simplify(Z));
end:

hsp1:=proc(S1,J1,S2,J2,i,F)
local result,X,Y,Z,ap;
X:=(-1)^ (J2+i+F)*sixj(J2,i,F,i,J1,1)
*sqrt(i*(i+1)*(2*i+1));
Y:=0;
if (S1=S2) then
Y:=(-1)^ (S1+J1)*sqrt(6*(2*J1+1)*(2*J2+1))
*sixj(J1,1,J2,1,S1,1);
end if;
Z:=(-1)^ (S1+1)*3*sqrt(5*(2*J1+1)*(2*J2+1)*(2*S1+1)
*(2*S2+1))*ck(1,1,2)
sixj(S1,1,S2,1/2,1/2,1/2)*ninej(S1,S2,1,1,1,2,J1,J2,1);
result:=combine(simplify(X*(Y+Z)))*ap;
end:

hjp:=proc(j1,J1,j2,J2,i,F)
local result,X,Y,Z,ap;
X:=sqrt(6*(2*j1+1)*(2*j2+1))*((-1)^ (j1+(1/2))
```

```
*sixj(j1,1,j2,1,1/2,1)
+3*ninej(1/2,1/2,1,1,1,2,j1,j2,1));
Y:=(-1)^ (j2+J1+(3/2))*sqrt((2*J1+1)*(2*J2+1))
*sixj(J1,1,J2,j2,1/2,j1)*X;
Z:=(-1)^ (J2+i+F)*sqrt(i*(i+1)*(2*i+1))
*sixj(J2,i,F,i,J1,1)*Y*ap;
result:=combine(simplify(Z));
end:

hjs:=proc(j1,J1,j2,J2,i,F)
local result,X,Y,Z,as;
if (j1<>j2) then result:=0;
else
result:=combine(simplify((-1)^ (i+F+j1+(3/2))
*sqrt(i*(i+1)*(2*i+1)*(2*J1+1)
(2*J2+1)*6)*sixj(J2,i,F,i,J1,1)
*sixj(J1,1,J2,1/2,j1,1/2)/2)*as);
end if;
end:

hsp:=proc(S1,J1,S2,J2,i,F)
local result;
result:=hsp1(S1,J1,S2,J2,i,F)+hss(S1,J1,S2,J2,i,F);
end;

hjj:=proc(j1,J1,j2,J2,i,F)
local result;
result:=hjp(j1,J1,j2,J2,i,F)+hjs(j1,J1,j2,J2,i,F);
end;
```

By careful inspection of the above procedures you should be able to reconstruct the tensor operator formulae used in the calculations.

Having established the forgoing results the next step would be to explore the mechanisms contributing to the observation of the highly forbidden transition observed in triply ionized nitrogen.

20.4   HYPERFINE INDUCED $f \longleftrightarrow f$ TRANSITIONS

From a theoretical point of view the spectroscopic properties of a given system are described by the line strengths of certain transitions, and in order to evaluate them there are two possible approaches. It is possible to include the physical mechanisms that are the most important for the proper description of the electronic structure of a system into the total hamiltonian and diagonalize it. As a result, in addition to the approximate energy of various states, also the wave functions describing these states are obtained. These functions, the best within the energy criterion, usually are used to evaluate the amplitude and eventually the line strength of a given transition. This procedure has been developed in chapters 11 and 12, illustrated in 13 and 14, 15, 16, and applied in the preceding section to describe the hyperfine induced transitions in NIV where the considerations are limited to the possible importance of nuclear magnetic dipole interactions. When following this way of determining the transition amplitude, some interactions via the operators representing certain physical mechanisms are not included. Evidence of this is clearly seen when analyzing the various contributions to the transition amplitude in the language of the perturbation approach as in the chapters 17, 18, and 19. In fact, the procedure described in these chapters presents the second way to a proper and more accurate description of the spectroscopic properties of the chosen systems. Indeed, since the $J - J$ mixing via crystal field potential is too weak ro explain the $f \longleftrightarrow f$ transitions in the lanthanides in crystals, the inter-shell interactions via V_{cryst} are taken into account as a perturbation, which modifies directly the transition amplitude. This is exactly the procedure that leads to the formulation of the Judd-Ofelt theory. Obviously not in all cases is it possible or even allowed to use perturbation theory and treat some interactions as perturbations that, in principle, just slightly modify the initial zero order value of energy. Fortunately the lanthanides are not this case.

As mentioned before, due to specific properties of the electronic structure of lanthanides, which originate from the screening of the open shell $4f$ electrons by the closed shells of the $5s$ and $5p$ symmetry, it is justified to use the perturbation approach to describe their properties. Thus, similarly as in chapter 17, continuing the discussion devoted to the theoretical reproduction and understanding of the $f \longleftrightarrow f$ electric dipole transitions, we start with the Hamiltonian in (17-16).

At this point of the discussion however we realize that the Hamiltonian in (17-16) should be extended by several new perturbations in order to take into account all possibly important mechanisms that might affect the transition amplitude. To summarize, the list of such contains:

1. V_{cryst} - the crystal field potential which is regarded as a forcing mechanism of electric dipole $f \longleftrightarrow f$ transitions by means of the Judd-Ofelt theory;
2. V_{corr} - non-central part of the Coulomb interaction that is responsible for the electron correlation effects playing a dominant role in the description of many electron systems;
3. V_{so} - spin orbit interaction which represents the relativistic part of interaction included within the non-relativistic approach;
4. V_{hfs} and V_{EM} - the new terms representing nuclear magnetic hyperfine interactions and electric multipole hyperfine interactions; the direct impact of these two mechanisms upon the transition amplitude is the subject of present discussion.[9]

Following the basic concept of the standard Judd-Ofelt theory we include here the impact due to the inter-shell interactions via the perturbing operators, similarly as when the crystal field potential was used in (17-1). Thus, the whole space is divided into two parts; one is spanned by the solutions of the zero order problem formulated for the ground configuration, P, and the remaining part distinguished by the projector Q is associated with the solutions obtained for the excited configurations. Keeping this formal distinction between the ground and excited configurations helps to avoid the errors caused by including certain contributions twice.

For the purpose of our analysis we define the electron part of the nuclear magnetic hyperfine interaction from (20-11) (without the Fermi contact term, since we describe the properties of the electrons from the $4f$ shell) in the terms of unit tensor operators,

$$V_{hfs} = \mathcal{A} r^{-3} \left(L^{(1)} - \sqrt{15} \langle \ell || C^{(2)} || \ell' \rangle W^{(12)1}(\ell \ell') \right) \cdot \mathbf{I}^{(1)} \tag{20-15}$$

where $\mathcal{A} = 2 \mu_B \mu_n g_I$ and $V_{hfs} \equiv P V_{hfs} Q + Q V_{hfs} P$. The symbol H_{hfs} is reserved to denote that part of the interactions that contribute to the energy, $P H_{hfs} P$. The nuclear part of (20-15), $I^{(1)}$, is regarded as any other angular momentum operator (a tensor operator of rank one), and therefore its reduced matrix element is easily evaluated by (14-6).

The electric multipole interactions, V_{EM}, are defined as in (6-33),

$$V_{EM} = \sum_k \frac{r_n^k}{r_e^{k+1}} \left(\mathbf{C}_e^{(k)} \cdot \mathbf{C}_n^{(k)} \right) \tag{20-16}$$

This general expansion contains even terms, but also the terms associated with odd-parity multipoles. The operator defined here is the inter-shell object and possesses the non-vanishing matrix elements between the functions that describe the states of the configurations of the same as well as opposite parities. Again, H_{EM} gives the contributions to the energy of the ground configuration, while $V_{EM} \equiv P V_{EM} Q + Q V_{EM} P$ affects directly the transition amplitude.

Instead of using the *multi - perturbation* approach in which all the perturbing operators listed above are included simultaneously, it is enough to apply its double version (as in section 17.2) and take into account only two different perturbations at the same time. There is the necessity to apply at least double perturbation theory in the case of magnetic hyperfine interactions, since their impact must be combined with the simultaneous interaction via the odd part of crystal field potential. Indeed, the parity of electric dipole transition operator is opposite to the parity of V_{hfs}, therefore to have non-vanishing contributions to the transition amplitude both impacts should be mediated by the V_{cryst} acting within the space of the excited configurations of opposite parities. The situation is different in the case of the perturbing influence of the electric multipole interactions, since they give rise to the non-zero contributions already at the second order (without the presence of the crystal field potential). Therefore their impact may be taken into account using a standard single perturbation approach. In order to establish the relative magnitude of various contributions, it is again required to include more than one mechanism at the same time.

20.5 NUCLEAR MAGNETIC HYPERFINE CONTRIBUTIONS

Apart from the nuclear part of the perturbing operator defined in (20-15), it is possible to distinguish two terms of different tensorial nature, namely,

$$h_{hfs}^1 = r^{-3} \sum_q L_q^{(1)} \tag{20-17}$$

and

$$h_{hfs}^2 = r^{-3} \langle \ell || C^{(2)} || \ell' \rangle W^{(12)1}(\ell \ell') \tag{20-18}$$

Already at first sight it might be concluded that the second-order contributions caused by V_{hfs} vanish. Inspection of the general expression for the second-order contributions defined in (17-6), with V_{cryst} replaced by V_{hfs}, leads to the conclusion that obviously the contributions from h_{hfs}^2 vanish. Indeed, the matrix elements $W^{(12)1}$ require functions of the same parity, and this condition is in contradiction with the parity requirements for the matrix elements of the electric dipole radiation operator. This situation is very clearly described by the reduced matrix elements of spherical tensors that are present in the expression for the second-order contributions. For the non-zero matrix element of $C^{(2)}$ in (20-18) the parity of ℓ' has to be the same as a parity of $\ell(\equiv f)$. At the same time, the non-zero reduced matrix element of $C^{(1)}$ (which is the angular part of the electric dipole transition operator present in (17-6)), requires the opposite parities of both one-electron states.

The first part of nuclear magnetic hyperfine interaction of (20-17) requires special attention, since it introduces a very interesting aspect of the general rules of the tensor operator algebra applied for the evaluation of matrix elements of various operators. The matrix elements of $L^{(1)}$, which is an orbital part of magnetic dipole radiation operator are non-zero only for the states of the same configuration. In the analysis presented here we are mainly interested in the perturbing impact of the inter-shell interactions, and therefore $V \equiv Qh_{hfs}^1 P + Ph_{hfs}^1 Q$. However, the angular momentum operator in h_{hfs}^1 of (20-17) is multiplied by the radial part, which may collect in an integral over the radial coordinates one-electron states that are different on both sides of a matrix element, within P and Q. The only condition to be satisfied is that both configurations differ by a single pair of one-electron states that are identified by different principal quantum numbers, but are of the same symmetry determined by the angular momentum. Thus, it is concluded that h_{hfs}^1 contributes to the transition amplitude when acting between the ground configuration $4f^N$ and singly excited configurations $4f^{N-1}n'f$. This means that the first part of the nuclear magnetic hyperfine interaction is capable of admixing to the functions of the ground configuration new components that are of the same parity. This of course violates the parity requirements for the non-vanishing matrix elements of the electric dipole radiation operator. It is concluded then, that h_{hfs}^1 indeed contributes to the transition amplitude but starting from the third order. At this order the contributions are determined by the triple products of matrix elements, and the parities of the intermediate configurations are properly adjusted via mediation of V_{cryst}, for example.

The third-order contributions represent the interplay of two physical mechanisms, the crystal field potential forcing the transition and the nuclear magnetic hyperfine interactions represented by $V_{hfs} = h_{hfs}^1 + h_{hfs}^2$. Without an appropriate energy denominator, which is omitted for the simplicity, we expect the third-order contributions to be determined by the following sequence of operators,

$$V_{cryst} D^{(1)} V_{hfs} \quad \text{and} \quad V_{hfs} D^{(1)} V_{cryst} \qquad (20\text{-}19)$$

It is possible also to include the intra-shell interactions within the Q space, since in general there are the third-order contributions determined by the other six different arrangements of the operators contributing to the transition amplitude at the third-order. Although this part of the investigation is beyond the main part of this model, for the completeness of the discussion those terms originating from $QV_{hfs}Q$ are presented in a separate section. However, it should be mentioned that in the search for new effective operators that might alleviate the strict selection rules and provide a reliable description of the unusual transitions, these terms introduce nothing new except new tensorial structures of the effective operators.

By following the same procedure, applying the same rules of coupling and re-coupling of angular momenta and also the tensor operators, and adopting the same approximations as in the previous discussion of $f \longleftrightarrow f$ electric dipole transitions presented in chapter 17, it is not difficult to derive the tensorial form of the effective operators that represent the new physical mechanism.

The first part of nuclear magnetic hyperfine interaction is expressed by the following unit tensor operator

$$h^1_{hfs} \Rightarrow \langle 4f|r^{-3}|n'f\rangle\langle f|\ell^{(1)}||f\rangle u_q^{(1)}(nf, n'f), \tag{20-20}$$

and the third-order correction, its electronic part, is determined by the operator that results from the contraction of operators in the sequence

$$\langle 4f^N\Psi_f|u_q^{(1)}(4f, n''f)u_\rho^{(1)}(n''f, n'\ell')u_p^{(t)}(n'\ell', 4f)|4f^N\Psi_i\rangle \tag{20-21}$$

Taking into account two terms with possible inter-shell interaction via the perturbing operators, the contribution to the transition amplitude is determined by the effective operator

$$^3\Gamma_{h^1_{hfs}}(h^1 DV_{CF} + V_{CF}Dh^1)$$

$$= \sum_{kq}\sum_{tp}(-1)^q B_p^t \overset{even}{\sum_{\ell'}\sum_{\ell''}}\delta(\ell'', \ell)\langle\varrho^t(4f \longrightarrow \ell')|r|\varrho^{-3}(4f \longrightarrow f)\rangle$$

$$\times \sqrt{\ell(\ell+1)(2\ell+1)}\langle\ell||\mathbf{C}^{(t)}||\ell'\rangle\langle\ell'||\mathbf{C}^{(1)}||\ell\rangle$$

$$\times \sum_{\lambda\mu}(-1)^{\lambda+p}[\lambda]\left(X^{t11}_{qp\rho}(\lambda\mu;\ell'\ell) + X^{t11}_{pq\rho}(\lambda\mu;\ell'\ell)\right)U_\mu^{(\lambda)}(\ell\ell) \tag{20-22}$$

where ρ denotes the components of the electric dipole radiation operator (the polarization), and the collection of angular momentum coupling coefficients for each sequence of operators is defined as follows,

$$X^{k_1k_2k_3}_{q_1q_2q_3}(\lambda\mu;\ell'\ell'') = \sum_{x\sigma}[x]\begin{pmatrix} k_1 & k_3 & x \\ q_1 & q_3 & -\sigma \end{pmatrix}\begin{pmatrix} x & k_2 & \lambda \\ \sigma & q_2 & -\mu \end{pmatrix}$$

$$\times \begin{Bmatrix} k_3 & x & k_1 \\ \ell & \ell' & \ell'' \end{Bmatrix}\begin{Bmatrix} k_2 & \lambda & x \\ \ell & \ell'' & \ell \end{Bmatrix} \tag{20-23}$$

When the summations over x, σ in (20-23) are performed, the product of these angular coupling coefficients is replaced by the product of two $3 - j$ symbols and one $9 - j$ symbol. These new coefficients are summed over a new index that is determined by the triangular conditions for the non-vanishing distinct terms. Thus, this manipulation does not provide any simplification of the tensorial structure of the effective operators, but leads to an alternative form.

The second part of the nuclear magnetic hyperfine interactions requires the contraction of the double tensor operators by means of the commutator presented in (18-17). In particular, the following sequence of the unit inter-shell tensor operators contributes at the third order (the first term of (20-19) with $V_{hfs} \Rightarrow h^2_{hfs}$),

$$\langle 4f^N \Psi_f | w^{(0t)t}(4f, n'\ell') w^{(01)1}(n'\ell', n''\ell'') w^{(12)1}(n''\ell'', 4f) | 4f^N \Psi_i \rangle \qquad (20\text{-}24)$$

where the first operator represents crystal field potential, the second is the electric dipole radiation operator, and finally the last one is from h^2_{hfs}. Taking into account also the sequence with the crystal field potential interchanged with the hyperfine operator, the sum of both contributions is determined by

$$\Gamma^{11}(h^2 DV + V Dh^2) = \sqrt{3} \sum_{\ell'} \sum_{\ell''} \sum_{tp} B^t_p \langle \varrho^t (4f \longrightarrow \ell') \mid r \mid \varrho^{-3}(4f \longrightarrow \ell'') \rangle$$

$$\times \langle \ell || \mathbf{C}^{(t)} || \ell' \rangle \langle \ell' || \mathbf{C}^{(1)} || \ell'' \rangle \langle \ell'' || \mathbf{C}^{(2)} || \ell \rangle$$

$$\times \sum_q (-1)^q \sum_y \sum_{\lambda,\mu} [y][\lambda]^{\frac{1}{2}} \sum_{x,\sigma} (-1)^x [x] \begin{pmatrix} t & 1 & x \\ p & \rho & -\sigma \end{pmatrix} \begin{pmatrix} x & 1 & \lambda \\ \sigma & q & -\mu \end{pmatrix}$$

$$\times \left[(-1)^{\lambda+y+1} A^{\lambda ty}_x(\ell\ell'\ell'') + B^{\lambda ty}_x(\ell\ell'\ell'') \right] W^{(1y)\lambda}_\mu(s\ell, s\ell) \qquad (20\text{-}25)$$

where

$$A^{\lambda ty}_x(\ell\ell'\ell'') = \sum_z [z] \begin{Bmatrix} 1 & z & 2 \\ \ell & \ell'' & \ell' \end{Bmatrix} \begin{Bmatrix} t & y & z \\ \ell & \ell' & \ell \end{Bmatrix} \begin{Bmatrix} 1 & z & 2 \\ \lambda & y & 1 \\ x & t & 1 \end{Bmatrix} \qquad (20\text{-}26)$$

and

$$B^{\lambda ty}_x(\ell\ell'\ell'') = \begin{Bmatrix} 1 & x & t \\ \ell & \ell' & \ell'' \end{Bmatrix} \begin{Bmatrix} 2 & y & x \\ \ell & \ell'' & \ell \end{Bmatrix} \begin{Bmatrix} 2 & 1 & 1 \\ \lambda & y & x \end{Bmatrix} \qquad (20\text{-}27)$$

with $\ell = 3$ for the $4f^N$ configuration of the lanthanides. The radial integrals of (20-25), similarly as in the case of (20-22), are defined by the perturbed functions, which are in fact not new or characteristic for this particular interaction. The perturbed function at the left-hand-side of the integral is the same as in the case of the second-order standard Judd-Ofelt theory, and it contains all the first order corrections due to the radial part of the crystal field potential; the second function in the integral is generated by the radial part of the perturbing operator h^2, which is the same as for the spin-orbit interaction in the case of the spherically symmetric potential.

Two different sequences of the initial operators in the product of matrix elements result in different angular parts of effective operators. They are represented here by the product of coupling and re-coupling angular momenta coefficients $A^{\lambda ty}_x(\ell\ell'\ell'')$ and $B^{\lambda ty}_x(\ell\ell'\ell'')$. However performing the summations over the $3j-$ of (20-25) and $6j-$symbols of each angular factor, it is possible to simplify the whole expression introducing the $12j-$symbol. The symmetry properties of the $12j-$symbols [10] show that both terms differ only by a phase factor $(-1)^{y+1}$, which means that y must be *odd*. Therefore the final expression for the third-order effective operator has a factor 2; its tensorial structure is as follows,

$$\Gamma^{11}(h^2 DV + V Dh^2) = 2\sqrt{3} \sum_{\ell'} \sum_{\ell''} \sum_{tp}^{odd} B_p^t \langle \varrho^t(4f \longrightarrow \ell') \mid r \mid \varrho^{-3}(4f \longrightarrow \ell'') \rangle$$

$$\times \langle \ell || \mathbf{C}^{(t)} || \ell' \rangle \langle \ell' || \mathbf{C}^{(1)} || \ell'' \rangle \langle \ell'' || \mathbf{C}^{(2)} || \ell \rangle \sum_q (-1)^q \sum_{y}^{odd} \sum_{\lambda,\mu}^{even} [y][\lambda]^{\frac{1}{2}} \sum_{x,\sigma} (-1)^{x+1} [x]$$

$$\times \begin{pmatrix} t & 1 & x \\ p & \rho & -\sigma \end{pmatrix} \begin{pmatrix} x & 1 & \lambda \\ \sigma & q & -\mu \end{pmatrix} \begin{Bmatrix} \ell'' & 2 & 1 & t \\ \ell & 1 & x & \ell' \\ \ell & y & \lambda & 1 \end{Bmatrix}$$

$$\times W_\mu^{(1y)\lambda}(s\ell, s\ell) \tag{20-28}$$

The limitation on y implies at the same time that the final rank of the double tensor operator must be *even*, if the hermiticity is required for the object contributing to the line strength. The fact that the effective operator is a double tensor operator with rank 1 with respect to the spin part of the space means that its contributions are not included when the transition amplitude is defined by the standard formulation of Judd and Ofelt in its semi-empirical realization. The contributions caused by the nuclear magnetic hyperfine interactions from (20-28) are included however within the relativistic parametrization scheme that is discussed in chapter 18. It is thus apparent that the one particle parametrization scheme based on double tensor operators, being extended for the spin space is more general, and includes many physical mechanisms in addition to those of the standard formulation. Indeed, for the ranks $\kappa = 0$ and $k = even$ in $W^{(\kappa k)\lambda}$ all the physical effects that are represented by the Judd-Ofelt parametrization are included; for $\kappa = 1$, we are taking into account the nuclear magnetic hyperfine interactions as in (20-28) among the other interactions within the spin-orbital space. It may then be concluded that at this point of our understanding of the nature of the $f \longleftrightarrow f$ transitions there is only a search for the physical interpretation of the contributions with $\kappa = 0$ and $\lambda = odd$.

The selection rules for the non-vanishing reduced matrix elements of spherical tensors in (20-28) indicate that the new effective operators include the perturbing influence of singly excited configurations described by $\ell' = even(d, g)$ and $\ell'' = odd(p, f)$. These parity requirements imply that the rank t of the crystal field potential is odd. In the light of all the difficulties with the evaluation of the odd rank crystal field parameters, this is not the easiest contribution to be treated in practice. However, from a physical point of view this means that here the same part of the crystal field potential perturbs the spherical symmetry of a free ion as in the case of the standard formulation of the Judd-Ofelt theory. Its perturbing influence, which plays a role of the forcing mechanism of the electric dipole transitions (in fact the only one known at this point of our discussion), creates the possibility for other physical effects, in particular the magnetic hyperfine interactions, to be included.

The main conclusion that is derived from the inspection of the tensorial structure of these new effective operators is that finally, in the non-relativistic approach, we have terms contributing to the transition amplitude, that change the standard selection rules. For example, for $\lambda = 0$ in (20-28), $y = 1$ and $x = 1$, and the effective operators have a simpler form, since the $12j$–symbol is reduced to two $6j$–symbols,

$$T_{0\longleftrightarrow0} = 6 \sum_{\ell'}^{even} \sum_{\ell''}^{odd} \sum_{tp}^{odd} B_p^t \langle \varrho^t(4f \longrightarrow \ell') \mid r \mid \varrho^{-3}(4f \longrightarrow \ell'') \rangle$$

$$\times \langle \ell || \mathbf{C}^{(t)} || \ell' \rangle \langle \ell' || \mathbf{C}^{(1)} || \ell'' \rangle \langle \ell'' || \mathbf{C}^{(2)} || \ell \rangle \sum_q (-1)^q \begin{pmatrix} t & 1 & 1 \\ p & \rho & -q \end{pmatrix}$$

$$\times \begin{Bmatrix} t & 1 & 1 \\ \ell'' & \ell & \ell' \end{Bmatrix} \begin{Bmatrix} 2 & 1 & 1 \\ \ell & \ell'' & \ell \end{Bmatrix} \langle 4f^6\,{}^7F_0| W_0^{(11)0}|4f^6\,{}^5D_0\rangle \qquad (20\text{-}29)$$

The matrix elements of these operators for $\ell' = d, g$, $\ell'' = p, f$, and $t = odd$ with its values limited by the symmetry of the crystal field, determine the transition amplitude of the $0 \longleftrightarrow 0$ in Eu^{+3}. Having the non-zero contributions of (20-29) we are now able to give a physical meaning to the intensity parameter Ω_0^{11} of the relativistic parametrization scheme discussed in section 18.4. It is interesting to mention that its presence in the set of intensity parameters is well justified by the effectively relativistic approach. Note that in fact, apart from the same unit tensor operators, in the case of the relativistic approach the radial integrals are defined by the relativistic components, while here the non-relativistic radial functions are used. At the same time, it is seen that this very effective operator, (20-29), also plays an important role in the non-relativistic description of the unusual transition for which it is the only term contributing to the amplitude. This is a comfortable and exceptional situation, since without performing any numerical calculations or even estimations, it is possible to judge the importance of an effective operator. Indeed, this is the most important term, since it is the only contribution to the amplitude of $0 \longleftrightarrow 0$ transition which exists.

20.6 ELECTRIC MULTIPOLE HYPERFINE CONTRIBUTIONS

Due to the parity requirements the electric dipole $f \longleftrightarrow f$ transitions are forbidden when the spherical symmetry of the system is assumed. This aspect of the spectroscopic properties of the lanthanides in crystals is discussed in detail in chapter 17, where the basic formulation of the standard Judd-Ofelt theory is presented. When the perturbing influence of the crystal field potential is taken into account, the first non vanishing terms contributing to the transition amplitude are of the second order. They are expressed as a product of two matrix elements of V_{cryst} and the actual transition operator $D^{(1)}$. These elements are connected via the intermediate states of the excited configurations. When the free ionic system approximation of the zero order problem is broken by the inclusion of the crystal field potential as a perturbation, all the other physical mechanisms that may affect the transition amplitude give rise to the contributions that are at least of the third order.

Here for the first time new second-order contributions are introduced. This means that there is another physical mechanism, different from the crystal field potential, which forces the $f \longleftrightarrow f$ transitions. In order to present these new terms the double perturbation approach is applied for the following hamiltonian

$$H = H_0 + \lambda(PV_{cryst}Q + QV_{cryst}P) + \mu(PV_{EM}Q + QV_{EM}P) \qquad (20\text{-}30)$$

where V_{EM} represents electric multipole hyperfine interactions defined in (20-16), and the crystal field potential is defined by (9-2). The contributions to the transition amplitude determined by the first order corrections to the wave functions, which are due to both perturbations, are defined in general in (17-22). The terms associated with the perturbing parameter λ, as before, are the Judd-Ofelt terms, while those multiplied by μ are new second-order contributions. They are determined by the following products of matrix elements

$$\Gamma_{EM}^{01} = \mu\{\langle \Psi_f^0 \mid D_\rho^{(1)} \mid \Psi_i^{01}\rangle + \langle \Psi_f^{01} \mid D_\rho^{(1)} \mid \Psi_i^0\rangle\} \qquad (20\text{-}31)$$

with the first order correction to the wave functions generated by the perturbing influence of V_{EM} expressed in a standard way

$$\Psi_i^{(01)} = \sum_{k\neq i}{}' \frac{\langle \Psi_k^0 \mid QV_{EM}P \mid \Psi_i^0\rangle}{E_i^0 - E_k^0} \Psi_k^0 \qquad (20\text{-}32)$$

Since the angular and nuclear moments are not coupled together, it is possible to express the second-order term as a product of the electronic part and a nuclear matrix element. The latter has the form of a nuclear moment

$$M_q^k(I) = \langle I \mid r_n^k \mathbf{C}_{q,n}^{(k)} \mid I\rangle \qquad (20\text{-}33)$$

with the quadrupole moment, $Q \equiv M_0^2(I)$ defined in (6-37), and the dipole moment for $k = 1$. The two sequences of operators in the matrix elements of the second-order terms, DV_{EM} and $V_{EM}D$, are expressed by the effective operators that have the same structure, but differ only by a phase factor. Similarly as in the case of the Judd-Ofelt effective operators, this is a source of the parity conditions that have to be satisfied for the non-vanishing contributions. In this particular case it is required that the parity of k, the rank of the multipole, is opposite to the parity of λ, which is the rank of unit tensor operator in the following expression for the effective operator,

$$\Gamma_{EM} = 2(-1)^\rho \sum_{k,q}^{odd} M_q^k(I) \sum_{\ell'}^{even}\sum_{\lambda}^{even} [\lambda]^{\frac{1}{2}} \begin{pmatrix} 1 & k & \lambda \\ \rho & q & -(\rho+q) \end{pmatrix}$$
$$\times A_k^\lambda(\ell') R_{JO}^{-k-1}(\ell') U_\mu^{(\lambda)}(\ell\ell) \qquad (20\text{-}34)$$

The similarity of these effective operators to those of the Judd-Ofelt theory is striking. Although both objects are derived from two different physical backgrounds, we have in (20-34) the angular part and the radial integral the same as in the case of the effective operator of Judd-Ofelt theory defined in (17-13), (17-14) and (17-39), respectively. In particular

$$A_k^\lambda(\ell') = [\lambda]^{1/2} \begin{Bmatrix} k & \lambda & 1 \\ f & \ell' & f \end{Bmatrix} \langle f\|C^{(1)}\|\ell'\rangle\langle \ell'\|C^{(k)}\|f\rangle$$

and

$$R_{JO}^{-k-1}(\ell') = \langle \varrho^{-k-1}(4f \longrightarrow \ell') \mid r \mid 4f \rangle$$

The non-vanishing reduced matrix elements above require that ℓ' is even and, as a consequence k is odd. These two limitations mean that first of all, similarly as in the case of the standard second-order approach, the effective operator in (20-34) represents the perturbing influence of all singly excited configurations in which an electron from the $4f$ shell is promoted to the one-electron state of $\ell' \equiv d, g$ symmetry. Since the radial integrals are defined in the terms of the perturbed functions discussed previously, all excitations of this type and for all the values of the principal quantum number, are taken into account (including the continuum). What is intriguing is the new forcing mechanism that leads to the non-vanishing second-order terms, which originate from the odd part of the electric multipole hyperfine interaction instead of the odd part of the crystal field potential of the standard Judd-Ofelt approach. The only formal difference is that in the expression (20-34) the odd crystal field parameters B_p^t (present in (17-12)) are replaced by the nuclear moment $M_q^k(I)$. In this way the reason, for which in practice it is impossible to calculate directly the transition amplitude that is defined within the standard formulation, is removed. Indeed, there is no reliable model for evaluation of the structural parameters available. It is possible only to determine the parameters with even ranks using a fitting procedure to adjust the energy expressed in the terms of parameters to its measured values, but unfortunately, they do not play important role in the $f \longleftrightarrow f$ transition theory.

Obviously, the similar tensorial structure of the two effective operators that are based on completely different physical mechanisms does not provide any information about their relative magnitude. The purely theoretical conclusions derived at this point show that although the physical origin of the Judd-Ofelt term and Γ_{EM} is different, they both do satisfy the same selection rules. Since k in (20-34) is odd, and λ is of opposite parity to k, then the effective operators are of even rank, as in the case of the Judd-Ofelt theory. This means that the standard intensity parameters Ω_λ from (17-15) represent in addition to all the other mechanisms discussed there the impact due to the electric dipole hyperfine interactions contributing to the transition amplitude at the second order.

The possibility of inclusion of interactions that are represented by an odd rank tensor operator other than the crystal field potential opens a new chapter in the investigations that are devoted to the theoretical description of the electric dipole $f \longleftrightarrow f$ transitions. Instead of the odd part of V_{cryst}, used as a forcing mechanism in (20-19), for example, the odd part of V_{EM} might be applied. Thus new terms $V_{EM} D^{(1)} V_{hfs}$ and $V_{hfs} D^{(1)} V_{EM}$ arise, which contribute to the transition amplitude at the third order. In particular these new terms give the chance for a better description of the hypersensitive transitions, which are formally defined as those mainly determined by the matrix element of the unit tensor operator $U^{(2)}$. As seen in (17-12), the description of hypersensitive transition, which is based on the standard Judd-Ofelt approach, requires contributions associated with the odd terms of crystal field potential for $t = 1, 3$. The majority of the hypersensitive transitions are observed for the systems with the symmetry for which the potential does not contain terms associated with $t = 1$.

The absence of this particular term was interpreted as the reason for the failure of a theoretical reproduction of hypersensitivity in all such cases. The situation is changed with the terms of (20-34), since they compensate for the lack of the contributions for $t = 1$. Through the term for $k = 1$ there is an additional contribution to the transition amplitude that is determined by the matrix element of $U^{(2)}$, regardless of the symmetry of the environment.

It is interesting to combine both interactions, electric multipole hyperfine and crystal field, and evaluate their simultaneous impact upon the transition amplitude at the third order. In general, both mechanisms are expressed by even and odd rank tensor operators as seen in (20-16). Therefore their role as an origin of the opposite parity components of the wave functions admixed to the states of $4f^N$ configuration may be interchanged. As always at the third order, there are two contributing expressions that differ by a sequence of the operators in matrix elements

$$^3\Gamma_{EM}(inter)$$

$$= \sum_{Xx} \sum_{Yy} \left\{ \frac{\langle \Psi_f^0 \mid PV_{EM}Q \mid Yy\rangle\langle Yy \mid D_\rho^{(1)} \mid Xx\rangle\langle Xx \mid QV_{cryst}P \mid \Psi_i^0\rangle}{(E_i^0 - E_{Xx}^0)(E_f^0 - E_{Yy}^0)} \right.$$

$$\left. + \frac{\langle \Psi_f^0 \mid PV_{cryst}Q \mid Xx\rangle\langle Xx \mid D_\rho^{(1)} \mid Yy\rangle\langle Yy \mid QV_{EM}P \mid \Psi_i^0\rangle}{(E_f^0 - E_{Xx}^0)(E_i^0 - E_{Yy}^0)} \right\} \quad (20\text{-}35)$$

where the interactions via perturbing operators are limited to those between the ground and excited configurations.

Applying the same approximations and the rules for the coupling of the inter-shell tensor operators as before, the final tensorial structure of the third-order contributions is as follows

$$^3\Gamma^{EM}(V_{EM}DV_{cryst} + V_{cryst}DV_{EM})$$

$$= \sum_{kq} \sum_{tp} (-1)^q B_p^t M_q^k(I) \sum_{\ell'\ell''} \langle \varrho^{-k-1}(4f \longrightarrow \ell')|r|\varrho^t(4f \longrightarrow \ell'')\rangle$$

$$\times \langle \ell||\mathbf{C}^{(k)}||\ell'\rangle\langle \ell'||\mathbf{C}^{(1)}||\ell''\rangle\langle \ell''||\mathbf{C}^{(t)}||\ell\rangle$$

$$\times \sum_{\lambda\mu} (-1)^{\lambda+p}[\lambda]\left(X_{q p \rho}^{tk1}(\lambda\mu; \ell'\ell'') + (-1)^{k-p} X_{p q \rho}^{tk1}(\lambda\mu; \ell''\ell') \right) U_\mu^{(\lambda)}(\ell\ell) \quad (20\text{-}36)$$

where the collection of angular momentum coupling coefficients for each sequence of operators is the same as for the terms caused by nuclear magnetic hyperfine interactions h_{hfs}^1, and they are defined in (20-23).

Similarly as before, the nuclear part of the electric multipole operator is uncoupled from the electronic part, and therefore it is regarded in (20-36) as a separate factor. The radial integral expressed by the perturbed functions is rather standard for the third-order terms, and again it is determined by the Judd-Ofelt perturbed function ϱ^t. The second perturbed function ϱ^{-k-1}, is the same as that of the radial integral of the second-order operator in (20-34). The middle reduced matrix element of the spherical tensor in (20-36) shows that the parities of one-electron excited states have to be opposite; this condition results in most interesting conclusions and provides new selection rules. Namely, if $\ell' = even(\equiv d, g)$, then rank $k = odd(= 1)$ in the definition of V_{EM} of (20-16), and consequently $\ell'' = odd(\equiv p, f)$; as a result the ranks of the tensor operators that define the crystal field potential t are *even*. This means that when the electric dipole hyperfine interactions are the forcing mechanism of the $f \longleftrightarrow f$ transitions, they are assisted by the interactions via *even* part of the crystal field potential. This result creates the first opportunity to perform *ab initio* -type calculations of the transition amplitude, since as mentioned above, it is a straightforward task to find the even rank crystal field parameters from the energy fitting procedure. When the parities of ℓ' and ℓ'' are interchanged, and the former is odd while the latter is even, then the interactions via the quadrupole moment, for $k = 2$, contribute to the transition amplitude; and since $t = odd$, the odd part of the crystal field potential forces the transition.

It is seen that the two parts, *even* and *odd* of both interactions, complement each other. In all cases however, there is no limitation on the values of the rank of the unit tensor operator $U^{(\lambda)}$. This means, that λ is determined only by the triangular conditions for the non-vanishing angular momentum coupling coefficients, and it has the values $1, 2, 3, 4, 5, 6$ in the case of $f-$ electrons. For even values of the rank λ nothing new is observed in the parametrization of $f-$spectra, since these are the very same terms as in the standard scheme. Thus the intensity parameters $\Omega_2, \Omega_4, \Omega_6$, when determined from the fitting procedure, potentially involve the impact due to the electric quadrupole hyperfine interactions. Of course we are unable to estimate their relative importance without numerical calculations. Note that any numerical analysis for the terms with the even crystal field potential is only a technical task and may be performed in an exact way.

The contributions with the *odd* ranks of $U^{(\lambda)}$ are new, and they are beyond the standard parametrization scheme. In particular, this means that the electric dipole hyperfine interactions (for $k = 1$) between the ground configuration $4f^N$ of the lanthanide ion and all the excited configurations of the opposite parity are not included within the standard parametrization scheme. Since these contributions do not have their counterparts defined within the standard approach to compare with, only the fact that they exist, and consequently they change the selection rules, make them important in the description of the $f \longleftrightarrow f$ transitions. Indeed, their presence at the third order relaxes the selections rules because of which in the standard scheme, for example, the transition $0 \longleftrightarrow 1$ observed in Eu^{3+} is completely forbidden. This conclusion shows that it is necessary at least to extend the list of intensity parameters by those with the odd ranks if the spectra are parameterized within the non-relativistic scheme.

20.7 SUMMARY

It is interesting to recall that the odd rank effective operators that contribute to the amplitude of $f \longleftrightarrow f$ transition have been introduced previously[11]. This alternative approach was based on the velocity form of the electric dipole radiation operator. Taking into account the non-locality of the potential of the zero order eigenvalue problem (the Hartree-Fock potential, in particular) it is evident that for many electron systems the transition amplitudes defined within the length and velocity forms differ from each other by the matrix element of the commutator of the non-local potential with the position operator, namely

$$\frac{i\hbar}{\mu} \langle \Phi_k \mid \vec{p} \mid \Phi_\ell \rangle = (E_k - E_\ell)\langle \Phi_k \mid \vec{r} \mid \Phi_\ell \rangle - \langle \Phi_k \mid [\vec{r}, V(\vec{r}, \vec{r}')] \mid \Phi_\ell \rangle \quad (20\text{-}37)$$

The last term in (20-37) exists only if the potential is non-local. This means that any discrepancy between the results of calculations of the transition amplitude with the length and velocity electric dipole radiation operators is not a measure of the quality of the functions used in the matrix elements (as is very often interpreted erroneously in the literature) but rather it indicates the non-local character of the potential included in the Hamiltonian.

As demonstrated previously[11] the matrix element of this commutator leads to new effective operators contributing to the transition amplitude defined with the velocity form. In this way, at the second order of analysis, in addition to the even rank operators of the standard formulation of the Judd-Ofelt theory, which originally is based on the length form, new effective operators of odd ranks are introduced. In fact, the matrix element of the commutator is expressed by one- and three-particle effective operators, however the results of numerical analysis demonstrated that the latter are relatively negligible. Thus, it is possible to conclude that by extending the parametrization scheme by the odd rank objects, not only are additional physical mechanisms taken into account, as discussed above, but the errors caused by the non-locality of the potentials used for solving the zero order eigenvalue problem are compensated.

The effective operator in (20-36) is expressed by the unit tensor operator U^λ that acts only within the orbital part of the space and which is related to the double tensor operator $[s]^{1/2} W^{(0\lambda)\lambda}$. Therefore, summarizing the discussion it is possible to conclude that

1. $W^{(0\lambda)\lambda}$ with $\lambda = even$ represent all the physical mechanisms that are described by the standard parametrization of Judd-Ofelt theory and its extension by the third-order electron correlation contributions. In addition these terms contain the impact due to the electric multipole hyperfine interactions defined by (20-34) and (20-36), at the second and at the third order, respectively;

2. $W^{(0\lambda)\lambda}$ with $\lambda = odd$ are the objects we were looking for at the end of the previous section; now we know that they represent third-order terms that originate from the electric multipole hyperfine interactions and are defined in (20-36);

3. $W^{(1k)\lambda}$ with $\lambda = even$ and odd represent in general all interactions within the spin-orbital space, and in particular the nuclear magnetic hyperfine interactions that are defined by the third-order effective operators in (20-28).

All these contributions are included by the parametrization scheme introduced in (18-36). Without the results of numerical calculations, we are unable to establish here a hierarchy of important mechanisms, especially in the case of those that are new and originate from the hyperfine interactions. However this general discussion and review of effective operators that are contributing to the transition amplitude show at least the direction for further theoretical analysis of spectroscopic properties of lanthanides in crystals.

In order to perform direct numerical calculations, similarly as in the case of the second and the third-order contributions to the transition amplitude discussed previously, it is a trivial task to evaluate the angular factors of the effective operators. Since they have the same values for all the lanthanide ions, once evaluated they may be used for analysis of any lanthanide system. The physical reality of each transition in a given system is represented by the matrix elements of the unit tensor operators of certain order between the states involved in the radiative process. If these are the unit tensor operators U^λ acting within the orbital space only, the evaluation of their matrix elements is done in the same manner as in the case of the Judd-Ofelt approach discussed in chapter 17. In the case of the double tensor operators $W^{(\kappa k)\lambda}$, the wave functions have to be presented in the same coupling scheme as for the former ones (for example in the intermediate coupling scheme), and a general expression that includes the genealogy of both states has to be taken into account. However, for $\kappa = 1$, as is the case of the non-relativistic approach presented in this chapter, it is possible to find the values of the matrix elements for the pure $L - S$ coupling in the Tables of Nielson and Koster[12].

The features of the electronic structure of each lanthanide ion are represented in the effective operator contributing to the transition amplitude by the radial integrals. As previously, we use here the perturbed function approach in order to include the impact of the complete radial basis sets of one-electron excited states of a given symmetry. In the case of the effective operators of the second and the third order that originate from the magnetic hyperfine interactions presented in (20-22) and (20-25) there are particular integrals: $\langle \varrho^t(4f \longrightarrow \ell') \mid r \mid \varrho^{-3}(4f \longrightarrow \ell'') \rangle$ for $t = 1, 3, 5, 7$, $\ell' = d, g$ and $\ell'' = p, f$. Thus in order to evaluate directly the impact of the magnetic nuclear interactions the following perturbed functions have to be generated : $\varrho^{1,3,5,7}(4f \longrightarrow d), \varrho^{1,3,5,7}(4f \longrightarrow g)$ (that are already known from the Judd-Ofelt approach of the second order), $\varrho^{-3}(4f \longrightarrow f)$, (the same as in the case of the influence of the spin-orbit interaction), and a new function present at the third order, $\varrho^{-3}(4f \longrightarrow p)$.

In the case of the dipole and quadrupole electric hyperfine interactions the radial integrals of the effective operators are rather similar, however new perturbed functions also have to be introduced. For the second-order effective operator in (20-34) the radial integral is of the same structure as in the case of the Judd-Ofelt effective operator. In particular, for the dipole interactions it has a form of $\langle \varrho^{-2}(4f \longrightarrow \ell') \mid r \mid 4f \rangle$ for $\ell' = d, g$ where a new perturbed function $\varrho^{-2}(4f \longrightarrow \ell')$ is involved. For the third-order effective operators, since in addition to the dipole also the quadrupole interactions are taken into account, we have the following integrals $\langle \varrho^{-2}(4f \longrightarrow \ell') \mid r \mid \varrho^{2,4,6}(4f \longrightarrow \ell'') \rangle$, for $\ell' = d, g$ and $\ell'' = p, f$, and $\langle \varrho^{-3}(4f \longrightarrow \ell') \mid r \mid \varrho^{1,3,5,7}(4f \longrightarrow \ell'') \rangle$, for $\ell' = p, f$ and $\ell'' = d, g$. Thus in addition to the functions already used in the previous analysis, the new ones are: $\varrho^{-2}(4f \longrightarrow d, g)$, and $\varrho^{2,4,6}(4f \longrightarrow p)$ and $\varrho^{2,4,6}(4f \longrightarrow f)$.

As an illustration of this approach, the numerical values of the radial integrals discussed here, and evaluated for some lanthanide ions are presented in chapter 21.

20.8   INTRA-SHELL INTERACTIONS

Although the intra-shell interactions via perturbing operators are beyond the concept of admixing to the wave functions of the ground configuration new components that contribute to the transition amplitude, as it is realized in the standard Judd-Ofelt theory, for the completeness of the presentation, the remaining third-order terms are introduced here. These additional contributions are determined by the matrix elements that contain the first order corrections to the wave functions that are due to both perturbations taken into account simultaneously,

$$\lambda\mu\left(\langle\Psi_f^0 \mid D^{(1)} \mid \Psi_i^{(11)}\rangle + \langle\Psi_f^{(11)} \mid D^{(1)} \mid \Psi_i^0\rangle\right) \tag{20-38}$$

where $\Psi_i^{(11)}$, the characteristic function for the double perturbation theory, as a consequence of the equation that it has to satisfy, is defined by the corrections previously introduced (see section 17.2, (17-23) and (17-24)),

$$\Psi_i^{(11)} = \sum_{Yy}\left\{\frac{\langle Yy \mid V_{cryst} \mid \Psi_i^{(01)}\rangle}{(E_i^0 - E_{Yy}^0)} + \frac{\langle Yy \mid V \mid \Psi_i^{(10)}\rangle}{(E_i^0 - E_{Yy}^0)}\right\} \mid Yy\rangle \tag{20-39}$$

In the case of the inter-shell interactions via $Qh_{hfs}^1 Q$ inserted into the first two terms of (17-25), the effective operator has the following form

$$^3\Gamma_{h_{hfs}^1}(Dh^1 V_{cryst} + V_{cryst}h^1 D)$$

$$= \sum_{tp} B_p^t \sum_{\ell'}\sum_{\ell''}^{even} \delta(\ell'', \ell')\langle\varrho^t(4f \longrightarrow \ell')\mid r\mid\varrho^{-3}(4f \longrightarrow \ell')\rangle$$

$$\times \sqrt{\ell'(\ell'+1)(2\ell'+1)}\langle\ell||\mathbf{C}^{(t)}||\ell'\rangle\langle\ell'||\mathbf{C}^{(1)}||\ell\rangle$$

$$\times \sum_{\lambda\mu}(-1)^{\lambda+p}[\lambda]\left(X_{ppq}^{t11}(\lambda\mu;\ell'\ell') + X_{ppq}^{1t1}(\lambda\mu;\ell'\ell')\right)U_\mu^{(\lambda)}(\ell\ell) \tag{20-40}$$

with the angular factors defined in (20-23). These are the third-order contributions that originate from the off diagonal configuration interaction between $4f^{N-1}n'\ell'$ and $4f^{N-1}n''\ell'$ for $\ell' = even(\equiv d, g)$ and for $n' \neq n''$. The diagonal terms for $n' = n''$ are more complex, and in order to derive their effective operator form some additional approximations have to be introduced. It is also possible to include the whole impact due to the nuclear magnetic hyperfine interactions in an approximate way using the measured value of the magnetic hyperfine structure constant. However such an approach would not provide information about the nature and sensitivity of the $f \longleftrightarrow f$ transitions to these interactions.

When $Qh_{hfs}^2 Q$ is used as a perturbation, its direct impact upon the transition amplitude is expressed by double tensor operators, as in the case of the inter-shell interactions. For example, for the sequence of the matrix elements $V h^2 D$, the effective operator has the structure

$$\Gamma^{11}(h^2 DV + V Dh^2) = \frac{1}{2}\sqrt{3} \sum_{\ell'} \sum_{\ell''} \sum_{tp}^{odd} B_p^t \langle \varrho^t(4f \longrightarrow \ell') \mid r^{-3} \mid \varrho^1(4f \longrightarrow \ell'') \rangle$$

$$\times \langle \ell ||\mathbf{C}^{(t)}||\ell'\rangle \langle \ell'||\mathbf{C}^{(2)}||\ell''\rangle \langle \ell''||\mathbf{C}^{(1)}||\ell\rangle \sum_q \sum_y \sum_{\lambda,\mu} [y][\lambda]^{\frac{1}{2}} \sum_{x,\sigma} (-1)^{x+y+t+\lambda-\mu} [x]$$

$$\times \begin{pmatrix} 1 & 1 & x \\ \rho & q & -\sigma \end{pmatrix} \begin{pmatrix} t & x & \lambda \\ p & \sigma & -\mu \end{pmatrix} \begin{Bmatrix} \ell & y & 1 & 2 \\ \ell & \lambda & 1 & \ell'' \\ \ell' & t & x & 1 \end{Bmatrix}$$

$$\times W_\mu^{(1y)\lambda}(s\ell, s\ell) \tag{20-41}$$

As is seen from this example, the effective operators that represent the intra-shell interactions have a similar tensorial structure to those which include the inter-shell interactions in (20-28). In general, when taking formally all possible sequences of operators in triple products of matrix elements (at the moment without their physical meaning), the derived effective operators contain the angular factors with various permutations of the indices that identify them. This is exactly what is observed comparing the structures of the intra-shell interactions via nuclear magnetic interactions V_{hfs}. The same conclusion is derived from the analysis of the effective operators with the intra-shell interactions included via the electric multipole hyperfine interactions V_{EM}. Indeed, the set of the effective operators defined in (20-36) is completed by the additional objects that are associated with the unit tensor operator $U^{(\lambda)}$ multiplied by the radial integrals with all possible assignments of the ranks of the perturbed functions and powers of the radial coordinate. The angular factors of these operators are determined by $X_{q_1q_2q_3}^{k_1k_2k_3}$, defined in (20-23), for all assignments of $(k_1, k_2, k_3) = (k, t, 1)$, with the appropriate order of their components q, p, ρ.

Thus, as expected, the intra-shell interactions do not provide effective operators that would change the selection rules for the non vanishing contributions to the transition amplitude established at the third order analysis of the objects representing the inter-shell interactions of various physical origin.

REFERENCES

1. http://www.users.cloud9.net/ cgseife/conway.html.
2. Wybourne B G, (1962) Nuclear Moments and Intermediate Coupling *J. Chem. Phys* **37** 1807.
3. Baker J M and Bleaney B, (1958) Paramagnetic Resonance in Some Lanthanon Ethyl Sulphates *Proc. R. Soc.,* **A245** 156.
4. Agladze N I and Popova M N, (1985) Hyperfine Structure in Optical Spectra of LiYF$_4$: Ho *Sol. St. Comm.,* **55** 1097.
5. Brage T, Judge P G and Profitt C R, (2002) Determination of Hyperfine-induced Transition Rates from Observations of a Planetary Nebula *Phys. Rev. lett.,* **89** 281101-1.

6. Huff L D and Houston W V, (1930) The Appearance of "Forbidden Lines" in Spectra *Phys. Rev.,* **36** 842.

7. Garstang R H, (1962) Hyperfine Structure and Intercombination Line Intensities in the Spectra of Magnesium, Zinc, Cadmium and Mercury *J. Opt. Soc. Am.,* **52** 845.

8. http://www.webelements.com/webelements/elements/text/N/isot.html.

9. Smentek L and Kędziorski A, (2006) Hyperfine Induced f-f Transitions *Spectr. Lett.,* **39** (in press).

10. Ord-Smith R, (1954) The Symmetry Relations of the 12*j* Symbol *Phys. Rev.,* **94** 1227.

11. Smentek-Mielczarek, L, (1995) An Alternative Formulation of the Judd-Ofelt Theory of $f \leftrightarrow f$ Transitions *J. Alloys Compd.,* **224** 81.

12. Nielson, C W and Koster, G F, (1963) *Spectroscopic Coefficients for the p^n, d^n, and f^n Configurations*, Cambridge: MIT Press.

21 Numerical Analysis of Radial Terms

Six hours a-day the young students were employed in this labour; and the professor showed me several volumes in large folio already collected, of broken sentences, which he intended to piece together; and out of those rich materials to give the world a compleat body of all the arts and sciences; ...

Jonathan Swift *Gulliver's Travels (1726)*

The numerical results and their graphical illustration presented here are the contribution of Andrzej Kędziorski, PhD candidate at Physics Department, Nicolaus Copernicus University in Toruń, Poland.

21.1 APPROXIMATIONS

In order to determine the accuracy and validity of the approximations that have to be introduced to derive the effective operators that contribute to the amplitude of $f \longleftrightarrow f$ transition, and as a consequence, to prepare the expressions for practical calculations, as an example the radial term of the Judd-Ofelt theory is analyzed. Thus the question is how to evaluate the radial integrals without great difficulty and within a satisfactory accuracy.

The second-order contributions to the transition amplitude have been defined in (17-6) by the perturbing expression,

$$\Gamma = \sum_{Xx} \left\{ \langle \Psi_f^0 \mid D_\rho^{(1)} \mid Xx \rangle \langle Xx \mid QV_{cryst}P \mid \Psi_i^0 \rangle / (E_i^0 - E_{Xx}^0) \right.$$
$$\left. + \langle \Psi_f^0 \mid PV_{cryst}Q \mid Xx \rangle \langle Xx \mid D_\rho^{(1)} \mid \Psi_i^0 \rangle / (E_f^0 - E_{Xx}^0) \right\}$$

where originally, as derived, there is a distinction between the energies and the wave functions of the initial and the final states of the process. While it is rather straightforward to evaluate the angular part of any of the effective operators that contribute to the transition amplitude, the main problem is to evaluate the radial integrals.

In general, the radial term of this second-order contribution, for each sequence of the operators, is expressed as a product of three terms. For example, the radial part for the sequence $PV_{cryst}QD^{(1)}$ is a product of the terms

$$\Gamma_1(rad) = \sum_{Xx} \langle \Psi_f^0 \mid r^t \mid Xx \rangle_r \langle Xx \mid r^1 \mid \Psi_i^0 \rangle_r / (E_i^0 - E_{Xx}^0)$$
$$\Rightarrow S(4f^{N-1}\Psi_f, 4f^{N-1}Xx)S(4f^{N-1}Xx, 4f^{N-1}\Psi_i)$$
$$\times \langle 4f_{\Psi_f} \mid r^t \mid (n'\ell')_x \rangle \langle (n'\ell')_x \mid r^1 \mid 4f_{\Psi_i} \rangle / (E_i^0 - E_{Xx}^0) \qquad (21-1)$$

where S denotes the overlap integrals between the $N - 1$ one-electron functions of the ground configuration $4f^N$ that are generated for initial and final states, and those that describe the excited configuration X in its state x. These one-electron functions obtained as solutions for the Hamiltonians that describe each case separately are not orthogonal to each other, and therefore integration over N radial variables results in the overlap integrals over $N - 1$ functions and the actual one-electron radial integral that contains the orbitals of the Nth electron. The function $n'\ell'$ is adjusted to the occupied core of the $4f^{N-1}$ configuration as the first excited one-electron state of ℓ' symmetry with the required number of nodes and orthogonality to the occupied orbitals of the same symmetry. For the second sequence of the operators, the second-order radial terms are analogous, except that the energy denominator is determined by the energy E_f^0 of the final state of the process instead of E_i^0 in (21-1).

How to avoid the problem of evaluation of the overlap integrals? The easiest way is to assume that a good approximation is to generate the excited one-electron states with frozen core orbitals, which are optimal, in particular in the sense of the Hartree-Fock model, for the ground configuration $4f^N$. At this point another question arises, namely for which state, the final or the initial of the radiative process, should the core be generated and subsequently frozen in the calculations for the excited configurations. The situation is even more complex, since in order to make the numerical calculations possible in practice several approximations for the energy denominators are necessary. Indeed, assuming only that the average energy of configuration may replace the energies of certain states in the energy denominators is it possible to perform the partial closure and derive the effective operators of the theory of $f \longleftrightarrow f$ transitions as presented in chapter 17.

In Table 21-1 the results of numerical Hartree-Fock calculations performed for the Eu^{3+} ion are presented to estimate the error caused by these approximations. All calculations have been performed by the numerical MCHF program of Charlotte

TABLE 21-1

Orbital and Total Energies[a] Evaluated by Numerical Hartree-Fock Method for Eu^{3+} Ion (in a.u.); All Calculations for the Excited Configurations Performed for Their Average Energy.[b]

	$(4f^6)_{AV}$	^7F	$^5D(3)$	$^5F(2)$	$^5I(2)$
$-\varepsilon_{4f}$	1.782	1.885	1.815	1.820	1.831
$-\varepsilon_{5d}$ [b]	1.230	1.228	1.229	1.229	1.229
$-\varepsilon_{5g}$ [b]	0.321	0.321	0.321	0.321	0.321
$\Delta E_{HF}(4f^6)$	0.358	0	0.243	0.225	0.188
$\Delta E_{HF}([4f^5]5d)$ [b]	0.910	0.898	0.906	0.905	0.904
$\Delta E_{HF}([4f^5]5g)$ [b]	1.820	1.805	1.814	1.813	1.812

[a] $\Delta E_{HF} = E_{HF} - E_{HF}^{7F}$, $E_{HF}^{7F} = 10422.029$ a.u.
[b] Orbital and total energies evaluated with the frozen core orbitals of $[Xe]4f^5$ that are optimal for the average energy of the $4f^6$ and for the energy states ^7F, $^5D(3)$, $^5F(2)$ and $^5I(2)$, respectively.

Froese Fischer[1]. The values of the total energies and the orbital energies evaluated for the average energy of the $4f^6$ configuration are compared in this table with those obtained from the calculations performed for certain energy states of the same configuration. As an example the following energy states have been chosen,

$$^7F \equiv (100)(10)^7F$$
$$^5D(3) \equiv (210)(21)^5D$$
$$^5F(2) \equiv (210)(21)^5F$$
$$^5I(2) \equiv (210)(20)^5I$$

The energies of the excited configurations $4f^55d$ and $4f^55g$, and the first one-electron excited states of d and g symmetries are evaluated with the frozen core of $[Xe]4f^5$, which is optimal for the appropriate energy of the ground configuration $4f^6$. It is seen from this table that the values obtained for the average energy of the ground configuration reproduce very well the values evaluated for various energy states. This means that at this point of analysis there is the possibility of replacing the total energies in the energy denominators of the radial terms by the orbital energies of appropriate one-electron states. To determine the error caused by this simplification, the results of calculations performed for the average energy of $4f^6$, $4f^55d$ and $4f^55g$ are collected in Table 21-2. If it is acceptable to use the functions generated for the average energy of the ground configuration instead of those for each particular excited configuration, the overlap integrals in (21-1) disappear.

The discrepancy between the values of the orbital energies evaluated within various schemes is the price we have to pay for the simplification of the radial term to the form treatable in practice. As a consequence, for further numerical analysis, the radial basis sets and appropriate energies are generated for the average energy of the ground configuration. In the particular case of the Eu^{3+} ion, these are the values in the first column of Table 21-1. In general this means that there is one common radial basis set of one-electron functions, and consequently the energy denominators of (21-1) are reduced to the difference between the orbital energies. At the same time, the overlap

---

TABLE 21-2
Orbital and Total Energies[a] Evaluated for the Average Energies of
$4f^6$, $4f^55d$, and $4f^55g$ Compared With the Values Obtained
With the Frozen Core $[Xe]4f^5$ (in a.u.)

	$(4f^6)_{AV}$	$(4f^55d)_{AV}$	$(4f^55g)_{AV}$
$-\varepsilon_{4f}$	1.782	2.104	2.416
$-\varepsilon_{5d}$	1.230	1.172	—
$-\varepsilon_{5g}$	0.321	—	0.321
ΔE_{HF} [a]	0.358	0.804	1.644
		(0.910) [b]	(1.820) [b]

[a] $\Delta E_{HF} = E_{HF} - E_{HF}^{7F}$, $E_{HF}^{7F} = 10422.029$ a.u.
[b] Evaluated for the frozen core of the average energy of $4f^6$ taken from Table 21.1

integrals in (21-1) are equal to unity, since all one-electron states are the solutions of the equations with the same Hamiltonian, and therefore they are orthonormal.

21.2 FUNCTIONS OF THE RADIAL BASIS SET

The quality of the radial functions generated for the average energy of the $4f^6$ configuration of the Eu^{3+} ion and for various energy states is presented in figure 21-1. In fact, following the practice of atomic spectroscopy, in this figure and in all presented here instead of the original radial functions $R(r)$ that are the solutions of the radial equations, the functions $P(r) = r R(r)$ are presented as those that define directly the radial probability (by an integral without a jacobian).

At the first sight the orbitals presented in Fig. 21-1 are identical no matter which procedure has been used for their generation. In the scale of this figure one-electron functions obtained for the average energy of $4f^6$ configuration overlap with those adjusted for the energies of various energy states of this configuration. This is the case of $4f$ orbital as well as the one-electron states of $5d$ and $5g$ symmetry.

In order to see the actual relation and differences between these functions, in Figures 21-2 and 21-3 the magnified plots are presented, each in three segments, which differ by a scale.

The most reliable way to evaluate the discrepancies between the functions obtained for the average energies of configurations, the ground and excited ones, and those from the calculations performed for the certain energy state of a given configuration is to compare the values of the radial integrals. In fact, not the functions by themselves are interesting for us, but rather the radial integrals that are the part of the effective

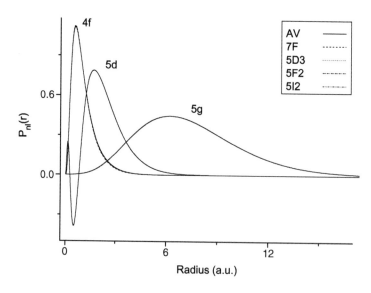

FIGURE 21-1 Atomic orbitals $4f$, $5d$ and $5g$ of Eu^{3+} generated for the average energy of $4f^6$ configuration and for several energy states.

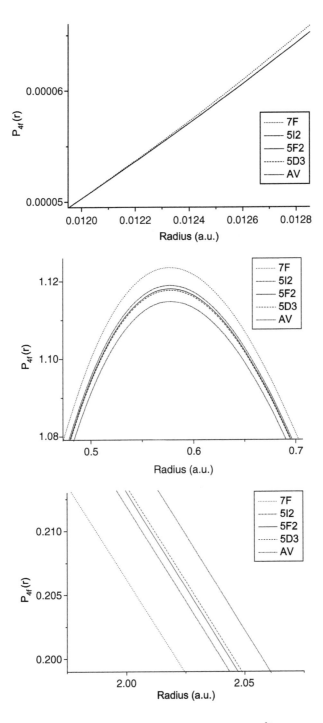

FIGURE 21-2 One-electron function of $4f$ symmetry for the Eu^{3+} ion generated for the average energy of $4f^6$ and for various energy states (enlargement of the plot from Fig. 21-1).

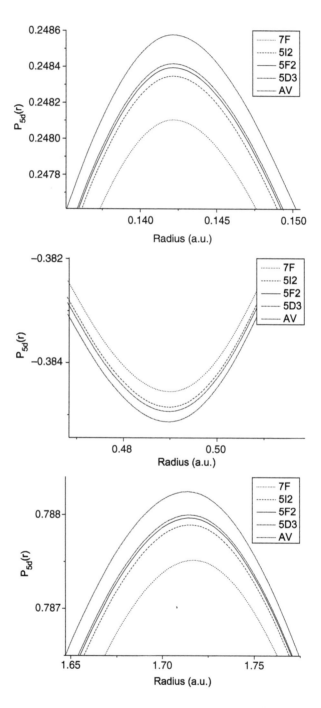

FIGURE 21-3 One-electron functions of $5d$ symmetry for the Eu^{3+} ion generated for the average energy of $4f^5 5d$ evaluated with the frozen core orbitals optimal for the average energy of $4f^6$ and for various energy states of this configuration (enlargement of the plot from Fig. 21-1).

TABLE 21-3
Radial Integrals $\langle nl|r^k|nl\rangle$ (in a.u.) Evaluated With the Hartree-Fock Functions for Eu^{3+}

$n\ell$	k	$(4f^6)_{AV}$	^7F	$^5D(3)$	$^5F(2)$	$^5I(2)$
4f	1	0.82	0.81	0.81	0.81	0.81
	2	0.84	0.81	0.83	0.83	0.82
	3	1.06	1.00	1.04	1.04	1.03
	4	1.62	1.49	1.58	1.57	1.56
	5	2.96	2.63	2.84	2.83	2.79
	6	6.28	5.39	5.97	5.93	5.84
	7	15.27	12.63	14.33	14.21	13.93
			Frozen Core			
5d	1	2.01	2.01	2.01	2.01	2.01
	2	4.61	4.63	4.62	4.62	4.62
	3	11.85	11.91	11.87	11.87	11.88
	4	33.77	33.20	33.84	33.85	33.88
	5	106.18	107.04	106.47	106.51	106.60
	6	366.30	369.80	367.48	367.64	367.99
	7	1378.16	1393.30	1383.27	1383.96	1385.48
5g	1	6.82	6.82	6.82	6.82	6.82
	2	50.86	50.86	50.86	50.86	50.86
	3	411.20	411.20	411.20	411.20	411.20
	4	3582.58	3582.64	3582.61	3582.62	3582.63
	5	33458.50	33459.24	33458.90	33458.95	33459.04
	6	333421.5	333429.6	333425.8	333426.4	333427.5
	7	3531185	3531280	3531237	3531243	3531256

operators. In Table 21-3 the expectation values of the radial distance in various powers evaluated with different functions are collected. These are the values obtained with the particular functions that are graphically illustrated in Fig. 21-2 and for which the appropriate energies are presented in Table 21-1.

In the first part of this table are the results obtained with the $4f$ functions; these are the particular radial integrals that determine the contributions to the transition amplitude, which are caused by the dynamic model (complementary to the static model of the Judd-Ofelt theory). In the second part of this table there are the values of radial integrals evaluated with one-electron functions of the first excited states of d and g symmetry. These in the first column again are generated for the average energy of $4f^6$ configuration, while in the following columns those obtained for chosen energy states of this configuration are presented.

The functions used to evaluate the integrals presented in Table 21-2 are illustrated graphically in Fig. 21-4 where the $4f$ orbitals of Eu^{3+}, which are generated for the average energies of $4f^6$, $4f^5 5d$ and $4f^5 5g$, respectively, are presented. In these calculations all the orbitals are varied to give an optimal energy. The magnified picture of their differences is presented in three sections in Fig. 21-5.

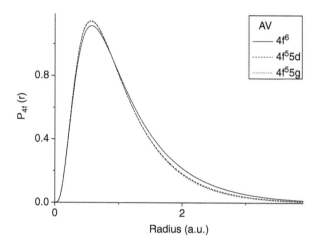

FIGURE 21-4 Radial parts of the orbital $4f$ of Eu^{3+} evaluated for the average energies of $4f^6$, $4f^55d$ and $4f^55g$; all the orbitals are varied to give an optimal energy for each electron configuration.

In support of the conclusions that are derived when analyzing the numerical results collected in Table 21-2, it is seen from the above figures that all functions are close enough to each other to be represented by those generated for the average energy of the ground configuration. A similar relation between the functions that result from various numerical procedures is found in the case of the excited one-electron states. For example in Fig. 21-6 the radial part of the orbital $5d$ generated for the average energy of $4f^55d$ configuration with the frozen core orbitals is compared with the resulting function from the procedure in which all orbitals are varied. Again, to demonstrate the differences which may give insight into the accuracy of the numerical procedure applied here, the same functions are presented in Fig. 21-7 in sections in various scales.

As before, the accuracy of the functions presented in these figures is easily measured through the comparison of the values of the radial integrals. In the first column of Table 21-4 the values of integrals obtained for the average energy of the ground configuration $4f^6$ and those with the frozen core for the excited configurations are collected. They are compared with integrals evaluated with the radial functions that are generated when all orbitals of $4f^55d$ and $4f^55g$ are varied. The values from the first column of this Table are used for further numerical analysis of the spectroscopic properties of the lanthanide ions. To be precise however it must be recalled that in fact when the perturbed function approach is applied, the atomic radial functions analyzed here are used in practical calculations as an initial radial basis set to build up the equations the solutions of which are the new functions (see section 17.4, and in particular equation (17-37)).

As a final characteristic of the one-electron functions used in the further numerical calculations, the radial densities of the $4f$, $5s$ and $5p$ states are presented in Fig. 21-8. The functions were generated for the average energy of the $4f^6$ configuration.

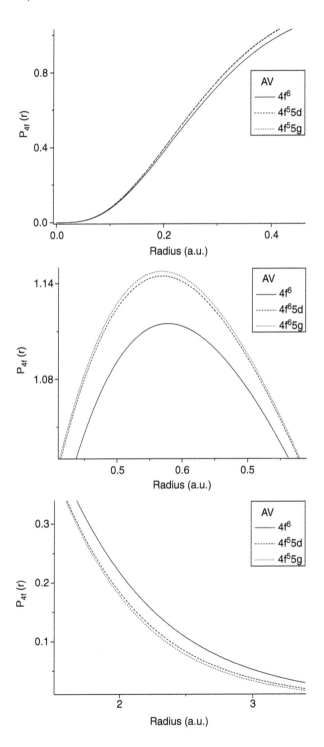

FIGURE 21-5 $4f$ orbitals from Fig. 21-4 presented in sections (note the change of the scales).

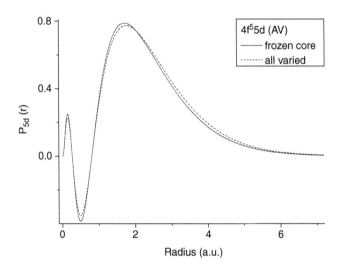

FIGURE 21-6 Radial parts of first excited one-electron state $5d$ of Eu^{3+} ion generated for the average energy of $4f^5 5d$ with the frozen core orbitals and with all orbitals varied.

The closed shells $5p$ and $5s$, although internal in the energy scale ($\varepsilon_{4f} = -1.87$, while $\varepsilon_{5p} = -2.06$ and $\varepsilon_{5s} = -2.85$, all values in a.u.), are in the space more distant than the $4f$ shell, as seen in this figure. Therefore the optically active electrons from the $4f$ shell are indeed screened by the stable closed shells $5p^6$ and $5s^2$ from the perturbing influence of the environment, if the ion is placed in a crystal, for example. It is interesting to mention that this property of the electronic structure of three positive lanthanide ions is visible, as reproduced here, already at the level of a simple

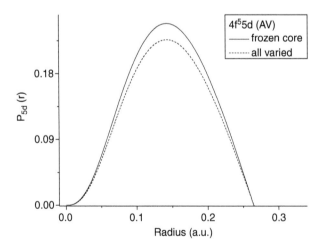

FIGURE 21-7 One electron functions $5d$ from the previous figure presented in sections with various scales.

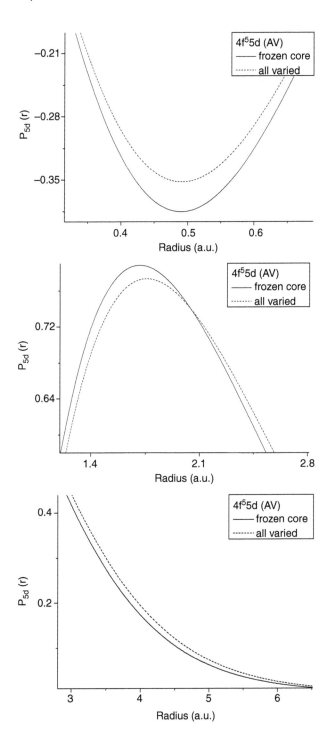

FIGURE 21-7 (*Continued*).

TABLE 21-4
Radial Integrals $\langle n\ell|r^k|n\ell\rangle$ Evaluated With the Hartree-Fock Functions Generated for the Average Energy of $4f^6$, $4f^5 5d$, and $4f^5 5g$ of Eu^{3+} Ion (in a.u.)

$n\ell$	k	$(4f^6)_{AV}$	$(4f^5 5d)_{AV}$	$(4f^5 5g)_{AV}$
$4f$	1	0.82	0.78	0.78
	2	0.84	0.75	0.74
	3	1.06	0.89	0.86
	4	1.62	1.25	1.19
	5	2.96	2.08	1.92
	6	6.28	4.00	3.56
	7	15.27	8.80	7.49
		Frozen Core		
$5d$	1	2.01	2.08	
	2	4.61	4.94	
	3	11.85	13.11	
	4	33.77	38.69	
	5	106.18	126.02	
	6	366.30	450.43	
	7	1378.16	1,755.56	
$5g$	1	6.82		6.84
	2	50.86		51.07
	3	411.20		413.49
	4	3582.58		3607.20
	5	33458.50		33727.35
	6	333421.5		336454.4
	7	3531185		3566784

model of Hartree-Fock calculations that are performed for the average energy of a configuration.

To continue the discussion on the evaluation of the radial term of the Judd-Ofelt effective operators it must be concluded that when the above approximations are adopted, the general definition of (21-1) is reduced to a much simpler integral

$$R^t_{JO}(\ell') = \sum_{n'} \frac{\langle 4f|r^t|n'\ell'\rangle\langle n'\ell'|r^1|4f\rangle}{(\varepsilon_{4f} - \varepsilon_{n'\ell'})} \qquad (21\text{-}2)$$

where the summation is performed over all excited one-electron states of ℓ' symmetry. In the description of the electric dipole $f \longleftrightarrow f$ transitions the most important are the excitations to the states of d symmetry, however those associated with the excitations to the $g-$ states are not negligible. The values of the radial integrals

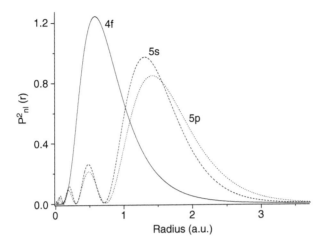

FIGURE 21-8 Screening of the $4f$ orbitals by the closed shells of the $5s$ and $5p$ symmetry; results obtained from the Hartree-Fock calculations performed for the average energy of $4f^6$ configuration of the Eu^{3+} ion.

that contribute to the term defined in (21-2) for the excitations to the $5d$ orbital are collected in Table 21-5. In the first column of this table there are the integrals obtained with the functions generated for the average energy of $4f^5 5d$, with all one-electron functions varied. In the second column there are integrals evaluated with the frozen core obtained for the average energy of ground configuration $4f^6$. The latter integrals, together with the appropriate energy denominators, are used to evaluate the radial part of the Judd-Ofelt effective operators, if the calculations are performed directly. For comparison in the two following columns of this table the values obtained for various energy states of $4f^6$ are also displayed.

TABLE 21-5
Radial Integrals $\langle 4f|r^k|5d \rangle$ (in a.u.) for Eu^{3+}, Evaluated With the Same Orbitals as Those Displayed in Table 21.4

t	$\langle 4f|r^t|5d \rangle_{AV}$	$\langle 4f|r^t|[4f^5]5d \rangle_{AV}$	$\langle 4f|r^t|[4f^5]5d \rangle^{7F}$	$\langle 4f|r^t|[4f^5]5d \rangle^{5F(2)}$
1	0.50	0.59	0.56	0.58
2	0.96	1.14	1.08	1.12
3	1.92	2.34	2.19	2.29
4	4.22	5.31	4.89	5.15
5	10.32	13.37	12.12	12.89
6	27.97	37.46	33.38	35.87
7	83.72	115.90	101.49	110.26

In accordance with the definition in (21-2), the radial integrals have to be divided by the appropriate energy denominators, and summed over all members of the radial basis set of one-electron excited states of ℓ' symmetry, discrete and the continuum.

In Table 21-6 various approximations of the energy denominators are presented. From all the numbers collected there the most important are those in the last column, since they are determined by the difference of the orbital energies of $4f$ and $5\ell'$ one-electron states generated for the average energy of $4f^6$ and with a frozen core. Indeed, as stated above, all practical calculations are performed for the average energy of the $4f^N$ configuration, and the excited functions are generated for the frozen core orbitals. The first two columns of Table 21-6 contain the results evaluated for the ground state ^7F of the ground configuration, and for the excited state $^5F(2)$ chosen arbitrarily as an example.

TABLE 21-6
Various Approximations for the Energy Denominators of $R^t_{j\,0}(\ell')$

ℓ'	$\Delta\varepsilon^{7F}(5\ell')$ [a]	$\Delta\varepsilon^{5F(2)}(5\ell')$ [a]	$\Delta\varepsilon(5\ell')$ [b]
d	0.657	0.591	0.552
g	1.564	1.499	1.461

[a] $\Delta\varepsilon^{7F}(5\ell') = \varepsilon^{7F}_{4f} - \varepsilon^{7F}_{5\ell'}$
[b] $\Delta\varepsilon(5\ell') = \varepsilon_{AV}(4f) - \varepsilon_{AV}(5\ell')$, where $\varepsilon_{AV}(4f)$ is obtained for the average energy of $4f^6$ and $\varepsilon_{AV}(5\ell')$ for the average of $4f^{N-1}5\ell'$ with frozen core of $4f^6$.

21.3 PERTURBED FUNCTIONS

Using the values collected in all tables presented up to this point we are able to evaluate the first component of the radial integral in (21-2). This particular term includes the impact of the first excitations from the $4f$ orbital to $5d$. The question is how many excitations to one-electron states of d symmetry should be taken into account in the summation over n'. Should the summation be limited to the discrete part of one-electron spectrum, or is the continuum part important enough to also be taken into account? Obviously the same questions concern the excitations to the states of g symmetry.

In Table 21-7 the completeness of the radial basis set of d symmetry is illustrated. The first column collects the values of the product of the radial integrals of the Judd-Ofelt term, while the next column presents the values obtained with the energy denominators included (this is why both values are of opposite signs). The latter are the contributions to the radial integrals of the Judd-Ofelt model, therefore the summation starts from $n' = 5$, which is the first excited (not occupied) one-electron state of d symmetry. The value of the summation performed over the terms up to the $n' = 9$ are displayed below each column. When they are compared with the results obtained for **all** one-electron states of d symmetry, it is clearly seen that more members than five of the excited set have to be taken into account. This conclusion

TABLE 21-7

Completeness of the Radial Basis Set Applied for Evaluation of the Judd-Ofelt Radial Term $R_{JO}^3(d)$ for Eu^{3+} Ion

n'	$\langle 4f|r^3|n'd\rangle\langle n'd|r|4f\rangle$	$R_{JO}^3(n'd)$
3	0.0036	—
4	0.3592	—
5	1.3822	−2.5040
6	−0.0494	0.0405
7	−0.0144	0.0100
8	−0.0064	0.0041
9	−0.0035	0.0022
	$\sum_{n'=3}^{9} 1.6713$	$\sum_{n'=5}^{9} -2.4472$
	$\langle 4f|r^4|4f\rangle = 1.6237$	$R_{JO}^3(d) = -1.7541$

is derived from the comparison of $\langle 4f|r^4|4f\rangle$, with the condition satisfied for the complete radial basis set $\sum_{n'}|n'd\rangle\langle n'd| = 1$, and the value of $R_{JO}^3(d)$ evaluated with the perturbed function. Indeed it is seen from this table that the results obtained for a few components (five) are far from those evaluated for the complete radial basis set. When the summation is performed up to $n' = 15$, the value of the product of the radial integrals is 1.3024, without the terms with the closed shells $3d$ and $4d$, and 1.6652, when they are included (these values have to be compared to 1.6237 that is evaluated for the complete radial basis set). The process of the convergence of the summation to the exact limit is slow, since six additional members of the set improve the result by only 0.0062. In fact this analysis shows the importance of the perturbed function approach as a tool for the evaluation of the radial integrals for all members of the radial basis sets, discrete and continuum. The results collected in this table demonstrate also the magnitude of the error that is made when only the first member of the set of d orbitals is taken into account, and the rest of the spectrum is simply neglected. Unfortunately this truncation is commonly applied in direct calculations reported in the literature. In order to make the numerical analysis of the radial integrals possible without the errors that are caused by the incompleteness of the basis sets, the values for all lanthanide ions that are evaluated with the perturbed functions are presented at the end of this chapter.

As demonstrated in the discussion of electric dipole $f \longleftrightarrow f$ transitions the task of evaluation of the radial integrals for the complete radial basis sets is simplified by application of the perturbed function approach.[2] In this approach instead of performing the summation in (21-2) in an explicit way, differential equations are solved for newly defined functions. In general, each perturbed function is defined by a typical perturbing expression (which is in fact a part of the Judd-Ofelt radial term)

$$\varrho^t(4f \longrightarrow \ell') = \sum_{n'} \frac{\langle n'\ell'|r^t|4f\rangle}{(\varepsilon_{4f} - \varepsilon_{n'\ell'})} P_{n'\ell'}$$

and consequently the radial term of the Judd-Ofelt effective operator is determined by a single radial integral of the form

$$R^t_{JO} = \langle \varrho^t(4f \longrightarrow \ell')|r^1|4f\rangle$$

In the particular case discussed here, the radial part of the crystal field potential, r^t, is the perturbation, and it is included in the definition of the perturbed function. In the case of other perturbing mechanisms taken into account in the description of the spectroscopic properties of the lanthanide ions, their radial parts should replace r^t in the above definition of ϱ (see chapter 17).

The initial basis sets of one-electron functions of a given symmetry that are occupied in the ground configuration of the lanthanide ion consist in each case of the orbitals that are generated for an average energy of a ground configuration. These are the only one-electron functions, together with their orbital energies, the knowledge of which is necessary for the construction of the differential equations (17-37) for the perturbed functions.

A graphical illustration of the perturbed functions generated for the Eu^{3+} ion is presented in Fig. 21-9. In this figure the perturbed functions $\varrho^t(4f \longrightarrow d)$ for $t = 3, 5$ are plotted on the background of the first excited state $5d$ and the high energy excited state $14d$. These atomic functions are included in this figure to demonstrate that the perturbed functions, although containing the impact due to all possible excitations to the states of d symmetry, including the continuum part of the spectrum, in fact have the nature of the first excited state $5d$. The dominant role of the first excited state is already seen in Table 21-7, where the largest contribution to the radial terms is due to the first excited orbital.

As seen from this figure, the perturbed functions are localized in the same region of space, and they possess the same number of nodes as the first excited state.

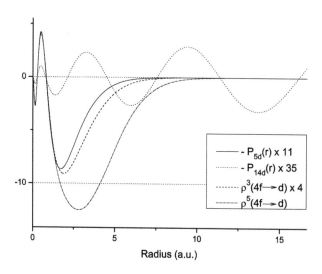

FIGURE 21-9 The perturbed functions $\varrho^t(4f \longrightarrow d)$ generated for the Eu^{3+} ion presented on the background of the atomic orbitals $5d$ and $14d$ generated for the average energy of the $4f^6$ configuration.

At the same time, since the $4f$ and $5d$ (and also $5g$) orbitals are localized in the same region (see Fig. 21-1), this means that the perturbed functions are localized also in the same region of space as $4f$. At first sight this result is surprising, especially when analyzing the graphical representation of the $14d$ orbital, which has the form of oscillating maxima and minima extended quite far from the origin, and consequently from the region of the atomic function $4f$. Thus the spatial behavior of the perturbed functions may be understood assuming that each function is a result of cancelation of the oscillating tails of the high energy one-electron states. As a consequence, the resulting perturbed functions are moved in towards the origin as observed in Fig. 21-9.

The spatial properties of the perturbed functions of the $f \longleftrightarrow f$ transition theory along with the simple concept of overlap and penetration of the wavefuctions allow one to treat the lanthanide ion as separated from the environment. Because of these properties, even in the case of complex chelates of lanthanides, it is possible to regard the wave function describing the whole system as a simple product of two distinct parts that are associated with each sub-system.

All the aspects of direct calculations of the radial integrals of the Judd-Ofelt effective operators are discussed here for the case of Eu^{3+}. Is this ion exceptional or are its properties and behavior representative of the remaining members of the lanthanide family? The answer to this question is presented graphically in Fig. 21-10 where the $4f$ atomic orbitals for all the lanthanide ions are displayed. It is interesting to mention that the variation of the properties or the change of the values of various integrals, for example $\langle 4f|r|4f \rangle$ presented in Fig. 21-11, are as smooth across the lanthanide series as is the change of the atomic orbitals presented in this figure. From the second part of this figure it is seen that with increasing atomic number across the series the $4f$ orbitals are moved in towards the origin; this property is known as **the lanthanide contraction**. The expectation value of the position operator in the energy state $4f$ for the ions across the lanthanide series demonstrates also this contraction. This property is illustrated in Fig. 21-11 where the values of the radial integrals for all ions evaluated with the Hartree Fock functions generated for the average energy of the ground configuration $4f^N$ are presented.

For the first time in 1941 Goeppert Mayer pointed out that the properties of lanthanide elements are determined by a sudden contraction and decrease in energy of the $4f$ electrons [3]. This collapse of the $4f$ orbitals results from the specific features of the effective potential for 4f electrons that has the form (in atomic units)

$$V_{eff} = \frac{\ell(\ell + 1)}{2r^2} + V(r)$$

where the first part is the centrifugal term, and $V(r)$ is the atomic central potential.

There are two potential wells separated by a potential barrier. For large r there is a shallow and very broad valley. For small r there is another minimum that is drawn into the interior and becomes deeper and narrower with increasing Z. The rivalry between these two wells determines the properties of f orbitals. When Z is small, the inner well is not deep enough, and therefore the lowest level of the whole system is that of the outer valley dominated by the Coulomb potential. Since the potential in this

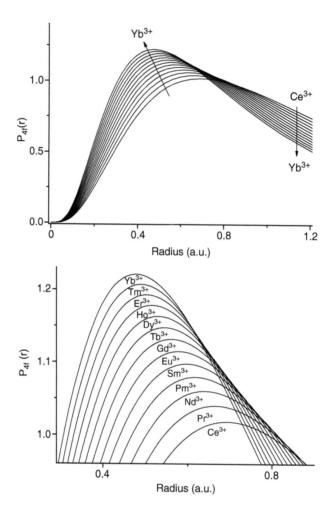

FIGURE 21-10 The $4f$ orbitals generated for the average energy of the ground configurations for all the lanthanide ions as graphical illustration of the lanthanide contraction.

region does not vary with Z, the binding energy of $4f$ electrons is almost the same as for the hydrogen atom. The first bound state of the inner well appears for elements at the beginning of the lanthanide series, for $Z = 58$. When the first electronic state in the inner well is below that of the outer well, the spatial shrinkage of the $4f$ function is observed. Since, at the same time the inner valley becomes deeper and narrower with increasing atomic number, the systematic shrinkage of $4f$ orbitals progresses across the lanthanide series.

In general, relativistic effects result in shrinkage and energy stabilization of the inner orbitals of s and p symmetries. At the same time this means that the screening of the nucleus is stronger in the relativistic description, and therefore the functions of the outer shells of d and f symmetries expand radially. This general trend makes

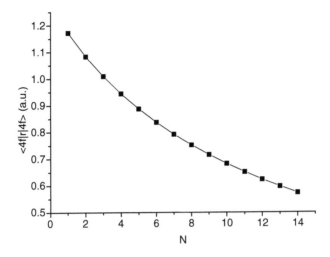

FIGURE 21-11 The values of the radial integrals $\langle 4f|r|4f \rangle$ evaluated with the Hartree-Fock orbitals generated for the average energy of the ground configuration $4f^N$ for ions across the lanthanide series.

the sudden shrinkage of the $4f$ orbitals for the lanthanides even more distinct, and therefore easily noticed.

Although the problem of the lanthanide contraction was addressed as early as 1941 with the pioneering work of Goeppert Mayer, still there is a broad interest in its consequences. Evidence of this may be found in the literature where an explosion of investigations is described. Indeed, there are discussions about the lanthanide contraction effect on grain growth in rare earth ions doped materials[4], on the surface characteristics of liquid rare earth metals[5], and on the structure of crystals[6-9]. The lanthanide contraction is observed when the elastic properties of ferrite garnets are analyzed[10], and also in the case of superconducting cuprates[11]. The lanthanide contraction is used to explain spectral line broadening[12], catalytic behavior of lanthanide compounds[13], metal-insulator transitions in rare earth materials[14], and vibronic transitions in various lattices with trivalent ions[15].

It has been demonstrated by Cheng and Froese Fischer [16] that the $4f$ and $5f$ wavefunctions of $Xe(54)$ are almost identical to the hydrogenic functions, while the $4f$ orbital for $Ba(56)$, $La(57)$ and for their ions are completely different, since they exhibit the collapse. The sudden collapse of $4f$ orbitals is very well illustrated by the shape of the functions of Ba and La obtained from the Hartree Fock calculations performed for the average energy of a configuration. At the same time for the lanthanides the $5f$ orbital has the nature of a hydrogenic $4f$ function. This means that, in the scheme with two potential wells, $5f$ orbitals are dominated by the outer well and its coulombic nature. The situation is again changed when the inner well becomes deeper with increasing atomic number, and the contraction of $5f$ orbital appears for the elements starting from the beginning of the actinide series.

The analysis of the results obtained for the ions across the lanthanide series support the statement that Eu^{3+} ion is not exceptional among the members of the series but is rather representative of the whole family. This conclusion is derived from inspection of the properties of atomic functions and the integrals evaluated with them. In Fig. 21-12 the Judd-Ofelt radial terms evaluated with the perturbed functions are presented for the various ranks of the crystal field potential $t = 1, 3, 5$ and excitations from the $4f$ shell to all one-electron states of the d symmetry. The curves presented there (note the reverse sign) are very regular and smooth across the lanthanide series. A similar regularity is observed in Fig. 21-13 where all the radial terms of the Judd-Ofelt

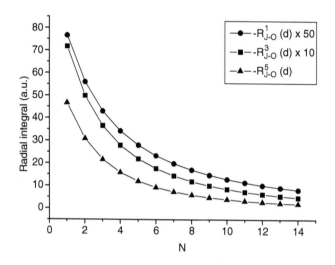

FIGURE 21-12 The values of the radial term $R^t_{JO}(d)$ evaluated with the perturbed functions for all ions across the lanthanide series.

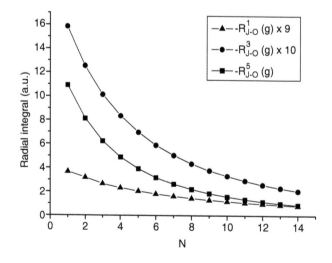

FIGURE 21-13 The values of the radial term $R^t_{JO}(g)$ evaluated with the perturbed functions for all ions across the lanthanide series.

approach with the excitations from the $4f$ shell to all the states with the g symmetry are displayed. Since the radial integrals reflect the electronic structure of a distinct ion (the angular parts of any effective operator have the same values for all ions), it is seen from these figures that it is not expected that the complementary systems have similar properties. The radial integrals change with the number of $4f$ electrons N (with the atomic number Z) monotonically, and if multiplied by the angular factors of any contributing effective operator (the same value for all ions), the particular contribution changes in the same way as the radial integrals.

21.4 VALUES OF RADIAL INTEGRALS FOR ALL LANTHANIDE IONS

In Table 21-8 the values of the radial integrals with r^k, for $k = 1, ...6$ are presented for all the lanthanide ions; these are the integrals which contribute at the second order to the amplitude of the $f \longleftrightarrow f$ transition defined within the dynamic model.

As discussed in chapter 17 it is important for the proper description of the spectroscopic properties of lanthanides in crystals to include electron correlation effects. They indeed have a very strong impact upon the transition amplitude, and the major part of this modification is determined by one particle effective operators of the third order. From a practical point of view, to perform the calculations beyond the single configuration approximation of the Judd-Ofelt model, it means to modify the radial integrals. Since the angular parts of the third-order electron correlation one particle operators are the same as those of the Judd-Ofelt terms, the transition

---

TABLE 21-8
Radial Integrals $\langle 4f|r^k|4f\rangle$ (in a.u.) for Ions Across the Lanthanide Series Evaluated With the Functions Generated for the Average Energy of Ground Configuration of Each Ion

$\langle 4f|r^k|4f\rangle$	1	2	3	4	5	6
Ce^{3+}	0.97	1.17	1.73	3.08	6.44	15.55
Pr^{3+}	0.93	1.08	1.55	2.65	5.36	12.53
Nd^{3+}	0.90	1.01	1.39	2.31	4.53	10.31
Pm^{3+}	0.87	0.94	1.26	2.04	3.89	8.63
Sm^{3+}	0.84	0.89	1.15	1.81	3.38	7.32
Eu^{3+}	0.82	0.84	1.06	1.62	2.96	6.28
Gd^{3+}	0.79	0.79	0.98	1.46	2.61	5.45
Tb^{3+}	0.77	0.75	0.91	1.33	2.33	4.76
Dy^{3+}	0.75	0.71	0.84	1.21	2.08	4.19
Ho^{3+}	0.74	0.68	0.79	1.11	1.87	3.71
Er^{3+}	0.72	0.65	0.74	1.02	1.69	3.31
Tm^{3+}	0.70	0.62	0.69	0.94	1.54	2.97
Yb^{3+}	0.69	0.60	0.65	0.87	1.40	2.67

amplitude defined within the extended model is determined by the matrix elements

$$
{}^3\Gamma = 2 \sum_{t,p}^{odd} B_p^t \sum_{k,q}^{even} \sum_{\ell'}^{even} (-1)^q [k]^{1/2} \begin{pmatrix} t & 1 & k \\ p & \rho & -q \end{pmatrix} A_t^k(\ell') R^t(\ell')
$$
$$
\times \langle 4f^N \Psi_f^0 \mid U_q^{(k)} \mid 4f^N \Psi_i^0 \rangle \tag{21-3}
$$

where the angular factor $A_t^k(\ell')$ is the same as in (17-12), and it is defined in (17-13), while the Judd-Ofelt radial integrals of second order are replaced by the radial term which contains the following integrals

$$
R^t(\ell') = R_{JO}^t(\ell') - R_{HF}^t(\ell' f) + \frac{1}{2}(N-1) R_t^0(\ell' f) \tag{21-4}
$$

and

$R_{JO}^t(\ell')$ is the original Judd-Ofelt radial term defined by (17-39), $R_{HF}^t(\ell' f) = {}^A R_{k_1 k_2}^{HF}(\ell' f) + {}^A R_{t1}^{HF}(\ell' f)$ is caused by the Hartree-Fock potential taken as a perturbation, and each ${}^A R_{1t}^{HF}$ is defined by (17-43), $R_t^0(\ell' f) = {}^A R_{k_1 k_2}^0(\ell' f) + {}^A R_{t1}^0(\ell' f)$ originate from the Coulomb interaction, and each ${}^A R_{1t}^s(\ell' f)$ is defined by the integral in (17-40).

The values of distinct contributions to $R^t(\ell')$ from (21-4) evaluated with the perturbed functions generated for all lanthanide ions are presented in Table (21-9).

Tables 21-10 to 21-12 below contain the values of the radial integrals of the effective operators that determine the contributions to the transition amplitude, which originate from the nuclear magnetic and multipole electric hyperfine interactions discussed in chapter 20. The appropriate integrals are evaluated with the perturbed functions, which are new and were never before generated. Here the results obtained for a few ions are chosen for this initial presentation.

TABLE 21-9
The Values of the Radial Integrals of Second Order Judd-Ofelt and Third- Order Electron Correlation Approach Evaluated With the Appropriate Perturbed Functions for Ions Across the Lanthanide Series (in a.u.)

Ion	t	ℓ'	$R_{JO}^t(\ell')$	$-R_{HF}^t(\ell' f)$	$\frac{1}{2}(N-1)R_t^0(\ell' f)$	$R^t(\ell')$
Ce^{3+}	1	d	−1.5318	—	—	−1.5318
		g	−0.4081	—	—	−0.4081
	3	d	−7.1698	—	—	−7.1698
		g	−1.5845	—	—	−1.5845
	5	d	−46.6154	—	—	−46.6154
		g	−10.9103	—	—	−10.9103

(Continued)

TABLE 21-9 (Continued)
The Values of the Radial Integrals of Second Order Judd-Ofelt and Third-Order Electron Correlation Approach Evaluated With the Appropriate Perturbed Functions for Ions Across the Lanthanide Series (in a.u.)

Ion	t	ℓ'	$R^t_{JO}(\ell')$	$-R^t_{HF}(\ell' f)$	$\frac{1}{2}(N-1)R^0_t(\ell' f)$	$R^t(\ell')$
Pr^{3+}	1	d	-1.1195	0.7626	-0.3914	-0.7483
		g	-0.3545	0.1746	-0.0896	-0.2695
	3	d	-4.9839	4.3395	-2.2275	-2.8719
		g	-1.2533	1.1682	-0.5996	-0.6847
	5	d	-30.8472	36.1334	-18.5438	-13.2576
		g	-8.1276	11.4283	-5.8643	-2.5636
Nd^{3+}	1	d	-0.8601	1.1464	-0.5884	-0.3021
		g	-0.2967	0.2888	-0.1482	-0.1561
	3	d	-3.6557	6.2482	-3.2072	-0.6147
		g	-1.0137	1.8296	-0.9391	-0.1232
	5	d	-21.6377	49.9231	-25.6202	2.6652
		g	-6.2348	17.0281	-8.7375	2.0558
Pm^{3+}	1	d	-0.6844	1.3413	-0.6886	-0.0317
		g	-0.2579	0.3646	-0.1872	-0.0805
	3	d	-2.7861	7.0242	-3.6056	0.6325
		g	-0.8345	2.1985	-1.1285	0.2355
	5	d	-15.8291	54.0401	-27.7322	10.4788
		g	-4.8953	19.5644	-10.0385	4.6306
Sm^{3+}	1	d	-0.5592	1.4346	-0.7364	0.1390
		g	-0.2264	0.4150	-0.2128	-0.0242
	3	d	-2.1858	7.2377	-3.7152	1.3367
		g	-0.6971	2.3921	-1.2278	0.4672
	5	d	-11.9573	53.7722	-27.5937	14.2212
		g	-3.9177	20.4391	-10.4869	6.0345
Eu^{3+}	1	d	-0.4662	1.4705	-0.7545	0.2498
		g	-0.2004	0.4480	-0.2300	0.0176
	3	d	-1.7541	7.1624	-3.6765	1.7318
		g	-0.5895	2.4779	-1.2718	0.6166
	5	d	-9.2644	51.5196	-26.4363	15.8189
		g	-3.1854	20.3996	-10.4658	6.7484
Gd^{3+}	1	d	-0.3952	1.4722	-0.7560	0.3210
		g	-0.1787	0.4688	-0.2406	0.0495
	3	d	-1.4339	6.9400	-3.5622	1.9439
		g	-0.5037	2.4968	-1.2813	0.7118
	5	d	-7.3294	48.4414	-24.8553	16.2567
		g	-2.6261	19.8651	-10.1907	7.0483

(Continued)

TABLE 21-9 (Continued)
The Values of the Radial Integrals of Second Order Judd-Ofelt and Third-Order Electron Correlation Approach Evaluated With the Appropriate Perturbed Functions for Ions Across the Lanthanide Series (in a.u.)

Ion	t	ℓ'	$R^t_{JO}(\ell')$	$-R^t_{HF}(\ell'f)$	$\frac{1}{2}(N-1)R^0_t(\ell'f)$	$R^t(\ell')$
Tb^{3+}	1	d	-0.3395	1.4542	-0.7462	0.3685
		g	-0.1602	0.4812	-0.2471	0.0739
	3	d	-1.1900	6.6453	-3.4108	2.0445
		g	-0.4343	2.4733	-1.2694	0.7696
	5	d	-5.8987	45.0969	-23.1378	16.0604
		g	-2.1901	19.0654	-9.7801	7.0952
Dy^{3+}	1	d	-0.2949	1.4244	-0.7312	0.3983
		g	-0.1445	0.4876	-0.2504	0.0927
	3	d	-1.0005	6.3209	-3.2663	2.0541
		g	-0.3773	2.4240	-1.2436	0.8031
	5	d	-4.8183	41.7815	-21.4352	15.5280
		g	-1.8457	18.1455	-9.3070	6.9928
Ho^{3+}	1	d	-0.2587	1.3878	-0.7128	0.4163
		g	-0.1309	0.4896	-0.2511	0.1076
	3	d	-0.8505	5.9899	-3.0740	2.0654
		g	-0.3303	2.3590	-1.2105	0.8182
	5	d	-3.9862	38.6281	-19.8162	14.8257
		g	-1.5698	17.1836	-8.8133	6.8005
Er^{3+}	1	d	-0.2288	1.3478	-0.6920	0.4270
		g	-0.1191	0.4882	-0.2510	0.1181
	3	d	-0.7299	5.6647	-2.9070	2.0278
		g	-0.2909	2.2751	-1.1725	0.8117
	5	d	-3.3345	35.6924	-18.3088	14.0491
		g	-1.3462	16.2256	-8.3215	6.5579
Tm^{3+}	1	d	-0.2038	1.3060	-0.6699	0.4323
		g	-0.1088	0.4846	-0.2486	0.1272
	3	d	-0.6319	5.3530	-2.7464	1.9747
		g	-0.2576	2.2068	-1.1325	0.8167
	5	d	-2.8172	32.9997	-16.9263	13.2562
		g	-1.1631	15.3002	-7.8463	6.2908
Yb^{3+}	1	d	-0.1828	1.2640	-0.6492	0.4320
		g	-0.0997	0.4794	-0.2460	0.1337
	3	d	-0.5511	5.0583	-2.5956	1.9116
		g	-0.2294	2.1270	-1.0914	0.8062
	5	d	-2.4012	30.5479	-15.6678	12.4789
		g	-1.0117	14.4215	-7.3950	6.0148
Lu^{3+}	1	d	-0.1648	1.2222	-0.6279	0.4295
		g	-0.0917	0.4728	-0.2431	0.1380
	3	d	-0.4837	4.7805	-2.4531	1.8437
		g	-0.2051	2.0471	-1.0504	0.7916
	5	d	-2.0622	28.3122	-14.5197	11.7303
		g	-0.8852	13.5901	-6.9680	5.7369

TABLE 21-10
Radial Integrals of Second-Order Effective Operators (20.34) that are Caused by the Electric Multiple Hyperfine Interactions

Integral	k	$Pr^{3+}(4f^2)$	$Eu^{3+}(4f^6)$	$Gd^{3+}(4f^7)$	$Tb^{3+}(4f^8)$
$\langle \rho^{-k-1}(d) \rangle$ [a]	1	0.1027	0.0606	0.0549	0.0501
	3	-19.3242	-14.5098	-13.9623	-13.5341
$\langle \rho^{-k-1}(g) \rangle$ [a]	1	-0.2184	-0.1850	-0.1786	-0.1727
	3	-0.5671	-0.6304	-0.6454	-0.6602

[a] $\langle \rho^{-k-1}(\ell') \rangle \equiv \langle \rho^{-k-1}(4f \rightarrow \ell') | r | 4f \rangle$.

TABLE 21-11
Radial Integrals of Third-Order Effective Operators Caused by Spin-Orbit Interaction, Nuclear Magnetic (20.28) and Electric Quadrupole (20.37) Hyperfine Interactions, Assisted by Interactions via the Crystal Field Potential

Integral	t	$Pr^{3+}(4f^2)$	$Eu^{3+}(4f^6)$	$Gd^{3+}(4f^7)$	$Tb^{3+}(4f^8)$
$\langle \rho_f^{-3} \rho_d^t \rangle$ [a]	1	5.5209	-2.4887	-2.1743	-1.9239
	3	-25.3936	-10.3939	-8.7955	-7.5500
	5	-175.6036	-62.8265	-51.6912	-43.2150
$\langle \rho_f^{-3} \rho_g^t \rangle$ [a]	1	-1.1839	-0.7553	-0.6911	-0.6365
	3	-6.5310	-3.4375	-3.0267	-2.6896
	5	-56.4390	-24.9861	-21.2636	-18.3095

[a] $\langle \rho_f^{-3} \rho_{\ell'}^t \rangle \equiv \langle \rho^{-3}(4f \rightarrow f) | r | \rho^t(4f \rightarrow \ell') \rangle$.

TABLE 21-12

Radial Integrals of the Third-Order Effective Operators Typical for the Electric Dipole Hyperfine Terms Defined in (20.37) for $k=$ odd, $t=$ even Evaluated for Several Lanthanide Ions

Integral	t	$Pr^{3+}(4f^2)$	$Eu^{3+}(4f^6)$	$Gd^{3+}(4f^7)$	$Tb^{3+}(4f^8)$
$\langle \rho_d^{-2}\rho_f^t \rangle$ [a]	2	−0.1897	−0.0676	−0.0551	−0.0455
	4	−1.3758	−0.3941	−0.3066	−0.2424
	6	−13.1092	−3.0684	−2.2897	−1.7397
$\langle \rho_d^{-4}\rho_f^t \rangle$ [a]	2	36.7692	16.5475	14.3065	12.5337
	4	273.4736	99.7468	82.5860	69.4691
	6	2676.	807.1281	643.0310	521.7883
$\langle \rho_g^{-2}\rho_f^t \rangle$ [a]	2	0.0806	0.0384	0.0330	0.0286
	4	0.5474	0.2139	0.1761	0.1466
	6	5.0303	1.6329	1.2950	1.0415
$\langle \rho_g^{-4}\rho_f^t \rangle$ [a]	2	−0.0189	−0.0158	−0.0152	−0.0147
	4	−0.0837	−0.0634	−0.0594	−0.0557
	6	−0.4426	−0.3393	−0.3139	−0.2901

[a] $\langle \rho_{l'}^{-k-1}\rho_{\ell''}^t \rangle \equiv \langle \rho^{-k-1}(4f \to \ell')|r|\rho^t(4f \to \ell'')\rangle$.

REFERENCES

1. Froese-Fischer C, (1977) *The Hartree-Fock Methods for Atoms*, Wiley - Interscience Publication, John Wiley and Sons Inc., New York.
2. Jankowski K, Smentek-Mielczarek L and Sokołowski A, (1986) Electron-correlation Third-order Contributions to the Electric Dipole Transition Amplitudes of Rare Earth Ions in Crystals I. Perturbed-function Approach *Mol. Phys.* **59** 1165.
3. Goeppert Mayer M, (1941) Rare-earth and Transuranic Elements *Phys. Rev.,* **60** 184.
4. Xue L A, Chen Y and Brook R J, (1988) The Effect of Lanthanide Contraction on Grain Growth in Lanthanide-doped $BaTiO_3$ *J. Mater. Sci. Lett.,* **7** 1163.
5. Sukhman A L and Gruverman S L, (1990) Surface Characteristics of Rare-earth Metals *Raspravy* **3** 124.
6. Ozeki T and Yamase T, (1994) Effect of Lanthanide Contraction on the Structures of the Decatungstolanthanoate Anions in $K_3Na_4H_2[LnW_{10}O_{36}] \cdot nH_2O$ (Ln = Pr, Nd, Sm, Gd, Tb, Dy) Crystals *Acta Crystallogr.,* **B 50** 128.
7. A. Chatterjee A Maslen E N and Watson K J, (1988) The Effect of the Lanthanoid Contraction on the Nonaaqualanthanoid(III) Tris(trifluoromethanesulfonates) *Acta Crystallogr.,* **B 44**, 381.
8. Wu Z, Xu Z, You X, Zhou X, Huang X and Chen J, (1994) Formation and Crystal Structures of (C5H5) 3Sm(THF) and (C5H5) 3Dy(THF) *Polyhedron,* **13** 379.
9. A. Dimos A, Tsaousis D Michaelides A, Skoulika S, Golhen S, Ouahab L, Didierjean C and Aubry A, (2002) Microporous Rare Earth Coordination Polymers: Effect of Lanthanide Contraction on Crystal Architecture and Porosity *Chem. Mater.,* **14** 2616.

10. Kvashnina O P, Kvashnina G M and Sorokina T P, (1992) Elastic Properties of Rare-earth Ferrite Garnets *Fiz. Tverd. Tela,* **34** 1306.

11. Tokiwa A, Syono Y, Oku T and Nagoshi M, (1991) Superconductivity and Crystal Structure of $Pb(Ba, Sr)_2 (Ln, Ca)Cu_3O_y$ and $Pb(Ba, Sr)_2(Ln, Ce)_2Cu3O_y$(Ln: Lanthanoid) with (Pb, Cu) Double Layers *Physica C,* **185-189** 619.

12. Ellens A, Andres H, ter Heerdt M L H, Wegh R T, Meijerink A and Blasse G, (1997) Spectral-line-broadening Study of the Trivalent Lanthanide-ion Series.II. The Variation of the Electron-phonon Coupling Strength through the Series *Phys. Rev.,* **B 55** 180.

13. Shiozaki R, Inagaki A, Ozaki A, Kominami H, Yamaguchi S, Ichihara J and Kera Y, (1997) Catalytic Behavior of a Series of Lanthanide Decatungstates $[Ln(III)W_{10}O_{36}^{9-}; Ln : La - Yb]$ for H_2O_2-oxidations of Alcohols and Olefins. Some Chemical Effects of the $4f^n$-electron in the Lanthanide(III) Ion on the Catalyses *J. Alloys Comp.,* **261** 132.

14. Sarma D D, Shanthi N, Barman S R and Mahadevan P, (1995) *Perspectives in Solid State Chemistry,* ed. K. J. Rao Narosa: New Delhi, India, p. 285.

15. Ellens A, Schenker S, Meijerink A and Blasse G, (1997) Vibronic Transitions of Tm^{3+} in Various Lattices. A Comparison with Pr^{3+}, Eu^{3+} and Gd^{3+} *J. Lumin.,* **72-74** 183.

16. Cheng K T and Froese Fischer C, (1983) Collapse of the 4f Orbital for *Xe*-like Ions *Phys. Rev.,* **A28** 2811.

22 Luminescence of Lanthanide Doped Materials

It is the mark of an instructed mind to rest satisfied with the degree of precision which the nature of the subject admits and not to seek exactness when only an approximation of the truth is possible.

Aristotle

The terms *laser* and *fiber optics* are modern and almost synonymous with modern technology. However, the phenomenon that defines their physical origin is known for a long time, since luminescence from inorganic materials was first observed in 1577 by Nicolas Monardes, who noticed a blue tint in water kept in a container made of a special kind of wood[1]. Three hundred years later, in 1852, fluorescence was introduced by Stokes as an emission of light. At that time the first spectroscopic principles were formulated[2], and subsequently used in 1854 as an analytical tool. Four hundred years after the first observation of luminescence, laser light was used to warm up and cut through metals, alloys and organic tissues. Almost 420 years passed before a sample was cooled with laser light and when in 1997 an optical refrigerator was constructed. In this experiment, due to luminescent cooling, the temperature of Yb^{3+}- doped fluoro-zirconate glass decreased by $16 K$!

Photoluminescence, fluorescence, phosphorescence, chemiluminescence, biolu-minescence, thermoluminescence, electroluminescence, radioluminescence, tribolu-minescence and *sonoluminescence* are the different phenomena observed in various materials and which result from different mechanisms of excitation. Each of these distinct terms is associated with a special experiment and an adjusted theoretical model.

In most cases inorganic crystals do not luminesce at room temperature with de-tectable efficiency. Therefore, from an experimental point of view, research is mainly directed to doped crystals. This is especially important when having in mind the po-tential preparation of new materials with well defined properties. As a consequence, luminescence in crystals is usually understood in the terms of luminescent centers that are localized at the impurities. In such cases the model of luminescence is simpler, since it is associated with a distinct ion. In particular when the rare earth ions are used as impurities, due to their special electronic structure, it is possible to separate the ion from its environment. As a result the ion is described as an isolated free sys-tem that is perturbed only by the surrounding crystal field. The basic concept of the description of the luminescence is thus the same as the assumptions of the standard Judd-Ofelt theory of electric dipole $f \longleftrightarrow f$ transitions (discussed in chapter 17).

Although the free ionic system approximation provides great simplification of a theoretical formalism, as Dexter stated in his paper in 1953[3], still the theoretical description of luminescence in crystals is a difficult task. This difficulty is caused by the complexity of the phenomena, since luminescence is an act which reflects the simultaneous and mutual interaction between matter, radiation, and very often the phonons. The problem is even more complicated by the fact that the accuracy of any theoretical description is very sensitive to the quality of the wave functions that describe the energy states of cooperating subsystems engaged in the process. This sensitivity is evident when analyzing the properties of materials with similar impurities that are doped in the same hosts, or those that are built of the same impurity ion placed in various hosts.

In the case of the sensitized luminescence the impurity ion radiates upon its excitation to a higher energetic state that results from the non-radiative transfer of the energy from another impurity ion or host. The two sources of the energy that is transferred to the activator distinguish the so-called impurity-sensitized[4−11] and host-sensitized[12−17] luminescence. The ion which is made to emit the radiation is called an *activator*, while the donator of the energy is the *sensitizer*. The process of energy transfer is realized by a virtual (not real) emission of energy by the donator, which is followed by reabsorption of this energy by the acceptor.

22.1   EXPERIMENTS

The simplest experiment concerns the impurity ion that is characterized by an absorption band suitable for the absorption of the energy from the radiating beam. If this condition is satisfied the impurity is excited and then returns to its ground state emitting a photon with the so-called Stokes' shift. Usually the emitted photon is less energetic than the light that excites the impurity ion, which is a luminescence center. If the radiating light does not match the scheme of the absorption spectra of a given impurity, or in the region of the energy of the light where only a forbidden transition is predicted, no excitation of the impurity takes place. It is possible however to excite such an ion indirectly using another impurity present in the host crystal. The latter absorbs the light and through a radiationless process it transfers the energy to the original impurity which when sensitized becomes a luminescence center. If the transfer of energy is of low probability, the second impurity in fact may itself become active and luminesce with a Stokes' shift. If these processes appear simultaneously, two bands are expected in the emission spectrum. This is a cooperative process, and due to the relaxation of the energy of two impurities, the line due to the sensitized luminescence is even less energetic than the previous one. The double Stokes' shift, as noted already in 1957 by Dexter[18] might very well be applied to obtain visible luminescence using for example a convenient source of ultraviolet light.

An experiment of Varsanyi and Dieke[19] illustrates the cooperative process in which one photon excites two centers simultaneously; this is the so-called down-conversions as presented in Fig. 22-1. The process of infrared - to - visible up-conversion, that is presented in this figure, also has a cooperative nature.[20] Due to the energy transfer between two centers, for example, a visible green emission arising from infrared excitation was observed for glass doped by $Yb^{3+} - Er^{3+}$.[21] Similarly, blue emission

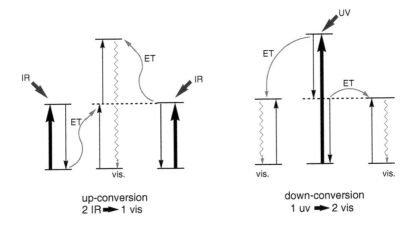

FIGURE 22-1 Cooperative processes of sensitized luminescence.

was obtained from the infrared radiation in the case of two impurities, Yb^{3+} and Tm^{3+}[22,23], to mention just a few pioneering experiments performed in the field, which has been rapidly developing over the years.

At the first sight the conversion process in which more highly energetic visible light is generated using a beam of lower energy seems to be impossible and simply has the symptoms of an experiment which violates the Stokes' law, the law which is in fact the realization of the energy conservation principle. It becomes clear that there is no violation of the law, if it is realized that more than one photon of the radiating beam is involved in the process (see for example Ref. 20 and 24).

There are two different processes of energy transfer that are distinguishable from an experimental point of view, resonant and non-resonant. The transfer via the resonant process depends on the concentration of the ions while the non-resonant case strongly depends on the structure of the crystal[25]. In the non-resonant case, when the excitation of the activator promotes it to the state with an energy smaller than the portion emitted by the sensitizer, a phonon is emitted, and the energy conservation principle is satisfied[26-28].

Many materials become fluorescent when doped with lanthanides[9,10,29-31]. In the case of phosphors the luminescent spectra of Ce^{3+} is in the far infrared, Gd^{3+} in the ultraviolet, while for the other ions of the lanthanide series the spectra is in the visible and near infrared regions. The lanthanide acceptors might be divided into categories characterized by the kind of their luminescence. The luminescence excitation of Ce^{3+}, Pr^{3+} and Tb^{3+} occurs in a broad $4f - 5d$ absorption band, in the case of Sm^{3+} and Eu^{3+} in a charge transfer band and Tm^{3+} in narrow bands due to $4f \longleftrightarrow 4f$ transitions, which in the majority of cases have a form of atomic-like spectroscopic lines. In fact, using appropriate rare earth ions and particular host lattices it is possible to select the light emission with the desired wavelength from the visible and near infrared regions of spectra. For example, Er^{3+} has its dominant emission at $1.5\mu m$ which is the preferred wavelength for long distance fiber-optic communication[32]. This property of rare earth ions has caused an increasing interest in the application of rare earth semiconductors as active media for light emitting diodes, injection lasers and other optoelectronic devices[33-35].

22.2   ELECTROSTATIC MODEL

What is the mechanism of the energy transfer between the sensitizer and activator?
What is the efficiency of the energy transfer and the sensitized luminescence?
The answers to these questions have been addressed in many publications. In some of them the energy transfer processes are regarded as a resonance between two allowed non-radiative (virtual) electric dipole transitions; this is the common case observed in organic systems. Through these two transitions the energy state of the sensitizer is lowered to elevate the activator, which after the excitation, radiates. This is the first step that has to be performed in order to sensitize the luminescence. There are also investigations devoted to the transition processes that result from the resonance of an allowed electric dipole transition in the sensitizer and forbidden transitions in the activator; this mechanism is observed in inorganic materials.

The first quantum mechanical investigations on luminescence were performed by Föster in 1948[36]. Only in the sixties Dexter developed the approach that defined the starting point for further theoretical analysis based on an electrostatic model of mutual interactions between the sensitizer and activator[3,18,37]. In this model the transfer rate depends on the distance between the cooperating centers[38].

The probability of energy transfer is proportional to the matrix element of the perturbing operator representing the interaction between the central ion and the ligands that is taken with the functions that describe the initial and final states of both subsystems[39]. Taking into account the multipole expansion of the Coulomb interaction between two centers (central ion and ligand, or in the nomenclature used here, sensitizer and activator) the dependence on the distance between them is very clear, namely

$$\mathcal{P}_{S \longrightarrow A} = \frac{T_6}{R^6} + \frac{T_8}{R^8} + \frac{T_{10}}{R^{10}} + \cdots \qquad (22\text{-}1)$$

where T_6, T_8 and T_{10} are the terms associated with the dipole-dipole, dipole-quadrupole and quadrupole-quadrupole interactions, respectively. In general, these terms are determined by appropriate matrix elements of the electric dipole, electric quadrupole, and higher multipole radiation operators with the functions of the energy states of activator and sensitizer.

In order to evaluate $\mathcal{P}_{S \longrightarrow A}$ the easiest would be to perform semi-empirical calculations based on the fitting of the numerators of the distinct terms in (22-1). In such a way all the problems with evaluation of appropriate matrix elements are avoided. However, in order to learn about the relative importance of various interactions and their role in the energy transfer, the expression in (22-1) has to be used in direct calculations. In such an approach new questions arise, namely which model to apply and which physical mechanisms should be included for the generation of the wave functions that are used for evaluation of appropriate matrix elements contributing to $\mathcal{P}_{S \longrightarrow A}$ of (22-1)? This issue is addressed in the next subsection.

In addition to the electrostatic model of the sensitized luminescence of the lanthanide doped materials, from a formal point of view, it is possible to include also the impact that is due to the electro-magnetic interactions[38]. However, it has been estimated that this type of interaction is negligible in comparison with those of an electrostatic nature. This relation was established for the case of the process of luminescence that consists of two allowed transitions, and it might be changed in the case of rare earth impurities for which the energy transfer is associated with the forbidden electric dipole transition. In both cases, in fact, the conclusions are based on estimations and speculations. Still within the main stream of theoretical investigations the energy transfer is regarded as a process that results from electrostatic interactions.

To outline the most important aspects of a direct evaluation of the amplitude of energy transfer it is assumed that the activator (the lanthanide ion) and the sensitizer (ligand) are well separated, and therefore their energy states are described independently; this is a case of host sensitized luminescence. Thus the rate of a transition from the initial state to the final state of a system is essentially determined by a product of two matrix elements

$$\langle f \mid V \mid i \rangle = \sum_{m\mu} \sum_{kq} [k, m, k+m]^{1/2} (4\pi)^{-3/2} \frac{C^{(k+m)}_{-q-\mu}(\Theta\Phi)}{R^{k+m+1}} B^{q\mu}_{km}$$
$$\times \langle 4f^N \Psi_f \mid D^{(k)}_q \mid 4f^N \Psi_i \rangle \langle \phi^L_f \mid D^{(m)}_\mu \mid \phi^L_i \rangle \tag{22-2}$$

where the common factor $C^{(k+m)}(\Theta\Phi)/R^{k+m+1}$ defines the geometry of the environment of the lanthanide ion. The radial part of this factor, R^{k+m+1}, with R being the radial coordinate of lattice particles defined within the static approximation, gives rise to the radial dependence of $P_{S \longrightarrow A}$ in (22-1). In the case of the host sensitized luminescence two multipoles $D^{(k)}_q$ and $D^{(m)}_\mu$ localized on two interacting centers, the activator and sensitizer, are coupled together by the coefficient $B^{q\mu}_{km}$. As before, also here the multipoles are defined in the terms of spherical tensors

$$D^{(k)}_q = \sum_i r^k_i C^{(k)}_{q,i}$$

The first matrix element in (22-2) describes the lanthanide ion; the second element is associated with the host and its evaluation depends on the particular choice of ligands[40].

22.3 EFFECTIVE OPERATOR FORMULATION

From a formal point of view the evaluation of the matrix element that describes the lanthanide ion in (22-2) is very similar to those that determine the amplitude of electric dipole $f \longleftrightarrow f$ transitions.[41-44] Therefore, in order to derive the effective operators of the rate of the energy transfer, it is necessary to undertake the following steps:

1. apply the perturbation approach and modify the wave functions, initially defined within the central field approximation for the free ionic system, by the corrections caused by the most important physical mechanisms,

especially those that occurred to be important in the description of the radiative processes observed in the lanthanides;

2. adopt the assumptions of the standard Judd-Ofelt theory of $f \longleftrightarrow f$ transitions on the energy denominators of the perturbing expressions (discussed in chapter 17); for the purpose of practical numerical calculations it means that the radial basis states of one-electron states are generated for the average energy of ground configuration $4f^N$; the energies of the excited configurations are also average and they are evaluated with the core orbitals frozen;

3. perform the partial closure in order to obtain the effective operator form of all contributions;

4. re-define the radial terms of all effective operators using the perturbed function approach to be able to take into account the excitations from the $4f$ shell to **all** one-electron excited states of a given symmetry (including the continuum);

5. perform numerical calculations to establish the hierarchy of important terms; this step is exactly the same as in the case of the description of the radiative transitions, and therefore it is fruitful to use here the experience gained in the previous chapters.

The first-order contributions to the energy transfer amplitude are determined by a simple matrix element

$$\Gamma^1 = \langle \Psi_f^0 \mid D_q^{(k)} \mid \Psi_i^0 \rangle \qquad (22\text{-}3)$$

where, due to the same parity of both wavefunctions describing the energy levels of $4f^N$ configuration, the rank of tensor operator k must be even. This means that the first-order contributions to the energy transfer, which are in fact defined for the free ionic system, contain the interaction of at least quadrupoles localized on the lanthanide ion. Thus the first-order terms contribute to the term T_8 of (22-1), in conjunction with the dipoles localized on the ligand, and to T_{10} of Dexter's expression with the quadrupoles that represent the ligand.

In the terms of the effective operators these contributions have a form of a single expression

$$\Omega^1 = \langle 4f \mid r^k \mid 4f \rangle \langle \ell \mid\mid C^{(k)} \mid\mid \ell \rangle U^{(k)}(\ell\ell) \qquad (22\text{-}4)$$

for $k = even$, with the values of k determined by the triangular condition for the non-vanishing matrix element of the spherical tensor.

In order to break the limitations of the free ionic system approximation of the zeroth order problem, most of all, the perturbing influence of the crystal field potential has to be taken into account. This physical necessity extends the amplitude of the energy transfer by the second-order terms that depends on the symmetry of the system via the crystal field parameters B_p^t,

$$^1\Omega^2_{cryst}(kq) = \sum_{tp} B_p^t \sum_{\ell'} \sum_{r\varrho} (-1)^{r-\varrho} [r]^{1/2} R_{CF}^{tk}(\ell') A_k^{rt}(\ell') \left[\epsilon_t^* \epsilon_k^* \epsilon_{\ell'} + \epsilon_t \epsilon_k \epsilon_{\ell'}^* \right]$$

$$\times \begin{pmatrix} t & k & r \\ p & q & -\varrho \end{pmatrix} U_\varrho^{(r)}(\ell\ell) \tag{22-5}$$

The parity symbols ϵ and ϵ^* select even and odd parts of the crystal field potential and the parity of the rank of the multipoles localized on the lanthanide ion; both are of the same parity. This means that since $\Omega_\lambda^2(kq)$ contains the impact due to the even and also odd parts of the crystal field potential, it contributes to all three terms of the Dexter's expression. However, from a formal point of view the terms for the higher multipoles, with $k > 2$, also contribute, and in direct calculations they should be taken into account. When the semiempirical procedure is applied, the adjusted parameters from the Dexter's expression that is limited to the three terms possibly compensate for the lack of the interactions of higher rank multipoles (and all the other physical mechanisms that are not taken into account explicitly).

The angular factor in (22-5) is the generalized Judd-Ofelt angular term, which has a similar form to that defined previously (the entry 1 in (17-13) is generalized here by an additional index),

$$A_k^{rt}(\ell') = [r]^{1/2} \begin{Bmatrix} k & r & t \\ \ell & \ell' & \ell \end{Bmatrix} \langle f \| C^{(t)} \| \ell' \rangle \langle \ell' \| C^{(k)} \| f \rangle \tag{22-6}$$

where ℓ' denotes the symmetry of one-electron state to which an electron from the shell $4f$ is promoted.

In the analysis of the electric dipole $f \longleftrightarrow f$ transitions it was verified numerically that electron correlation effects have a very strong impact upon the amplitude. Thus, here the expression for the amplitude of energy transfer should also be extended by the terms that go beyond the single configuration approximation. As before, the operator representing electron correlation effects, as defined as a non-central part of the Coulomb interactions, has two parts. Namely, the central field approximation potential (in practice the Hartree-Fock potential) and the original two particle Coulomb interactions. These two origins of new contributions to the rate of the energy transfer are distinctly seen in the final form of the second-order electron correlation effective operator

$$^1\Omega^2_{corr}(kq) = \epsilon_k \langle \ell \| C^{(k)} \| \ell \rangle \sum_{\ell''} \epsilon_{\ell''}^* \left[\delta(\ell'', f)\left(R_{HF}^k(\ell'') + \frac{7}{2}(N-1)R^{0k}(\ell'')\right) \right.$$

$$\left. + \frac{1}{2}[k]^{-1}(N-1)\langle \ell \| C^{(k)} \| \ell'' \rangle^2 R^{kk}(\ell'') \right] U^{(k)}(\ell\ell) \tag{22-7}$$

It is seen from this expression that in order to include the perturbing influence of electron correlation effects it is sufficient to modify the radial integrals by appropriate terms. The first integral in (22-7) is caused by the Hartree-Fock potential, while the remaining two terms are associated with the Coulomb operator. In all cases k is even, and therefore electron correlation effects contribute to the interaction of even rank multipoles.

Due to the two particle character of the Coulomb interaction operator there are additional terms of the second order that are two particle objects (in the case of the radiative electric dipole transitions the two particle electron correlation effective operators are of the third order). The general form of such objects is as follows

$$^2\Omega^2_{corr}(kq) = \epsilon_k[k]^{-1/2}\sum_{\ell''}\epsilon^*_{\ell''}\sum_{s>0}\epsilon_s R^{sk}(\ell'')\langle\ell\ ||\ C^{(s)}\ ||\ \ell\rangle$$

$$\times \sum_{x>0}[x]^{1/2}A^{xs}_k(\ell'')\sum_{i<j}\left[u^{(s)}_i(\ell\ell)\times u^{(x)}_j(\ell\ell)\right]^{(k)}\qquad(22\text{-}8)$$

where the angular part of this operator is defined by (22-6).

There are also two particle effective operators that are of the third order when the crystal field potential and electron correlation effects are taken into account simultaneously. For example, for the triple product of matrix elements with the sequence of operators $V_{CF}D^{(k)}V_{corr}$, the effective operator has the following form

$$^2\Omega^3(sing) = \sum_{tp}B^t_p\sum_{s>0}\sum_{kq}\sum_{\ell'}\sum_{\ell''}\epsilon_{\ell'}\epsilon^*_{\ell''}\epsilon_s\epsilon^*_k\epsilon^*_t P_1(tks,\ell'\ell'')W_1(tks,\ell'\ell'')$$

$$\times\sum_y\sum_{x>0}[y]^{1/2}[x](-1)^{x+y-\eta}\begin{pmatrix}t & k & y\\ p & q & -\eta\end{pmatrix}\begin{Bmatrix}k & y & t\\ \ell & \ell' & \ell''\end{Bmatrix}$$

$$\times\begin{Bmatrix}s & x & y\\ \ell & \ell'' & \ell\end{Bmatrix}\sum_{i<j}\left[u^{(s)}_i(\ell\ell)\times u^{(x)}_j(\ell\ell)\right]^{(y)}_\eta\qquad(22\text{-}9)$$

where $P_1(tks,\ell'\ell'')$ is the appropriate radial term, and the angular factors consist of the product of the reduced matrix elements associated with the distinct operators of the perturbing sequence,

$$W_1(tks,\ell'\ell'') = \langle\ell\ ||\ C^{(t)}\ ||\ \ell'\rangle\langle\ell'\ ||\ C^{(k)}\ ||\ \ell''\rangle\langle\ell''\ ||\ C^{(s)}\ ||\ \ell\rangle\langle\ell\ ||\ C^{(s)}\ ||\ \ell\rangle$$

$$(22\text{-}10)$$

As previously, this sequence of the reduced matrix elements determines the selection rules for the non-vanishing contributions and defines the parities and values of all the ranks of the operators and the symmetry of the excited states. All these conditions are represented in $^2\Omega^3(sing)$ by the parity symbols. In particular, the contributions defined in (22-9) contain the perturbing influence of singly excited configurations of even and odd parities, in comparison to the parity of $4f^N$ configuration. This influence is included via the odd part of the crystal field potential, V_{cryst}, for $t = odd$. The rank of the effective operator s is limited to even values (s is the rank of the spherical tensors in the multipole expansion of the Coulomb interaction within the lanthanide ion), and the whole expression modifies the interaction of the odd rank multipoles ($k = odd$).

The two particle character of the Coulomb interaction allows the inclusion at the third order the perturbing influence of doubly excitations. In such a case, when the perturbing influence of the crystal field potential is included simultaneously, the effective operator has the form

$$^2\Omega^3(doub) = \sum_{tp} B_p^t \sum_s \sum_{kq} \sum_{\ell'} \sum_{\ell''} \epsilon_{\ell'}\epsilon_{\ell''}\epsilon_s^*\epsilon_k^*\epsilon_t^* P_4(tks, \ell'\ell'')W_3(tks, \ell'\ell'')$$

$$\times \sum_y \sum_{x>0} \sum_{r>0} [y]^{1/2}[x,r](-1)^{r+k-\eta} \begin{pmatrix} t & k & y \\ p & q & -\eta \end{pmatrix} \begin{Bmatrix} s & x & t \\ \ell & \ell' & \ell \end{Bmatrix}$$

$$\times \begin{Bmatrix} k & r & s \\ \ell & \ell'' & \ell \end{Bmatrix} \begin{Bmatrix} s & x & t \\ y & k & r \end{Bmatrix} \sum_{i<j} [u_i^{(r)}(\ell\ell) \times u_j^{(x)}(\ell\ell)]_\eta^{(y)} \quad (22\text{-}11)$$

with the angular factor of the form

$$W_3(tks, \ell'\ell'') = \langle \ell \| C^{(t)} \| \ell' \rangle \langle \ell' \| C^{(s)} \| \ell \rangle \langle \ell \| C^{(s)} \| \ell'' \rangle \langle \ell'' \| C^{(k)} \| \ell \rangle \quad (22\text{-}12)$$

The structure of the effective operators defined by (22-11) is more complex than in the previous case, since there is no limitation on the ranks of unit tensor operators r and x. Indeed, these ranks are the final values that result from the coupling of the appropriate angular momentum, and they are not present in the reduced matrix elements of W_3. At the same time, here the ranks of spherical tensors in the expansion of Coulomb potential s are odd. The ranks k and t are also odd. This means that these effective operators contribute to the terms that represent the interactions of the odd-rank multipoles that are assisted by the odd part of the crystal field potential.

The effective operators that contribute to the rate of the energy transfer presented here as an example are defined within the non-relativistic approach. This means that in order to include relativistic effects, one has to repeat a similar analysis as in the case of the radiative processes. Consequently the description of the energy transfer would be defined by the objects acting within the spin-orbital space. Following the same rules as described in the previous chapters, it is rather straightforward to derive new effective operators that are in all cases associated with double tensor operators. For example we have the terms of the second order

$$^{rel}\Gamma_{cryst}^2(k) = 2[k]^{1/2} \sum_{tp} B_p^t[t]^{1/2} \sum_{\kappa_1 k_1} \sum_{\kappa_2 k_2} \sum_{\ell'} A_{k_1 k_2}^{\kappa_1 \kappa_2}(kt\ell') \sum_{\kappa_3 k_3} a$$

$$\times \sum_{z,\zeta} [z]^{1/2}(-1)^{\kappa_3+k_3+1-\zeta} [\kappa_3, k_3]$$

$$\times \begin{pmatrix} k & t & z \\ q & p & -\zeta \end{pmatrix} \begin{Bmatrix} \kappa_2 & \kappa_3 & \kappa_1 \\ \frac{1}{2} & \frac{1}{2} & \frac{1}{2} \end{Bmatrix} \begin{Bmatrix} k_2 & k_3 & k_1 \\ \ell & \ell' & \ell \end{Bmatrix}$$

$$\times \begin{Bmatrix} \kappa_1 & k_1 & k \\ \kappa_2 & k_2 & t \\ \kappa_3 & k_3 & z \end{Bmatrix} W_\zeta^{(\kappa_3 k_3)z} \quad (22\text{-}13)$$

where $A^{\kappa_1\kappa_2}_{k_1k_2}(kt\ell')$ is defined by the angular and radial terms in the following way

$$A^{\kappa_1\kappa_2}_{k_1k_2}(kt\ell') = \sum_{i_1,i_2}^{2} \beta^{k\ell'\ell}_{\kappa_1 k_1}(j_{i_1} j'_{i_2})\beta^{t\ell\ell'}_{\kappa_2 k_2}(j'_{i_2} j_{i_1})$$

$$\times R^k(j_{i_1} j'_{i_2})R^t(j'_{i_2} j_{i_1})$$

$$\times \langle j_{i_1}||C^{(k)}||j'_{i_2}\rangle\langle j'_{i_2}||C^{(t)}||j_{i_1}\rangle \qquad (22\text{-}14)$$

and i_1 and i_2 number $j_\pm$ and $j'_\pm$; the angular factors and the radial integrals are defined as before, namely

$$\beta^{x\ell'\ell}_{\kappa_1 k_1}(j_{i_1} j'_{i_2}) = (-1)^{\kappa_1+k_1+x}[j_{i_1}, x, j'_{i_2}]^{1/2}\begin{Bmatrix} \ell' & \ell & k_1 \\ s & s & \kappa_1 \\ j'_{i_2} & j_{i_1} & x \end{Bmatrix}, \qquad (22\text{-}15)$$

The radial integrals, as relativistic counterparts of those from the non-relativistic approach, contain the *large* and *small* components, and they are defined as follows

$$R^x(j_{i_1}, j'_{i_2}) = \langle P^{j_{i_1}}|r^x|P^{j'_{i_2}}\rangle + \langle Q^{j_{i_1}}|r^x|Q^{j'_{i_2}}\rangle. \qquad (22\text{-}16)$$

The reduced matrix element of the spherical tensor is the generalization of the intra-shell case

$$\langle j_{i_1}||C^{(x)}||j'_{i_2}\rangle = (-1)^{j_{i_1}+1/2}[j_{i_1}, j'_{i_2}]^{1/2}\varepsilon(\ell + x + \ell')\begin{pmatrix} j_{i_1} & x & j'_{i_2} \\ \frac{1}{2} & 0 & \frac{1}{2} \end{pmatrix}. \qquad (22\text{-}17)$$

In addition, the factor a in (22-13) is equal 2 when the parity of appropriate ranks of operators is the same, otherwise it vanishes, namely,

$$a = 2 \quad \text{if } p(\kappa_1 + k_1 + 1 + \kappa_2 + k_2 + t) = p(\kappa_3 + k_3)$$

$$0 \quad \text{otherwise} \qquad (22\text{-}18)$$

When the electron correlation is included via the Coulomb interaction taken as a perturbation, the two particle effective operator has a rather complex structure,

$$^{rel}T^2_{corr}(k) = -2\sum_{s}^{even}[s]^{1/2}\sum_{\ell'}\sum_{\kappa_1,k_1}\sum_{\kappa_3,k_3}\sum_{\kappa_4,k_4}A^{\kappa_1\kappa_3\kappa_4}_{k_1k_3k_4}(ks\ell')$$

$$\times \sum_{x}[x]^{1/2}\sum_{\kappa_2,k_2}\varepsilon(\kappa_2 + k_2)[\kappa_2, k_2]$$

$$\times \begin{Bmatrix} \kappa_4 & \kappa_2 & \kappa_3 \\ \frac{1}{2} & \frac{1}{2} & \frac{1}{2} \end{Bmatrix}\begin{Bmatrix} k_4 & k_2 & k_3 \\ \ell & \ell' & \ell \end{Bmatrix}\begin{Bmatrix} \kappa_1 & k_1 & k \\ \kappa_4 & k_4 & s \\ \kappa_2 & k_2 & x \end{Bmatrix}$$

$$\times \sum_{i<j}\left[w_i^{(\kappa_3k_3)s} \times w_j^{(\kappa_6k_6)x}\right]^{(\lambda)} \qquad (22\text{-}19)$$

where $\mathcal{A}_{k_1 k_3 k_4}^{K_1 K_3 K_4}(ks\ell')$ is defined by the angular and radial terms in the following way

$$\mathcal{A}_{k_1 k_3 k_4}^{K_1 K_3 K_4}(ks\ell') = \sum_{i_1, i_2, i_3} \beta_{K_1 k_1}^{k\ell'\ell}(j_{i_1} j_{i_2}') \beta_{K_4 k_4}^{s\ell\ell'}(j_{i_2}' j_{i_1}) \beta_{K_3 k_3}^{s\ell\ell}(j_{i_3} j_{i_3}) R^k(j_{i_1} j_{i_2}') R^s(j_{i_3} j_{i_2}' j_{i_3} j_{i_1})$$

$$\times \langle j_{i_1} || C^{(k)} || j_{i_2}' \rangle \langle j_{i_2}' || C^{(s)} || j_{i_1} \rangle \langle j_{i_3} || C^{(s)} || j_{i_3} \rangle \langle j_{i_2}' || C^{(t)} || j_{i_1} \rangle,$$

$$(22\text{-}20)$$

and i_1, i_2 and i_3 number $j_\pm$ and $j_\pm'$, and the coefficients β, radial integrals and reduced matrix elements are defined by (22-15)-(22-17), respectively.

In order to make this presentation complete it should be realized that additional effective operators that originate from the exchange interactions should also be taken into account. Now the situation becomes even more complex than in the case of the direct interactions presented above. In the case of the exchange terms there are the integrals with the functions that are not referred to the same coordination frame, and therefore a certain transformation has to be used prior to their evaluation. Thus, the task now is to evaluate the integral which is not any longer expressed as a simple product of separate parts associated with the lanthanide and the ligand, namely

$$\langle \Psi_f(1)\phi_f^L(2) \mid \frac{e^2}{r_{12}} P_{12} \mid \Psi_i(1)\phi_i^L(2) \rangle \tag{22-21}$$

where P_{12} permutes the coordinates of two interacting electrons. Using the Dirac identity for the permutation operator, and after transforming the functions to a common coordination frame, the simplest effective operator that contributes to the energy transfer at the first order, and which is due to the exchange interactions, has the form

$$^s X = \sum_{k_1=0}^{2\ell} \sum_{k_2=|l_f^L - l_i^L|}^{l_f^L + l_i^L} \sum_{k=|k_1-k_2|}^{k_1+k_2} \Pi_{k_1 k_2 k}\left(n\ell n_f^L \ell_f^L, n\ell n_i^L \ell_i^L\right)$$

$$\times \langle \Psi_f(1)\phi_f^L(2) \mid \hat{X}_s(k_1 k_2 k; 12) \mid \Psi_i(1)\phi_i^L(2) \rangle \tag{22-22}$$

where $n\ell$, $n^L \ell^L$ are one-electron functions of the lanthanide ion and ligand, respectively. The operator in the matrix element is the tensorial product of two double tensor operators,

$$\hat{X}_s(k_1 k_2 k; 12) = -2[k]^{-1/2} \sum_q \sum_{xy} [x, y]^{1/2}(-1)^{q+x+k_2} C_{-q}^{(k)}(R) \begin{Bmatrix} x & y & k \\ k_2 & k_1 & 1 \end{Bmatrix}$$

$$\times \left[w^{(1k_1)x}(1) \times w^{(1k_2)y}(2) \right]_q^{(k)} \tag{22-23}$$

and Π is the generalized version of the exchange radial integral G^k which is defined as follows[45],

$$\Pi_{k_1 k_2 k}\left(n\ell n_f^L \ell_f^L, n\ell n_i^L \ell_i^L\right) = \sum_{\kappa}^{\min(2\ell,\,\ell_f^L+\ell_i^L)} \langle \ell \parallel C^{(\kappa)} \parallel \ell \rangle \langle \ell_f^L \parallel C^{(\kappa)} \parallel \ell_i^L \rangle$$

$$\times A \sum_{\tau\tau'}^{\infty} \sum_{\lambda=|\ell_i^L-\tau|}^{\ell_i^L+\tau} \sum_{\lambda'=|\ell_f^L+\tau'|}^{\ell_f^L-\tau'} (-1)^{\kappa+\phi}[\tau,\tau',\lambda,\lambda']\begin{pmatrix}\ell & \tau & \kappa \\ 0 & 0 & 0\end{pmatrix}\begin{pmatrix}\ell & \tau' & \kappa \\ 0 & 0 & 0\end{pmatrix}$$

$$\times \begin{pmatrix}\ell_i^L & \tau & \lambda \\ 0 & 0 & 0\end{pmatrix}\begin{pmatrix}\ell_f^L & \tau' & \lambda' \\ 0 & 0 & 0\end{pmatrix}\begin{pmatrix}\lambda & \lambda' & k \\ 0 & 0 & 0\end{pmatrix}\begin{Bmatrix}\ell & \ell & k_1 \\ \tau & \tau' & \kappa\end{Bmatrix}\begin{Bmatrix}k & k_2 & k_1 \\ \lambda & \ell_i^L & \tau \\ \lambda' & \ell_f^L & \tau'\end{Bmatrix}$$

$$\times \int_0^\infty \int_0^\infty \frac{r_<^k}{r_>^{k+1}} R_{n\ell}(r_1)\vartheta_{\tau\lambda n_i^L \ell_i^L}(r_1, R)R_{n\ell}(r_2)\vartheta_{\tau'\lambda' n_f^L \ell_f^L}(r_2, R)r_1^2 r_2^2 dr_1 dr_2$$

$$(22\text{-}24)$$

$R_{n\ell}$ are the radial functions of the $n\ell^N (\equiv 4f^N)$ configuration of the lanthanide ion, and ϑ are the transformed radial functions of the ligand electrons; the latter are defined as follows

$$\vartheta_{\tau\lambda n\ell}(r, R) = \frac{1}{\pi}\int_{-\infty}^{\infty} j_\lambda(kR)j_\tau(kr)\bar{R}(k)k^2 dk \qquad (22\text{-}25)$$

and

$$\bar{R}(k) = 4\pi \int_0^\infty R(r)j_L(kr)r^2 dr \qquad (22\text{-}26)$$

where $j_\lambda(kR)$ is the spherical Bessel function.

Should we improve the representation of the exchange interactions by higher order terms that are caused by the perturbing influence of the crystal field potential and electron correlation effects? What about the relativistic version of the impact of exchange interactions upon the energy transfer rate?... We have not yet presented here any expression that determines the amplitude of the energy transfer in the case of the impurity sensitized luminescence[46] ...

Instead of continuing the presentation of the explosion of various terms we have to admit that, as Albert Einstein noticed,

> Quantum mechanics is very impressive. But an inner voice tells me that it is not yet the real thing. The theory yields a lot, but it hardly brings us any closer to the secret of the Old One. In any case I am convinced that He doesn't play dice.

These words bring our theoretical discussion down to earth. If the presentation here would be devoted to Racah algebra and to its logic and beauty, in order to illustrate this powerful tool of atomic spectroscopy, we would derive without much difficulty more expressions for new effective operators. In the context of the spectroscopic properties of the materials doped by the lanthanides, however, such a presentation would be just an art for the sake of art. Therefore at this point we should rather address the issue of the relation between purely theoretical expressions, as those presented above, and physical reality. At the same time, in order not to discredit the theory and basic research...

> We must not forget that when radium was discovered no one knew that it would prove useful in hospitals. The work was one of pure science. And this is a proof that scientific work must not be considered from the point of view of the direct usefulness of it. It must be done for itself, for the beauty of science, and then there is always the chance that a scientific discovery may become like the radium a benefit for humanity.

Maria Skłodowska-Curie, Lecture at Vassar College (May 14, 1921)

22.4 CONFRONTATION WITH NATURE: TISSUE SELECTIVE LANTHANIDE CHELATES

The rare earth doped materials have not only revolutionized modern techniques and technology, but also clinical and diagnostic medicine, the pharmaceutical industry, molecular biology, and the understanding of the mystery of life, not to mention the purely scientific value of their applications in astronomy or astrophysics. Painless contrast agents to diagnose brain hemorrhages, laser surgical scalpels to remove tumors, endoscopic resections and laparoscopy, probes that follow the pathway of iron in the blood stream or monitor a transport of calcium, and at the same time modern technology based on liquid crystals - these are only a few examples of the applications of rare earth doped materials. Although lasers are used practically in almost all possible aspects of life, their impact upon the state of the art of surgery is one of its most spectacular applications since the first laser beam was generated in 1960.

Materials that contain the rare earth ions are recognized not only as good laser media but are also widely used in newly developed technologies such as semiconductor light-emitting devices and as dopants in optical fibers. Due to their favorable spectroscopic properties, rare-earth-doped fiber lasers offer rich possibilities in telecommunication applications, atmospheric pollution measurements, noninvasive medical diagnostic procedures and industrial monitoring and control.

There are very impressive applications of the rare earth doped materials in bioinorganic chemistry. The importance of these materials has increased, since it was discovered in 1970 that rare earth ions are ideal probes in the investigation of the binding sites in proteins. The attractive role of the lanthanides is connected with the research on calcium binding in proteins. Calcium, an important element in biochemistry, as an inert system, is very difficult to examine directly. The chemical similarity of rare earth ions to calcium, and their ability to luminesce at room temperature, are the main reasons why lanthanides are so commonly used as probes. Substitution of Ca by a rare earth ion has enabled one to establish the position of calcium in trypsin, an enzyme of the pancreatic juice. It was also found that admixing Tb^{3+} to another protein, pancreatic elastase, results in strong luminescence of Tb, and it was also verified that binding of calcium, magnesium and zinc under such conditions is possible.

Magnetic resonance imagining (MRI) of biological structures is an alternative technique to light-based microscopy. The efficiency of this technique is not limited by the scattering of light to the layers of cells on the surface, and in addition it does not produce toxic substances as do the methods that use dyes and fluorochromes. This is a useful method to monitor cellular Ca^{2+} and its role in the physiology of a cell and internal biochemical processes. Recently, the lanthanide ion Gd^{3+} has been used as a contrast agent in MRI. This ion is characterized by the half-filled shell of equivalent $4f$ electrons, and it belongs to the group of the so-called S-state ions (their ground state is $^8S_{7/2}$); it is a good contrast agent due to its high magnetic moment. To avoid the toxicity of the aqueous ion, all but one of the coordination sites of Gd^{3+} are bound by a chelate, which usually forms a cage encapsulating the ion. The remaining site is left for the water molecule that produces the signal in the imaging experiments.

In order to comprehend the scale and variety of known applications of the lanthanide ions it is sufficient to explore the Internet to easily find information about the extensive role they play. Here we cite the description found at just one address, *pubs.acs.org/cen/80th/lantanides.html*, where the role of the lanthanides is described in catalytic converters and alloys for very stable permanent magnets used in antilock braking systems, air bags, and electric motors, and in their traditional use for staining glass and ceramic. In fact, as mentioned at this address, the lanthanide materials are so commonly present in everyday life that...the red, green and blue luminescence of Eu^{3+} ion upon uv light is used to protect against counterfeit euro banknotes, introduced in Europe in 2002.

These are but a few examples that demonstrate the wide range of applications of the lanthanide ions in various disciplines and show the motivation for the theoretical research even if at the first sight there is no direct connection between basic research and the practical use of its results.

There is a large group of organic chelates doped with the lanthanides that play a crucial role in investigations devoted to the mastering of the spectroscopic tools of the early detection of cancer.[47-50] Presently extensive experimental work in vivo is being performed to monitor, to recognize and to measure tumors in various tissues. These systems are also important for the applications in radioimmunotherapy as local media, which via the radiation of the radioactive isotope encapsulated inside the cage destroy the cancerous cells. The architecture of the cage, its symmetry and stability play a dominant role in the effectiveness of the tissue selective lanthanide chelates applied as probes and radiopharmaceuticals. Although many aspects of the luminescence of chelates doped by the lanthanides are evidently observed in laboratories, still the mechanism of the energy transfer responsible for the host sensitized luminescence is unknown, or at least unclear. Indeed, the main problem (and at the same time the most intriguing part of the spectroscopic properties of these materials) is to understand the process of energy transfer between the sensitizer, the host that is usually built as cage by a chelate, and the activator, the role of which is always played by the lanthanide ion. Therefore crucial to the further development of the theoretical model and the experimental (most of all clinical) skills are detailed numerical

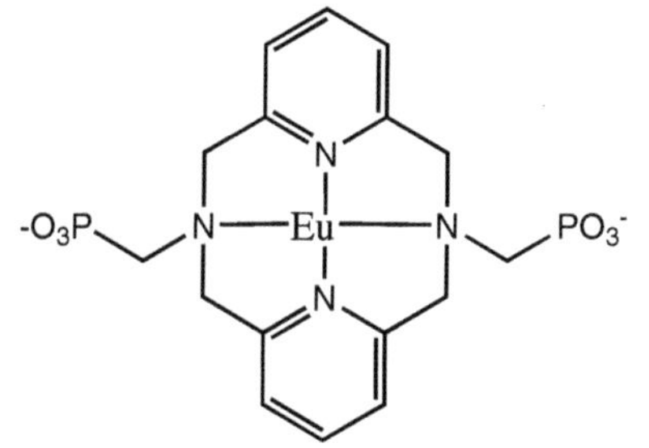

BP2P

FIGURE 22-2 Structure of a cage built of the chelate BP2P.

calculations on the impact of various physical mechanisms upon the efficiency of the luminescence in order to understand its origin.

With respect to the way the energy is harvested and then subsequently transferred to the lanthanide ion, all chelates are divided into two classes. In one class the whole cage absorbs the energy from the external beam, and then as a sensitizer it donates this energy to the lanthanide ion to excite it. An example of such a chelate is presented in Fig. 22-2; BP2P is the abbreviation for $N, N' - bis(methylenephosphonicacid) - 2, 11 - diaza[3.3] - (2, 6)pyridinophane$.[51,52] The lanthanide ion is an active part of the probe, and it is responsible for the signals monitoring the cancerous growth. Thus, for this particular kind of chelate, the cage plays a double role as a sensitizer of luminescence (donor of energy) and as a perturbation to the lanthanide ion that forces the radiative transitions. At the same time, the attraction between the arms built of the negatively charged groups PO_3^- increases the uptake of the agent by the skeletal system due to the electrostatic attraction of the positive ions of Ca^{+2} that are present in the bone tissue. There is also electrostatic interaction between these arms and the three positive lanthanide ions, which results in the folding of the arms to form a cage.

In the case of the second kind of chelates with lanthanide ions it is possible to distinguish three subsystems that take part in the process. Namely, there is a cage which is the host that perturbs the central ion causing the forced electric dipole transitions observed as signals. The second and optically active part is the central ion responsible for the luminescence that monitors the cancerous tissues. The third part is the antenna, which harvests the energy from the external beam and plays the role of a sensitizer of the luminescence. In Fig. 22-3. several examples of chelates are presented schematically. Here the side arms of the parent systems of DOTA (abbreviation for $1, 4, 7, 10 - tetraazacyclododecane - 1, 4, 7, 10 - tetraaceticacid$) and DOTP (abbreviation for $1, 4, 7, 10 - tetraazacyclododecane - 1, 4, 7, 10 - tetraphosphonicacid$) play the role of the antenna. In the modified cages of DOTA and DOTP presented below the parent systems, one of the side arms is replaced by a chromophore in order to reinforce the process of harvesting the light and increasing the efficiency of the luminescence of the agent. In DOTA-1 and DOTP-1 a benzene ring is tethered to the cage via the amide linkage. In DOTA-2 and DOTP-2 the role of

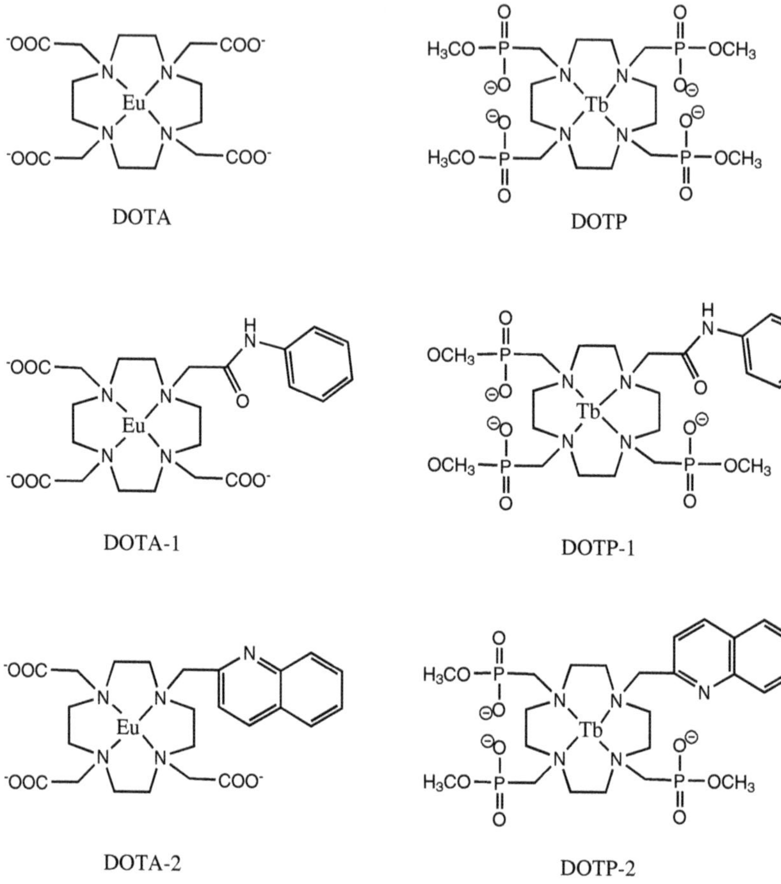

FIGURE 22-3 Parent systems of DOTA and DOTP, and their structures modified by the attached antennas, DOTA-1, DOTA-2 and DOTP-1, DOTP-2.

the antenna is played by the quinoline attached to the main cage via a single methylene bridge.

The parent chelate DOTA exists in two interconverting diastereomeric structures of a capped square antiprism. The basal plane of the cage is built of four amine nitrogens of the macrocycle, and the capped plane is occupied by four oxygen atoms that are part of the acetate sidearms. The second isomeric form was predicted by the observation of two peaks in the emission spectra of the chelated Eu ion, which were assigned to the electric dipole transitions $^7F_0 \longleftrightarrow {}^5D_0$. Since the final and initial states of this transition are non-generate, the only possible explanation for these two lines was to have two different, simultaneously existing isomers. This conclusion, deduced from the experimental data, was supported by the NMR investigations, and it was found that the two forms of DOTA differ by the twist angle between the basal and the capped planes.[53] In both forms there is eight-fold coordination with the lanthanide ion, and the ninth position is occupied by the water molecule, which via the hydrogen bonding caps the cage. In the parent DOTA chelate the acetate arms harvest the energy to transfer it directly to the central ion. From a theoretical point of view it means that the second matrix element of (22-2), which contributes to the amplitude of the energy transfer, describes only the arms of the chelate. At the same time, the cage into which the lanthanide ion is inserted disturbs the symmetry of its closest environment, and this perturbation is responsible for the luminescence resulting from the electric dipole $f \longleftrightarrow f$ transitions.

DOTA complexes with various encapsulated lanthanides are used widely as diagnostic probes for the early detection of cancerous cells in various tissues. As reported in all published experimental investigations, the cancerous tissue is monitored by the luminescence of the lanthanide ion, which is transferred as a part of the absorbed agent. It is very interesting, although not understood yet, that the cancerous cells absorb several times more of the appropriate agent than the healthy part of the same tissue.

In the particular case of the chelate with the Tb ion, the electric dipole $f \longleftrightarrow f$ transitions between the energy levels $^5D_4 \rightarrow \ ^7F_J$, for J= 0-6, are responsible for the observed atomic-like emission lines that monitors the presence of the agent in the properly addressed tissue. In the case of Eu ion the transitions $^5D_0 \rightarrow \ ^7F_J$, for various J, are observed. Among the transitions allowed by the second-order Judd-Ofelt theory (although forbidden by the Laporte parity selection rules) in this particular group there is the highly forbidden transition $^5D_0 \longleftrightarrow \ ^7F_0$ that was used in the process of identification of the geometry of DOTA.

The parent system of DOTP in Fig. 22-3, where the acetate sidearms of DOTA are replaced by methylenephoshonate sidearms, is assumed to exist in only one conformation. The complex is eight coordinated and again the basal plane is built of four amine nitrogens. The charge of the whole cage is -5 and therefore it is expected that it plays even a more important role in the diagnosis of cancer. Since the surface of the bone tissue is covered by the positively charged ions of calcium, the probe being more negative is there strongly attracted. The same mechanism is expected in the case of the cancerous cells in soft tissues, since they are usually manifested by the calcifications present in the vicinity of the tumor.

The efficiency of the process of the light harvesting is increased by replacing one of the arms of the parent system by a specially chosen chromophore. In Fig. 22-3 such cases are presented below the original cages. Although such substitutions in principle destroy the axial symmetry of the cages, there is evidence, based on the results of numerical calculations, that from a theoretical point of view it is possible to assume that the cage surrounding the lanthanide ion is practically not altered. If the structure of the cage is stable and this assumption is well satisfied, even in the presence of the antenna, only the cage with its preserved axial symmetry has a direct impact upon the luminescence of the lanthanide ion. This means that the antenna is separated from the central ion well enough to be treated, as in the case of the parent chelate with actual axial symmetry, as the part which is responsible only for the harvesting and subsequent transfer of the energy.

In order to verify this concept and to gain insight into the mechanisms of the observed phenomena, which consequently would help establish a reliable theoretical model of the host sensitized luminescence in the cages doped by the lanthanides, density functional calculations (DFT) have been performed for all the structures presented in Fig. 22-3.[53]

In Fig. 22-4 the optimized geometry of DOTA chelates is presented. It is vividly seen that indeed the organic chelates have the structure of the cages with an encapsulated central ion. In the case of the parent system the acetate sidearms are folded to close the cage by the oxygen atoms (numbered in the figure), which are coordinated with the central ion. The situation is different in the case of the agents modified by certain antennas, although the cage is still closed. The fourth carboxylate oxygen which is coordinated with the central ion in DOTA, is replaced in DOTA-1 by the oxygen from the amide. The distance of the latter from the central ion is comparable to the distances of the remaining oxygens that form the capping plane. In the structure of DOTA-2, in addition to three oxygens from the original acetate arms, the nitrogen atom from the quinoline is coordinated. In both cases the chromophores attached to the cages, and coordinated with the central ion, are sticking out as though prepared to harvest the light. It is seen from these optimized geometries that indeed the chromophores play a role of antenna and do not participate in the subsequent step of sending the spectroscopic signals from the selected tissue.

From among all results of the DFT calculations presented here the most surprising is the finding that the substitution of one acetate arm of the parent system by an antenna does not change the structure and also the size of the cage! As a consequence, in all three cases presented in Fig. 22-4 the lanthanide ion *feels* the same immediate crystal field. In all cases the ion is embedded in the same crystal field potential, which in fact is responsible for the electric dipole $f \longleftrightarrow f$ transitions. This means that the cage by itself plays the role of the forcing mechanism of the radiation that is detected as signals monitoring the presence of cancer. These numerical results support the observations, while at the first sight difficult to understand, that indeed the emission spectra of the same chelate, but with various antennas, are almost the same.

The results of the DFT calculations performed for DOTP and its modification by two different antennas are similar. It is seen from Fig. 22-5 that again the chelates have the structure of cages, and the original triple coordination with the lanthanide ion on the capping plane is completed by the fourth oxygen atom from the amide linkage, and by the nitrogen atom in the case of the quinoline attached as a chromophore.

The analysis presented here provides support for the basic assumption that the theoretical model of the luminescence may be divided into two steps, as anticipated above. Namely, there is the first step of energy absorption by the side arms or antenna and the second one, which due to the perturbing influence of the cage, forces the electric dipole radiation of the central ion.

The results of numerical calculations performed for various ions across the lanthanide series provide information about the stability of the structure of each chelate. In addition they demonstrate that there are no major structural changes caused by inserting various lanthanides into the same cage. This particular result supports the practice in which *Gd* ion, for example, is replaced by the radioactive *Sm* ion in order to measure the uptake of the chelate by a certain kind of tissue. Prior to the DFT calculations[53], it was simply assumed that there are no changes of the initial geometry and the results of measurements are realistic. This stability of the structure and application of the radioactive isotopes give hope for the therapeutic role of the

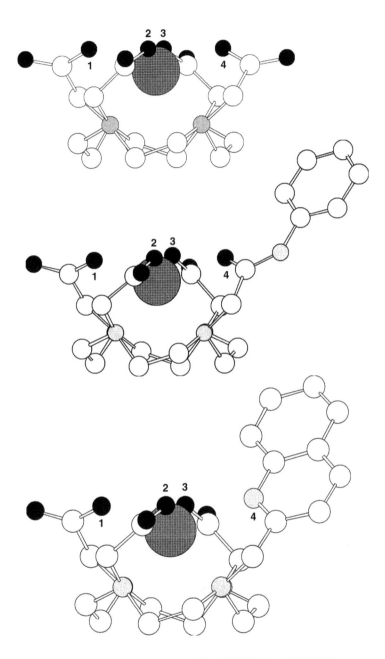

FIGURE 22-4 Optimized geometry (DFT)of DOTA, DOTA-1 and DOTA-2.

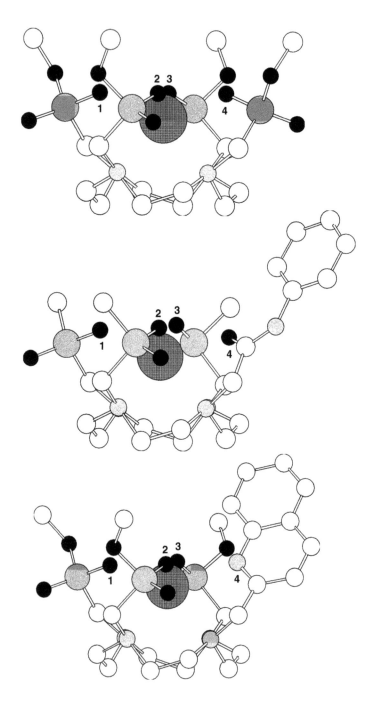

FIGURE 22-5 Optimized geometry (DFT) of DOTP, DOTP-1 and DOTP-2.

chelates and development of modern, effective, non-invasive and free of dangerous side effects pharmacology.

In fact, to some extent this hope has already materialized in clinical medicine. The most striking result is described in the publication *Radiolabeled somatostatin analog [$^{177}Lu - DOTA^0, Tyr^3$]octreotate in patients with endocrine gastroemteropancreatic tumors"* by D. J. Kwekkeboom et al, published in 2005.[54] This report was followed by a question *Does $^{177}Lu-$ labeled octreotate improve the rate of remission of endocrine gastroenteropancreatic tumors?*, by A. N. Eberle and C. Beglinger, published in Nature Clinical Practice Endocrinology and Metabolism.[55] In the latter paper very optimistic and almost realistic promises for all who suffer are announced: *Chemiotherapy has not been satisfactory and prolonged treatment is limited because of potential toxic effect. The use of radiolabeled somatostatin* (peptide hormone) *analogs represents an important new therapeutic option with promising efficacy and low toxicity*. This means that the new therapeutic tools have already left the testing stage and are being used in patients!

In October 2004 a new drug, *Fosrenol* was approved for the treatment of hyperphosphatemia, the disorder in which the level of phosphorus in the blood is too high. Although phosphorus plays a critical role in bone mineralization, cellular structure, genetic coding and energy metabolism, its high level can lead to pathological calcifications of soft tissues, which may cause dangerous clinical consequences, especially when localized in vital organs. What is so special about this new medicine? Fosrenol is the carbonate salt of **lanthanum**. It dissolves easily in the digestion acid producing a free lanthanum ion, which possesses the ability of binding the dietary phosphate extracted from the food during digestion. The lanthanum phosphate complexes are insoluble, and therefore the level of the phosphate in the serum is lowered.

Lanthanides also play an important role in homeopathy to cure migraine, dyslexia, multiple sclerosis, AIDS, arthritis, and other health problems. This is the most enigmatic component of their attractiveness to which the book *Secret Lanthanides* by Jan Scholten is devoted. In fact homeopathy by itself is mysterious and its curing power is based on the belief rather than on scientific merit (but from my personal experience I know that it works!), however *The beginning of knowledge is the discovery of something we do not understand* (Frank Herbert). Thus, being open to new challenges provided by the experimental work and observations, we should listen to hear, and watch to see, trying to understand and learn more about nature. I have to confess at the same time that

I am among those who think that science has great beauty. A scientist in his laboratory is not only a technician: he is also a child placed before natural phenomena which impress him like a fairy tale.

Maria Skłodowska-Curie

REFERENCES

1. Monardes N (1577), Joyfull News out of the New Founde Worlde, 1 London, reprinted by Knopf, New York, **1**, 1925.
2. Stokes GG (1852), On the change of refrangibility of light, *Philos. Trans.,* **142**, 463.

3. Dexter DL (1953), Theory of sensitized luminescence in solids, *J. Chem. Phys.*, **21**, 836.

4. Axe JD and Weller PF (1964), Fluorescence and energy transfer in $Y_2O_3 : Eu^{3+}$, *J. Chem. Phys.*, **40**, 3066.

5. Sommerdijk JL (1973), Influence of host lattice on the infrared-excited visible luminescence in Yb^{3+}, Er^{3+}-doped fluorides, *J. Lumin.*, **6**, 61.

6. van der Ziel JP and Van Uitert LG (1969), Optical emission spectrum of Cr^{3+}-Eu^{3+} pairs in europium gallium garnet, *Phys. Rev.*, **186**, 332.

7. Kouyate D, Ronfard-Haret JC, and Kossanyi J (1993), Electroluminescence of rare earths-doped semiconducting zinc oxide electrodes: kinetic aspects of the energy transfer between Sm^{3+} and Eu^{3+}, *J.Lumin.*, **55**, 209.

8. Zhong G, Feng Y, Yang K, and Zhu G (1996), Enhanced luminescence of Eu(III) by La(III), Gd(III), Tb(III) and Y(III) in langmuir-blodgett films of mixed rare earth complexes, *Chem. Lett.*, **25**, 775.

9. Georgobani AN, Tagiev BG, Tagiev OB, and Izzatov BM (1995), Photoluminescence of rare earths in $CaGa_2S_4$, *Inorg. Mater.*, **31**, 16.

10. Viney IVR, Ray B, Arterton BW, and Brightwell JW (1992), Electroluminescence, Proc. Int. Workshop 6th, (Ed. Singh V P and McClure J C), Cinco Puntos Press, El Paso, TEX., p. 128.

11. Sun J-Y, Du H-U, Pang W-Q, and Shi C-S (1994), Photoluminescence properties of divalent europium ion doped in zeolites, *Gaodeng Xuexiao Huaxue Xuebao*, **15**, 961.

12. Auzel FE (1973), Materials and devices using double-pumped-phosphors with energy transfer, *Proc. IEEE*, **61**, 758.

13. Van Uitert LG and Soden RR (1962), Enhancement of Eu^{3+} emission by Tb^{3+}, *J. Chem. Phys.*, 36, 1289; Van Uitert LG Soden RR, and Linares RC (1962), Enhancement of rare-earth ion fluorescence by lattice processes in oxides, **36**, 1793.

14. Petoud S, Bünzli J-CB, Schenk KJ, and Piguet C (1997), Luminescent properties of lanthanide nitrate complexes with substituted bis(benzimidazolyl)pyridines, *Inorg. Chem.*, **36**, 1345.

15. Panadero S, Gomez-Hens A, and Perez-Bendito D (1997), Kinetic determination of carminic acid by its inhibition of lanthanide-sensitized luminescence, Fresenius,' *J. Anal. Chem.*, **357**, 80.

16. Schrott W, Buechner H and Steiner UE (1996), Dual mechanism of energy transfer in benzoate sensitized luminescence of Eu^{3+} and Tb^{3+}, *J. Infor. Recor.*, **23**, 79.

17. Mayolet A, Zhang W, Martin P, Chassigneux B, and Krupa J-C (1996), The intermediate role of the Gd^{3+} ions in the efficiency of the host-sensitized luminescence in the lanthanum metaborate host lattice, *J. Electrochem. Soc.*, **143**, 330.

18. Dexter DL (1957), Possibility of luminescent quantum yields greater than unity, *Phys. Rev.*, **108**, 630.

19. Varsanyi F and Dieke GH (1961), Ion-pair resonance mechanism of energy transfer in rare earth crystal fluorescence, *Phys. Rev. Lett.*, **7**, 442.

20. Johnson LF, Guggenheim HJ, Rich TC, and Ostermayer FW (1972), Infrared-to-visible conversion by rare-earth ions in crystals, *J. Appl. Phys.*, **43**, 1125.

21. Alexeyev AN and Roshchupkin DV (1996), Diffraction of surface acoustic waves on the zigzag domain wall in a Gd (MoO) crystal, *J. Appl. Phys.*, **68**, 159.

22. Auzel F (1966), Quantum counter by energy transfer between two rare earth ions in a mixed tungstate and in a glass, *C. R. Acad. Sci. B (France)*, **262**, 1016.

23. Auzel F (1967), Quantum counter by transfer of energy from Yb^{3+} to Tm^{3+} in a mixed tungstate and in a germanate glass, *C. R. Acad. Sci. B (France)*, **263**, 819.

24. Kuroda H, Shionoya S, and Kushida T (1972), Mechanism and controlling factors of infrared-to-visible conversion process in Er^{3+} and Yb^{3+}-doped phosphors, *J. Phys. Soc. Japan,* **33**, 125.

25. Van Uitert LG and Johnson LF (1966), Energy transfer between rare-earth ions, *J. Chem. Phys.,* **44**, 3514.

26. Johnson LF, Van Uitert LG, Rubin JJ, and Thomas RA (1964), Energy transfer from Er^{3+} to Tm^{3+} and Ho^{3+} ions in crystals, *Phys. Rev.,* **A133**, 494.

27. Yamada N, Shionoya S, and Kushida T (1971), Phonon-assisted energy transfer — comparison between theory and experiment, *J. Phys. Soc. Jon,* **30**, 1507.

28. Yamada N , Shinoya S, and Kushida T (1972), Phonon-assisted energy transfer between trivalent rare earth ions, *J.Phys. Soc. Japan,* **32**, 1577.

29. Miura N, Sasaki T, Matsumoto H, and Nakano R (1992), Band gap energy dependence of emission spectra in rare earth doped $Zn_{1-x}Cd_x S$ thin film electroluminescent devices, *Jpn. J. Appl. Phys. Part 1,* **31**, 295.

30. Feng J, Shan G, Maquieira A, Koivunen ME, Guo B, Hammock BD, and Kennedy IM (2003), Functional europium oxide nanoparticles used as a fluorescent label in an immunoassay for atrazine, *Anal. Chem.,* **75**, 5282.

31. Zhu G, Si Z, Jiang W, Li W and Li J (1992), Study of sensitized luminescence of rare earths. Fluorescence enhancement of the europium-terbium (or Lutetium)-β-diketone–synergistic agent complex system and its application, *Spectrochim. Acta,* **48A**, 1009.

32. Tiedje T, Colbow KM, Gao Y, Dahn JR, Reimers JN, and Houghton DC (1992), Role of coulomb repulsion in 4f orbitals in electrical excitation of rare-earth impurities in semiconductors, *Appl. Phys. Lett.,* **61**, 1296.

33. Ennen H, Pomrenke G, Axmann A, Eisele K, Haydle W, and Schneider J (1985), 1.54-μm electroluminescence of erbium-doped silicon grown by molecular beam epitaxy, *Appl. Phys. Lett.,* **46**, 381.

34. Eagleshman DJ, Michel J, Fitzgerald EA, Jacobson DC, Poate JM, Benton JL, Polman A, Xie Y-H, and Kimerling LC (1991), Microstructure of erbium-implanted Si, *Appl. Phys. Lett.,* **58**, 2797.

35. Isshiki H, Kobayashi H, Yugo S, Kimura T, and Ikoma T (1991), Impact excitation of the erbium-related 1.54 μm luminescence peak in erbium-doped InP, *Appl. Phys. Lett.,* **58**, 484.

36. Förster T (1948), Intermolecular energy transference and fluorescence, *Ann. Phys.,* **2**, 55.

37. Dexter DL (1962), Cooperative optical absorption in solids, *Phys. Rev.,* **126**, 1962.

38. Di Bartolo B (1984), *Energy transfer processes in condensed matter, NATO ASI Series B,* (Eds. Di Bartolo B and Karipidou A), Plenum Press, New York, **114**, 103.

39. Altarelli M and Dexter DL (1970), Cooperative energy transfer and photon absorption, *Opt. Commun.,* **2**, 36.

40. Malta O (1997), Ligand-rare-earth ion energy transfer in coordination compounds. A theoretical approach, *J. Lumin.,* **71**, 229.

41. Smentek L and Hess BA Jr. (2000), Theory of host sensitized luminescence in rare earth doped materials. I. Parity considerations, *J. Alloys and Comp.,* **300–301**, 165.

42. Smentek L and Hess BA Jr. (2000), Theory of host sensitized luminescence in rare earth doped materials. II. Effective operator formulation, *J. Alloys and Comp.,* **300–301**, 165.

43. Smentek L and Hess BA Jr. (2002), Theory of host sensitized luminescence in rare earth doped materials. III. Numerical illustration, *J. Alloys and Comp.,* **336**, 56.

44. Smentek L, Wybourne BG, and Hess Jr. BA (2002), Theory of host sensitized luminescence in rare earth doped materials. IV. Relativistic approach, *J. Alloys Comp.,* **341**, 67.

45. Smentek L (2002), Two-center exchange interactions in rare earth doped materials, *Int. J. Quant. Chem.*, **90**, 1206.

46. Smentek L (2005), Impurity sensitized luminescence in rare earth doped materials, *J. Sol. St. Chem.,* **178**, 470.

47. Bornhop DJ, Hubbard DS, Houlne MP, Adair C, Kiefer GE, Pence BC, and Morgan DL (1999), Fluorescent tissue site-selective lanthanide chelate, Tb-PCTMB for enhanced imaging of cancer, *Anal. Chem.,* **71**, 2607.

48. Griffin JMM, Skwierawska AM, Manning HC, Marx JN, and Bornhop DJ (2001), Simple, high yielding synthesis of trifunctional fluorescent lanthanide chelates, *Tetrahedron Lett.,* **42**, 3823.

49. Bornhop DJ, Griffin JMM, Goebel TS, Sudduth MR, Bell B, and Motamedi M (2003), Luminescent lanthanide chelate contrast agents and detection of lesions in the hamster oral cancer model, *Appl. Spectrosc.,* **57**, 1216.

50. Bell BA, Shilagard T, Goebel T, Bornhop DJ, and Motamedi M (2004), Spectroscopic detection of oral cancer using lanthanide chelate complex as topical contrast agent, *Lasers Surg. Med.,* **34**, 50, Suppl. S16.

51. Houlne MP, Agent TS, Kiefer GE, McMillan K, and Bornhop DJ (1996), Spectroscopic characterization and tissue imaging using site-selective polyazacyclic terbium(III), *Appl. Spectrosc.,* **50**, 1221.

52. Hubbard DS, Houlne MP, Kiefer GE, McMillan K, and Bornhop DD (1998), Endoscopic fluorescence imaging of tissue selective lanthanide chelates, *Bioimaging,* **6**, 63.

53. Smentek L, Hess, BA, Jr. Cross JP, Manning HC, and Bornhop DJ (2005), Density-functional theory structures of 1,4,7,10-tetraazacyclododecane-1,4,7,10-tetraacetic acid complexes for ions across the lanthanide series, *J. Chem. Phys.,* **123**, 244302.

54. Kwekkeboom DJ, Teunissen JJ, Bakker WH, Kooij PP, de Herder WW, Feelders RA, van Eijick CH, Esser J-P, Kam BL, and Krenning EP (2005), Radiolabeled Somatostatin Analog [177Lu-DOTAO, Tyr3]Octreotate in Patients with Endocrine Gastroenteropancreatic Tumors, *J. Clin. Oncol.,* **23**, 2754.

55. Eberle AN and Beglinger C (2005), Does ^{177}Lu-labeled octreotate improve the rate of remission of endocrine gastroenteropancreatic tumors? *Nature Clinical Practice Endocrinology & Metabolism,* **1**, 20–21.

Index